Springer Series in Wood Science

Editor: T. E. Timell

Martin H. Zimmermann
Xylem Structure and the Ascent of Sap (1983)

John F. Siau
Transport Processes in Wood (1984)

Robert R. Archer
Growth Stresses and Strains in Trees (1986)

W. E. Hillis
Heartwood and Tree Exudates (1987)

W. E. Hillis

Heartwood and Tree Exudates

With 47 Figures

Springer-Verlag
Berlin Heidelberg New York
London Paris Tokyo

Dr. W. E. Hillis
CSIRO (retired)
Division of Chemical and Wood Technology
Bayview Avenue
Clayton, Victoria 3168, Australia

Series Editor:
T. E. Timell
State University of New York
College of Environmental
Science and Forestry
Syracuse, NY 13210, USA

Cover: Transverse section of *Pinus lambertiana* wood. Courtesy of Dr. Carl de Zeeuw, SUNY College of Environmental Science and Forestry, Syracuse, New York

ISBN-13: 978-3-642-72536-4 e-ISBN-13: 978-3-642-72534-0
DOI: 10.1007/978-3-642-72534-0

Library of Congress Cataloging in Publication Data. Hillis, W. E. Heartwood and tree exudates. (Springer series in wood science) Includes bibliography and index. 1. Heartwood. 2. Plant exudates. 3. Forest products. I. Title. II. Series. QK647.H55 1987 582.16'010425 87-4832

Softcover reprint of the hardcover 1st edition 1987

2131/3130-543210

Preface

Why prepare a treatise on Heartwood and Tree Exudates? Why consider both topics together? What approach should be taken in their treatment?

The exudates were one of the earliest items of trade between family, tribal, and racial groupings in prehistoric times. They became used extensively as items for the manufacture of implements and as commercial goods for illumination, for cosmetic, religious and magical purposes. Later heartwood from various trees entered intra- and international trade for prestigious and religious buildings (when cedars were used), for the furniture (e.g., ebony, mahogany) of the nobility, for boats and vehicles. Consideration of their relevance to anthropology, and to the origin of technological developments in different primitive cultures, would satisfy a personal life-long interest. Attention to such a topic is urgently needed now that the development and destruction of land and forests is increasing and wiping out the traces of earlier people to meet the demands of the rapidly enlarging populations of today. The latter represents an even more urgent need. Increasingly, mankind will depend on renewable resources produced at low energy cost. Forest products are one of these and the greater demands for them will require their growth and utilization with reduced loss and waste.

One of the major products of trees is heartwood, and one of the main features of heartwood is the presence of extractives. The formation of extractives – not only in heartwood but also in other tissues – and of exudates can assist the resistance of trees to disease and insect attack. Greater understanding of their formation and their role in tree protection is needed if our remaining forests and the new plantations are to be maintained in a healthy and productive condition. On the other hand, lower amounts of extractives than normal would improve the use of wood for some purposes. The choice of fast-growing species for intensively managed forests or plantations, of management strategies and shorter growth cycles is resulting in greatly increasing proportions of sapwood being provided commercially. It is desirable, and in some cases necessary, to control that proportion and this depends on an improved understanding of the factors regulating heartwood formation.

Eucalypts and pines are being planted on increasing areas throughout the world as fast-growing sources of fiber and wood. In their native habitat of Australia, the most commonly mentioned cause of degrade in eucalypts is that of kino veins or pockets. Accordingly, because of the increasing importance of the *Eucalyptus* genus, information on the occurrence, etiology, and structure of kino veins and their effect on wood quality should be readily available. On the other hand, the features controlling the formation of resin in conifers is important, as these extractives and exudates are the source of chemicals for value-added products. These and related features have been considered in this book.

Studies concerning the formation of extractives are complex and interrelated, and to assist work in these areas this treatise considers different facets of established knowledge that appear relevant. It will be seen that even in closely related fields, the same basic principles have sometimes been studied in one area and conclusions drawn without recognition of the contemporaneous and similar developments in the other fields. Knowledge of a broad picture would provide mutual support for studies in basic sciences and enable specialists to perceive the overall relevance of their conclusions. The different segments have been classified so as to assist future inclusion of information that will develop total understanding. The details of most facets are available elsewhere; the major aspects of their relevance are discussed here.

Chemical studies alone will not provide the answer to heartwood formation and neither will anatomical, cytological, or other studies. An interdisciplinary concept of this complex field is needed, but this has its own problems. It has been attempted on the advice of Schrödinger (1945; quoted by Berlyn 1970) "...the spread, both in width and depth, of the multifarious branches of knowledge during the last hundred odd years has confronted us with a queer dilemma. ...I can see no escape from this dilemma (lest our true aim be lost forever) than that some of us should venture to embark on a synthesis of fact and theories, albeit with second hand and incomplete knowledge of some of them – and at the risk of making fools of ourselves."

Although the extractives of heartwood (and bark) usually differ in chemical composition from those of the exudates from the same tree, they are formed by the same basic processes. It was the study of extractives by stalwarts in different countries that laid the foundations of organic chemistry and biochemistry, and a summary of that development has been presented elsewhere (Hillis 1986). The tree can arrange its physiology and biochemistry to produce, with astonishing selectivity, single compounds of considerable complexity in great purity or a suite of particular compounds for reasons that are not yet understood. Examples of the variations and aberrations of heartwood and exudate formation to illustrate these remarkable events were gathered in various centers. These sample collections are now being abandoned and with the disappearance of virgin and over-mature forests will never again be available. Consequently, the general applicability of theories developed from young trees of a few genera cannot be tested.

Some of the complex compounds from trees could serve as intermediates in the manufacture of industrial and commercial needs and be obtained without the frequent waste and pollution which accompanies chemical methods. The establishment of chemotaxonomy enables the prediction of suitable sources of these complex compounds from extractives or exudates. Their collection on a scale similar to that practised many years ago could provide employment for rural communities in some societies if supplies can be maintained and quality assured. Greater understanding of their biosynthesis should result in the enhancement of the consistent production of required compounds in the tree. The development of chemomorphology will enable the study of the relevant tissues to discern and define the physiological apparatus with which specific compounds are formed. Perhaps in the future some technological societies will develop this knowledge so that the required compounds are synthesized selectively in "biotrons" from simple carbohydrates by means of enzymes and with little pollution. This book should provide a step in that direction.

My scientific interest in exudates, heartwood and extractives was aroused in CSIRO and mainly in the Division of Forest Products during its spectacular existence. I am indebted particularly to Stan Clarke, Eric Dadswell and Margaret Chattaway for their encouragement during my early development of studies; I thank Warren Hewertson for his support and the use of CSIRO facilities at the end. I have benefited from discussions with numerous scientists, too many to list, both in my own and in many other countries. I thank them for their contributions to this field of study. I acknowledge the help and encouragement of Tore Timell, the Editor of this Series, and the staff of Springer-Verlag during the preparation of this book.

Melbourne, Australia W. E. Hillis
Spring 1987

Contents

Chapter 1

Introduction

1.1 Prehistoric and Ancient Use

The earliest evidence of the synthesis of resin during the evolution of the plant kingdom is among the gymnospermous plants. Only those belonging to the Coniferales are known to have produced resin (Langenheim 1969). Samples of these resins, which exist now in the form of amber, have been found in the Early Cretaceous and Tertiary strata (70 million to 2 million years old) and later geological deposits. They apparently arose from species of the families of Araucariaceae, Podocarpaceae, Pinaceae and the Cupressaceae – Taxodiaceae complex. The large deposits of amber that have been collected from some regions suggest that the early species may have formed resin more freely than species in recent forests. Amber was probably the first product of the forest to be admired, used, and traded over great distances.

The earliest written records and archeological studies show that forests were much more extensive 10 000 years ago than at present. They consisted of mixtures of species, particularly in tropical regions, which were controlled solely by the forest environment, and many trees were large in size and probably over-mature. Forests provided many of the needs of the small groups of the human race scattered throughout the world. With the elementary implements available to them, only small-sized objects from the forest could be used. The range of exudates from trees and plants were used for various purposes such as the provision of torches, caulking for boats, bulking material to aid the attachment of sharp stones to wooden handles, for coating various articles, for poisons, medicines, and jewellery. There is evidence of trade in exudates in ancient times, and doubtless various tribes gradually learned how to initiate the formation of exudates.

With the development of cutting tools, logs could be cut for construction and other purposes, and knowledge of the desirable features of heartwood was accumulated particularly with regard to appearance and durability. Queen Hatshepsut and other Egyptian royalty and nobles in the 15th century B.C. had furniture incorporating ebony and ivory overlaid with gold (Meiggs 1982). Ebony was used increasingly in high-class furniture in eastern Mediterranean countries, Persia imposed an annual tribute of 200 ebony logs when she conquered Egypt. Later Ptolemy Philadelphus (286–247 B.C.) staged a spectacular procession in Alexandria, Egypt, when Ethiopian gift-bearers carried 600 ivory tusks and 2000 ebony logs. Apparently at that time there were good supplies of ebony in Ethiopia and southern India. About 100 years later Pliny described the black heartwood as "close-grained", heavy and long lasting; and he also recognized the durability of cedar and the preservative properties of cedar oil (Meiggs 1982). Cedar from Lebanon was used to build the temple of Solomon in Jerusalem in the 10th century B.C. (II Chronicles, Chapter 2, verse 16).

Appreciation of the durability of heartwood was developing at the same time in oriental countries. The Horuji Temple was built in Japan during the Nara period (A.D. 710–784) from the durable wood of *Chamaecyparis obtusa* and it still stands (Kohara 1972). The Todaiji constructed in the same period was later to become the largest wooden building in the world. In China large logs of *Persea nan-mu* were used in the columns of Imperial palaces and buildings and some of those erected in the 15th century A.D. still exist.

1.2 Changing Uses of Forests

With the increase in refinements in living, and in material requirements, international trade in timber and exudates increased. At the beginning a major part of those exports were highly colored timbers or extracts of them, for the provision of dyestuffs. Dark-colored or figured heartwoods of tropical species were popular for luxury furniture. Exudates and wood and bark extracts from a wide variety of sources ("minor forest products") became an important facet of trade. Gradually the specialized timbers to provide masts and the hulls of large ships gave way to the more utilitarian needs of construction where low cost, assured supplies and sometimes quality were needed. Increasing amounts of wood were used for heating or for industrial purposes and for the provision of fiber for pulp and paper.

Expanding populations and industrial activities in some regions resulted in the destruction of prime forests, so that far-sighted people began to control their use and in the 16th century foresters began activities to manage and develop forests on a sustained yield basis in Europe and eventually elsewhere. The increased use of forests in tropical and semi-tropical countries yielded many examples of defects, such as pockets and shakes containing resins and compounds with a wide range of compositions, exudates of various types and heartwoods with characteristics different from those encounterd in the cooler climates of Europe and America. The study of components of extractives hastened the development of organic chemistry and then biochemistry and the disciplines which resulted from them (Hillis 1986). The use of wood extractives and exudates has recently declined considerably in favor of alternatives or synthetic materials which are cheaper, more effective, and more uniform in quality. Also in the period of low-cost energy and ample supplies, utilization problems caused by some heartwoods and by defects were avoided by technological developments such as pulping of resinous species to fiber by the kraft process or by forming large beams of required strength after cutting out defects and finger-jointing the short pieces followed by glue-lamination.

Increasing populations have resulted in inadequate supplies of wood products in an increasing number of regions. International trade in forest products exceeds US $50 billion and is increasing. More and more countries which were formerly large exporters are now importing forest products. Even with the current methods of improved forest utilization, there will be inadequate supplies to meet the demands of the rapidly increasing global population. The increasing demands must be met from increasing areas of plantations. Economic and environmental aspects will require many of these plantations to be established from fast-grown, exotic species to which forest science must give more attention, not only to their growth rate and to the form and habit of the tree, but also to heartwood and chemical aspects. The decreasing area of land per capita demands greater attention to increasing volume production of wood free from decay and defects but also to the production of wood most suited for processing into products in the greatest yield to meet consumers' needs.

The increasing need to use larger proportions of plantation trees as well as secondary species from tropical forests has led to changing qualities which introduce utilization problems due to heartwood, extractives, and exudates. For example, the impermeable nature of heartwood reduces the rate of penetration of preservatives and pulping liquors. Similarly, extractives can lower the quality of adhesive bonds or the quality of pulp and paper. On the other hand, heartwood can have the color, durability and stability required for high-quality products.

The following chapters consolidate much of the observations gained from different resources, many of which no longer exist, in order to provide a basis to assist the growth and utilization of future resources. It is not possible to present in this book all the previously observed details on heartwood and exudates which have been frequently minor observations in broader studies. Many of the details on exudates are in inaccessible journals or have been collected by Howes (1949, 1950) Tschirch and Stock (1935), Mantell et al. (1942). The details presented here are to clarify or substantiate various aspects leading to the scientific basis of the formation of heartwood and exudates.

Chapter 2

Definitions and Descriptions

2.1 Introduction

In cross-sections of freshly felled mature trees a number of zones can usually be observed, a pale-colored outer zone – the sapwood – and a dark-colored inner core that is often situated concentrically – the heartwood. Both these zones change in appearance on storage and always become darker, especially with some of the heartwoods of hardwood species. Between these major zones there may be a third zone, depending on the species, either an intermediate zone of wood of varying colors and width, or a transition zone paler in color than the surrounding tissues and narrow in width.

Cavities of a wide range of sizes may be present in the wood some of which may contain extracellular material. This material may also appear on the bark of the tree as an exudate.

2.2 Sapwood

In addition to the vertically oriented, long, thin fibers formed in the cambium, there are a number of other structural xylem elements. The rays extend from the outside of the stem inwards towards the center of the tree on the xylem side and outwards into the phloem on the bark side. The individual cells of these rays are bricklike in shape, very much shorter than the fibers, and are arranged end-to-end. Many species also contain vertical parenchyma cells of a similar structure. In the thin wall of each of the parenchyma cells are openings by which there can be contact between the individual ray cells and also between these cells and the neighboring fibers, vertical parenchyma cells, and in hardwoods the vessels. These openings or "pits" can have a variety of structures and sizes.

2.2.1 Definitions

2.2.1.1 Sapwood

In the living tree, the fibrous elements die soon after their development has been completed and a certain amount of cytoplasmic debris is left on the lumen wall. On the other hand, many of the ray and vertical parenchma cells remain alive for a considerable time in the outer, pale-colored annulus or the sapwood, which has been defined as "the portion of the wood that, in the living tree, contains living cells and reserve materials (e.g. starch)" (Anonymous 1957). In addition to reserve materials (which disappear at the heartwood periphery), other types of cell contents, both inorganic and organic, can be observed in the sapwood as well as the heartwood (see Chapter 5.2, 5.3, 5.4, 5.6).

Ziegler (1968) has drawn attention to the difference between the living and the physiologically active cells in sapwood. For example, there were water-conducting cells in the two outermost growth rings only of a 45-year-old *Quercus phellos* whereas there were living cells in 21 growth rings from the cambium. Part of the sapwood has a conducting function and part a storage function.

2.2.1.2 Included Sapwood

Included sapwood describes "wood included in the heartwood, having the appearance and properties of sapwood. Living cells are no longer present as in normal sapwood but reserve materials may remain" (Anonymous 1957). If an injury is of sufficient severity to destroy a portion of the cambium and to interrupt the files of ray parenchyma with the cambial zone in the living tree, then the sapwood situated posterior to that tissue will not be transformed to heartwood. The affected region is eventually overgrown. In surrounding regions, heartwood is formed, leaving an isolated area of dead sapwood cells (Fig. 2.1). Included sapwood has been noticed in a wide range of trees, and particularly in dark-colored hardwoods, damaged by mechanical injury, fire or by insects (e.g., Erdtman and Rennerfelt 1944, Dadswell and Hillis 1962, Hillis 1962b, McGinnes et al. 1969, McCaw 1983, Perry et al. 1985). Included sapwood has been noticed on the pith side of resin pockets in *Larix decidua*, *Picea* species, and *Juniperis virginiana* (Knuchel 1947, Hillis 1962b) and of eucalypt kino pockets. Included sapwood in the heartwood of *Schinopsis quebracho-colorado* contained the polyphenols of normal sapwood (Roux and Evelyn 1960). No evidence has been found of migration of heartwood polyphenols into the included sapwood of several species. The target ring pattern (Section 2.5.2.5.1) is a related phenomenon to included sapwood.

2.2.2 Area of Sapwood in Log Cross-Sections

The sapwood has three important functions in the living tree. "It conducts large quantities of sap, with gases and salts in solution, from the roots to the cambium and leaves, provides a strong and rigid stem which lifts the foliage to a position where it can secure the desired amounts of air and light, and serves as an important reservoir for the storage of food substances" (Hook et al. 1972).

The area of sapwood before it is transformed into heartwood varies considerably between families, genera, and species. For example, the average sapwood areas, as percentages of cross-sectional areas of trees with diameters between 64–77 cm, of *Larix occidentalis*, *Thuja plicata*, *Picea engelmannii*, *Pseudotsuga menziesii*, *Pinus contorta*, and *P. ponderosa* were 12, 16, 28, 29, 50, and 61 respectively (Dobie and Wright 1978). The number of growth rings of sapwood can be characteristic of some species after the tree has grown out of the juvenile stage and the number is often more consistent than the width, and for example in *Pinus*

a

b

Fig. 2.1.a. Cross-section of a *Eucalyptus* species showing included sapwood which resulted after patches of the cambium had been destroyed. The tissues which grew over the patches were subsequently transformed into heartwood in the normal manner (scale 5 cm). (Hillis 1962b). **b** Sector of *Callitris intratropica* with included sapwood, commencement of zone of included sapwood, and a trace of sandarac (*S*) at an injured region (scale 3 cm)

banksiana and *Larix laricina* there is no relationship between the two (K.C. Yang et al. 1985). In some species the number is dependent on the age of the tree, the environmental conditions and unknown causes. Trees with rapid growth rates can have the same number of rings of sapwood but greater width than slow-growing and suppressed trees. Open or moist sites as a result often support trees with wider sapwood than sites with crowded stands or drier soils.

There are tree species, relatively few in number compared with the total number of timber-producing genera, which do not form colored heartwood. Although they can grow to a considerable size, even up to one meter in diameter, they contain starch and living cells at the center of the tree. *Brachychiton acerifolius*, *Alstonia scholaris* (Chattaway 1952), *Alnus glutinosa*, *A. incana*, *Populus tremuloides*, *Aesculus hippocastanum*, *A. rubicanda*, *Carpinus betulus*, *Corylus avellana*, *Sambucus nigra* and *S. racemosa* (Büsgen and Münch 1929), and *Terminalia superba* (Bauch 1980) are examples.

However, all the trees in the one genus are not necessarily sapwood trees. *Betula alleghaniensis* and *B. lenta* form colored heartwoods but not *B. verrucosa*, *B. pubescens*, and *B. alba; Acer tartaricum* and *A. dasycarpum* contain true heartwood but not *A. pseudoplatanus*, *A. platanoides*, and *A. negundo* (Büsgen and Münch 1929, Trendelenburg and Mayer-Wegelin 1955).

Büsgen and Münch (1929) have listed species which do not normally form heartwood, but contain so-called "ripewood", which differs from sapwood in that it has a lower moisture content and that it contains small amounts of colored extractives (see Intermediate Wood, Section 2.3)

2.2.3 Variation of Area of Sapwood

2.2.3.1 In a Species

In some species the area and age of sapwood can vary considerably between trees. Several trees of old-growth *Sequoia sempervirens* had sapwoods containing a mean number of 137 growth rings and a mean width of 8.6 cm when the average growth ring width of the outer heartwood was 0.62 mm. Younger trees of the same species with ages between 61 and 88 years had about the same mean width (8.1 cm) of sapwood but a much smaller number of growth rings (43) when the average growth ring width of the outer heartwood was 2.53 mm (Anonymous 1981a). Heartwood begins to form in *Pinus sylvestris* of 25, 40, and 70 years age in southern, central, and northern Sweden respectively (Hägglund 1951). A close relationship was reputedly found between the age of heartwood inception and the mean maximum temperatures associated with these locations. A 230-year-old tree of the same species grown in Norway had a sapwood width of 3.5 cm containing 100 growth rings with the average ring width for the tree being 0.18 mm. A 48-year-old tree grown in Germany had a sapwood width of 6 cm containing 35 growth rings with average ring width being 0.46 mm (Bauch 1980). The number of sapwood rings in mature *Quercus robur* grown in different parts of Europe during the past 800 years ranged between 7 to 24 with the means (14 to 18) being different for different regions and lowest in Finland (14) (Baillie et al. 1985). Eckstein et al. (1986) obtained similar values for oaks from North Poland. The range of number of sapwood rings may be due to differences in tree age (see Section 2.2.3.3, Fletcher 1986).

The percentage of sapwood rings in a cross-section usually increases gradually as the top of the tree is approached. *Larix decidua* grown in Prussia and at widely different altitudes in Switzerland, and also *Picea abies* show this behavior (Trendelenburg and Mayer-Wegelin 1955). On the other hand the percentage of sapwood rings in several *Pinus* species of Europe and North America is less at 3–5 m above ground level than at any other level in the tree (see Section 2.5.2.2), and the rate of growth does not appear to affect this position. The minimum percentage area of sapwood occurs at a lower height in young trees than in old trees or trees grown on poorer quality sites. In 34 mature trees of *Abies grandis*, which contained about 40 growth rings of sapwood, the proportion of sapwood increased with height when the years of sapwood were expressed as a percentage of the age of the cross-sectional disc. When the sapwood area was expressed as a percentage of disc area, the proportion of sapwood was not affected by height (Puritch 1977).

It has long been known that the boundary between sapwood and heartwood in many species does not follow a particular annual ring either at different heights in the tree or in the same cross-section (Büsgen and Münch 1929). The sapwood width at breast height in *Pseudotsuga menziesii* increased with stem diameter, and this width remained relatively constant up the stem, although the number of annual rings in it decreased (Brix and Mitchell 1983). In a group of 54 *P. menziesii* trees with a range of sizes and average age of 85 years, the average sapwood characteristics were 21 growth rings and 3.3 cm width. However, the range of the number of growth rings in the sapwood of this group was between 5 and 44 (J.H.G. Smith et al. 1966).The number of rings of sapwood in *Juglans nigra* of a wide variety of ages, size, and growth rates was between 3 and 23 years and the width between 4 and 80 mm. In *Prunus serotina*, the corresponding figures were 3 to 34 and 4 to 44 respectively. There was a wide variation between and within sites, and the sapwood was more narrow with fewer growth rings in the higher positions than near the butt in trees of *J. nigra* and *P. serotina* (Nelson 1976). In another study of *J. nigra* (Szopa et al. 1980) the trees growing on the intermediate sites contained the lowest percentage overall of sapwood. Within sites the number of sapwood rings, sapwood width, and percentage sapwood were all significantly correlated with height position in the tree. However, the trees from the good site had faster growth and wider sapwood than trees on other sites, and at 0.3 m and 3 m height the percentage of sapwood was significantly greater in trees from good sites. As expected, the percentage number of sapwood rings in *Quercus* species increased as the height above ground level increased (Trendelenburg and Mayer-Wegelin 1955), although the actual number of sapwood rings remained about the same. The number of sapwood rings decreased slightly with increasing height in *Q. alba*, but wounding retarded heartwood formation (McGinnes and Shigo 1975). The widest part of the sapwood was found at the base of *Thuja plicata* (Wellwood and Jurazs 1968), *Pseudotsuga menziesii* (Wellwood 1955) *Pinus banksiana* and *Larix laricina* (K.C. Yang et al. 1985). The narrowest part was between 2 and 10 m. The average differences between these two parts in the latter two species were 84 percent and 58 percent respectively (K.C. Yang et al. 1985).

2.2.3.2 In Different Species

The amount of sapwood in mature trees of the same genus and of different genera can vary considerably.

The width of sapwood in trees belonging to the same genus is not always uniform. In the *Pinus* genus, *P. palustris* usually has a narrow sapwood, *P. monticola* and *P. echinata* wide sapwoods, and *P. caribaea*, *P. taeda*, and particularly *P. ponderosa*, have very wide sapwoods. *P.radiata* usually has 14 growth rings of sapwood, whereas *P.banksiana* has 33 on the average. At the same age, *P.sylvestris* has more sapwood rings than *Picea abies*, which in turn has more than *Larix decidua* (Trendelenburg and Mayer-Wegelin 1955) and *L.laricina* has on the average 19 sapwood rings (K.C. Yang et al. 1985).

There is also a wide variation in the number of sapwood rings in the hardwoods. For example, the number ranges from 1–2 in *Catalpa speciosa*, 3–4 in *Robinia pseudoacacia* and in *Castanea dentata*, 14 in *Quercus robur* and *Prunus avium*, 25–30 in *Magnolia acuminata*, 30–40 in *Acer saccharum*, and over 40 in *A.pseudoplantanus*, over 40 in *Carpinus betulus*, 30–50 in *Fraxinus excelsior*, 60–80 in *Betula lenta* and *Magnolia grandiflora*, and 80–100 in *Nyssa sylvatica* (Sargent 1926 and various sources). The width of sapwood in many *Eucalyptus* species grown in south-eastern Australia is about 1.5–3.5 cm and about 5 growth rings are present. However, there are exceptions between situations and species. More than 15 growth rings have been observed in 2.5 cm of sapwood in a sample of *E.delegatensis*, and *E. maculata* has a sapwood width of 4 cm or more, but eucalypts growing in semi-arid regions, such as *E. salubris*, have a width of 0.5 cm sapwood (see Section 2.2.3.4).

2.2.3.3 Due to Age of Tree

The width of sapwood in *Eucalyptus* species over 15 years old remains practically unchanged, as do the number of growth rings, throughout the rest of the life of the tree. The number of sapwood rings has been positively correlated with tree age in *Juglans nigra*, *Prunus serotina* (Nelson 1976), *Pinus banksiana*, *Larix laricina* (K.C. Yang et al. 1985), and *Pseudotsuga menziesii* (J.H.G. Smith et al. 1966). A curvilinear relationship has been found between the number of sapwood rings and tree age, but the width of the sapwood of these species is more or less constant at all tree ages. For example, the sapwoods in 9-year-old *J. nigra* averaged 4.2 growth rings with widths between 4.5 and 13.6 mm (Nelson et al. 1981).

The relationship for *Prunus serotina* has been expressed as $\log_{10}Y = 0.1414 + 0.5962 \log_{10}X$ ($r^2 = 0.68$) and for *Juglans nigra* $\log_{10}Y = 0.3023 + 0.5254 \log_{10}X$ ($r^2 = 0.64$) (Nelson 1976), where X represents tree age and Y the number of sapwood rings.

On the other hand, it has been found that the percentage volume of sapwood and number of growth rings decreases with age (e.g., Trendelenburg and Mayer-Wegelin 1955, Harris and Orman 1958), although the actual number of rings in the sapwood, and its width, may increase. This increase has been observed with dominant *Quercus* and *Picea* species (Trendelenburg and Mayer-Wegelin 1955) and *Pinus sylvestris* (Werberg 1929 as quoted by Dadswell and Hilis 1962), and has been attributed to the increase in the size of the crown of the tree (see also Section 2.2.3.4). Hollstein (see Trenard 1982) found the number of sapwood rings in the European *Quercus* species to vary from 16 to 26 according to the age of the trees from 50 to over 200 years.

2.2.3.4 Due to Rate of Growth

The width and volume of sapwood of several *Pinus* species and of a *Picea* species were greater in rapidly grown than in slow-grown trees (Hillis 1962b, Harris 1954a, MacKinney and Chaiken 1935). High correlations have been found between sapwood width and growth rate of *Pinus radiata* (Carrodus 1972, Bamber and Fukazawa 1985). Relationships between growth rate and sapwood width have been shown with *Picea abies*, *Abies alba*, and *Quercus sessiliflora* (Büsgen and Münch 1929). On the other hand, dominant *P.ponderosa* trees (grown in New Zealand) have a smaller volume (Harris and Kripas 1959), and dominant *Quercus* species have half the number of growth rings (Trendelenburg and Mayer-Wegelin 1955) of sapwood than suppressed trees of the same age. Still further contrasts between species are shown with *Pseudotsuga menziesii* (J.H.G. Smith et al. 1966, Brix and Mitchell 1983), *Thuja plicata* (Wellwood and Jurazs 1968), *Juglans nigra*, and *Prunus serotina* (Nelson 1976), where sapwood growth rate was found to vary more closely with sapwood width than with the number of sapwood rings, as in the behavior of *Pinus banksiana* and *Larix laricina*. With these last two species the growth rate was not reflected in the sapwood width, but the number of sapwood growth rings was strongly correlated with the sapwood growth rate (K.C. Yang et al. 1985). Some of these contrasts in behavior may be due to differences in the levels in the tree where the samples were measured.

Fast-grown eucalypts can have sapwoods of greater width than normal, but in several cases the number of growth rings present was still 4–6. The sapwood area in rapidly grown (1.5 rings cm^{-1}) 18-20-year-old *E. camaldulensis* in Israel increased from 31 to 47 percent of the cross-sectional area with increasing height, in comparison with an increase in values in slow-grown (3.2 rings cm^{-1}) trees of from 16 to 26 percent (Tischler 1976). In these and other specimens (Chudnoff 1961), the width of sapwood in a tree remained about the same, irrespective of the height above ground. The width (about 6.6 cm) of the sapwood in the faster-grown specimens was greater than that (about 2.5 cm) in the slow- grown ones, but the number of growth rings in both was about the same, namely 5 (Tischler 1976). *E.tereticornis* fast-grown in India showed a similar behavior (Purkayastha et al. 1980). Fast-grown *E. nitens* in Australia had about 3 growth rings and 3–4 cm width of sapwood (R.J. McKimm, personal communication). A sample of *E. dalrympleana* slow-grown at high altitudes had 29 growth rings in the sapwood. The sapwood in a number of *E.viminalis* and *E.globulus* that were unhealthy and had very small crowns and few leaves, was much thinner than normal (M.P. Dawson personal communication). On the other hand, the percentage of heartwood rings (relative to sapwood) is greater in fast-growing, low-altitude provenances

Table 2.1. Variations among crown classes in width of sapwood at different stem heights of *Pseudotsuga mensiesii*

Crown class	Average diameter at breast height (cm)	Mean width of sapwood (cm)		
		Top	Middle	Base
Dominant	29.0	3.81	3.12	4.39
Codominant	20.3	2.51	2.18	2.69
Intermediate	18.3	1.85	1.75	2.16

of *Larix decidua* than in slow-growing provenances in the high altitudes of Switzerland (Leibundgut 1983).

Crown class can influence the width of the sapwood with dominant trees having wider sapwood than lower crown classes, as shown in Table 2.1 (Wellwood 1955). This aspect is further discussed in the next section.

2.2.3.5 Due to Environment

As would be expected from the observations that a more rapid growth rate results in trees with a greater volume of sapwood, European workers have found that the more favorable the climate and the site, the greater is the percentage of sapwood rings in *Pinus sylvestris* (Trendelenburg and Mayer-Wegelin 1955) and *Pseudotsuga menziesii* (Chalk 1951).

In a critical assessment of the sapwood-heartwood relationships in second-growth *Pseudotsuga menziesii* (59–78 years old), Wellwood (1955) found that the quality of the site did not have a strong effect on the *width* of the sapwood band. However, the *areas* of sapwood in the middle and basal sections of trees grown on average sites were significantly greater than those in sections of trees on good sites, due to the smaller diameter of the former group of trees. Trees of the same crown class possessed the same width of sapwood for the two different site groups. In a further study (J.H.G Smith et al. 1966) on this species, diameter breast height, crown class, and crown width were found to be the most important tree variables affecting sapwood width, but tree crown class does not directly govern the thickness of sapwood. The association of crown class with sapwood thickness must derive primarily from the increased bole size, bark thickness, and growth rate, and bole size was found to be the most useful. It was found that the relationship between sapwood thickness (T) and diameter outside bark at breast height (D) and the rate of growth in the last decade (G) in inches could be expressed as $T = 0.21 + 0.070\,D + 0.73\,G$ (J.H.G Smith et al. 1966). Another relationship along the same lines has been found for the heartwood diameter in *Pinus radiata* (Hillis and Ditchburne 1974) (see Section 2.5.2.3.1)

Pilz (1907) stated that "the greater the water uptake of the tree, the lower the tendency of the living cells to form heartwood, as they are still needed to provide the crown with sap." This opinion is supported by the observations of Paul (1952), who found that the sapwood of several *Pinus* species was wider in open-grown trees with large crowns and rapid growth than in trees of the same species and same diameter but with smaller crowns and slower growth. The larger sapwood volumes in *Pinus*, *Abies*, and *Picea* species have also been correlated with larger quantities of foliage (Büsgen and Münch 1929). With *Pseudotsuga menziesii* Wellwood (1955) found that, for a particular site, those trees with a dominant crown class have a wider sapwood than the codominants and much wider than the intermediates. Wellwood considered that tree vigor was expressed more strongly by crown class than by site index.

Benic (1956) was able to derive a relationship between crown volume and sapwood area of *Fraxinus excelsior* by taking into consideration site class, crown class, and the height of the section. In his opinion, the age at which sapwood changes into heartwood depends on the crown volume/stem diameter ratio. With *Picea sitchensis* and *Pinus contorta*, evidence was found to support the view that the relationship between foliage area and sapwood area depends on the permeability of the sapwood and the local climate through its influence on transpiration rate (Whitehead et al. 1984). Close relationships have been found between sapwood cross-sectional area and foliage mass in *Abies amabilis, A.balsamea, A.lasiocarpa, A.procera, Acer pseudoplatanus, Picea engelmannii, P.rubens, Pinus contorta, P.ponderosa, Populus tremuloides, Pseudotsuga menziesii, Quercus alba, Q.coccinea, Q.velutina*, and *Tsuga heterophylla* (Dixon 1971, Grier and Waring 1974, Waring et al. 1977, 1982, Snell and Brown 1978, Rogers and Hinckley 1979, Long et al. 1981, Kaufmann and Troendle 1981, Marchand 1984) and to a lesser extent with *Eucalyptus viminalis, E.radiata*, and *E.obliqua* (Brack et al. 1985). The slopes of the relationships differ with site (Whitehead 1978) and species (Waring et al. 1982). The sapwood area (or width) appears to be best measured at breast height with *Pseudotsuga menziesii*. Proportional increases in foliage area and sapwood area of this species took place in the trees remaining after thinning so that the balance between demand and supply of water could be maintained, although different portions of the crown may have different influences (Brix and Mitchell 1983). A model based on data from a large number of *Pinus sylvestris* has shown the mean annual ring width of the sapwood varied with the needle biomass per unit of sapwood basal area (Albrektson 1984).

The greater width of sapwood in one of three *Pseudotsuga menziesii* trees was attributed to proximity to a river (Chalk 1951). Later Polge (1982) found with several provenances of the same species that irrigation resulted in a highly significant wider sapwood spread over more years. The greater amount of solar radiation received by one side of trees of *Pinus banksiana* and *Larix laricina* apparently resulted in sapwood on that side being 17 percent wider in the first species and 20 percent wider in the second, but in both cases there were fewer sapwood rings (K.C. Yang et al. 1985). However, with *Pseudotsuga menziesii* this relationship between number of rings and width of sapwood existed with dominant but not other trees (Polge 1982).

The width and proportional amount of sapwood in *Picea abies* affected by environmental pollution decrease with increasing incidence of disease. Furthermore, the moisture content of the sapwood in diseased trees was less than that in healthy trees (Frühwald et al. 1984).

2.2.4 Sapwood Contents

Primary metabolites are accumulated in the sapwood to maintain respiration and other processes when photosynthates cannot be supplied directly. Carbohydrates are the chief storage materials in many hardwoods, whereas "fats" or "resins" are the most common storage form in softwoods (see Table 2.2).

There is less information available on nitrogenous materials or sugars which may also be present. The amounts of all storage materials in sapwood can vary according to the season, the production of flowers and seeds, damage, and environmental effects (see Chapters 5.3 and 5.4). Polyphenolic materials are often present in the ray parenchyma of sapwood but in small amounts that increase towards the pith or heartwood boundary.

The accumulation of inorganic compounds can commence in sapwood. Calcium oxalate crystals occur in specially developed crystalliferous cells in both the vertical and ray parenchyma of sapwood and heartwood or they occur loosely in the parenchyma cells themselves. Calcium oxalate crystals are relatively rare in softwoods, but in hardwoods they can occur in a range of sizes from small particles to comparatively large crystals (Chattaway 1955a, 1956). Berry-like aggregates of silica are formed midway in the sapwood of *Syncarpia glomulifera* (Hillis and de Silva 1979).

Table 2.2. Classification of tree genera based on types of food reserve in pith and wood of young twigs and branches during winter[a]

Predominantly "fat"	Both starch and "fat"	Predominantly starch
Hardwoods		
Aesculus	*Betula* (some species)	*Acer*
Betula (some species)	*Juglans* (some species)	*Carpinus*
Catalpa	*Populus* (some species)	*Carya*
Cornus (some species)	*Prunus*	*Castanea*
Juglans (some species)	*Rhus* (most species)	*Celtis*
Populus (most species)	*Robinia*	*Cladrastis*
Tilia	*Salix*	*Cornus* (some species)
Rhus (some species)	*Sambucus*	*Crataegus*
Viburnum (some species)	*Viburnum* (some species)	*Diospyros*
		Fagus
Softwoods		*Fraxinus*
		Gleditsia
Pinus	*Abies*	*Hamamelis*
Picea	*Ginkgo*	*Ilex*
Pseudotsuga		*Liquidambar*
Tsuga		*Liriodendron*
Taxus		*Magnolia*
		Nyssa
		Plantanus
		Quercus
		Syringa
		Ulmus
		Zanthoxylum

[a] The data of Sinnott (1918) as quoted in Kramer and Kozlowski 1960

2.2.5 Discolored Sapwood

Living sapwood usually shows a dynamic response to injury, and discolored wood containing extractives is formed in a zone several mm wide around the area containing microorganisms. Heartwood, on the other hand, shows a passive response (Shigo and Hillis 1973). If the discolored wood formed in this Stage I is unable to resist invasion, the infection will spread further by successive microorganisms which will result in discoloration, and if the situation is suitable the processes will go to Stage III and decay. Discussion of the large and important studies of discoloration, discolored wood, and wetwood are outside the scope of the present work; references to those studies may be found in Shigo and Hillis (1973), Hillis (1977), and Bauch and Baas (1984).

2.3 Intermediate Wood

2.3.1 Description

Intermediate wood has been defined as the inner layers of the sapwood that are transitional between sapwood and heartwood in color and general character (Anonymous 1957) (see Fig. 2.2).

The existence of a zone ("ripewood") of different color between the sapwood and heartwood of an *Ulmus* species was recorded by Nördlinger in 1860. When present, as in *Picea abies*, *Acer campestre*, and *Abies alba*, the zone of intermediate wood is of variable width of about 5 to 10 cm in a log cross-section, and with a moisture content that is lower than that of the sapwood and is similar to that of the heartwood. As the amount of extractives is low, the intermediate wood is often paler in color than the adjacent zones. It can be confused with the narrow transition zone (q.v.) and the terms have been used interchangeably. The "transition zone" in *Tsuga heterophylla* (Barton and Gardner 1966) is probably intermediate wood. It was about 35 growth rings wide in one specimen, but less in others, and the site of brown stain formation on storage. Both intermediate wood and transition zone can be present, as in *Eucalyptus sideroxylon*. The "ripewood" which Büsgen and Münch (1929) described in different species that did not form dry heartwood normally and with moisture contents lower than the sapwood is probably the same as intermediate wood. The descriptions of these woods are so variable that a general definition is not possible.

It is uncertain whether all intermediate woods or ripewoods contain living cells. Chattaway (1952) found them in *Sloanea woollsii* and *Diospyros pentamera* but they are not present in European ripewoods (Frey-Wyssling and Bosshard 1959). Starch is absent in the intermediate wood of *Nothofagus cunninghamii* (Chattaway 1952).

The very gradual color change in cross-sections of *Acer saccharum* prevents the demarcation of heartwood or other zones. Furthermore, living cells containing starch begin to decrease gradually from 10–15 years after formation in the cambium and living cells can still be found in wood 115 years old (Good et al.

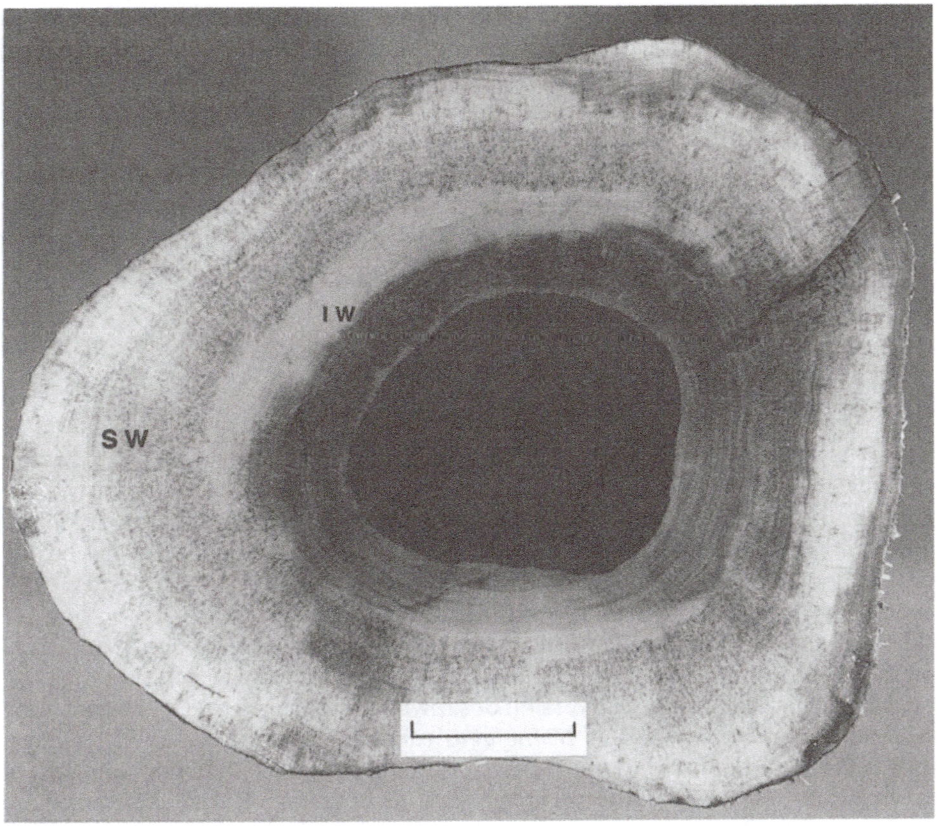

Fig. 2.2. Cross-section of a *Diospyros* species from Sulawesi one week after cutting. The sapwood (*SW*) is speckled with fungal growth. The wide intermediate wood (*IW*) has become pink to red, particularly in the inner portions adjacent to the intense black of the typical ebony heartwood (scale 5 cm)

1955). In one specimen of this species the moisture and mineral contents and pH were the same in wood 10–15 and 60–65 years after formation. The gradual increase in color occurred together with the death of the parenchyma cells and deposition of colored material.

2.3.2 Occurrence

Ring porous hardwoods can have wide intermediate woods (Yazawa and Ishida 1965b) and they have been observed in *Fraxinus*, *Ulmus*, *Kalopanax*, *Phellodendron*, *Morus*, *Tilia*, *Carpinus*, *Rhamnus*, *Salix*, and *Diospyros* species, *Abies alba*, *Acer campestre*, *Crataegus oxyacantha*, *Picea abies*, *Sorbus terminalis*, *Sloanea woollsii*, *Nothofagus cunninghamii*, and *Euonymus europaeus* (Büsgen and Münch 1929, Frey-Wyssling and Bosshard 1959, Chattaway 1952, Yazawa and Ishida

1965b). The ripewood zone in young *Larix leptolepis* is wider at ground level than elsewhere, and its width decreases up to a height of less than 0.5 cm at 0.3 m, where heartwood is formed (Hirai 1952).

The clearly delineated "intermediate wood" in *Tectona grandis* could be more accurately defined as "transition zone". In 15-year-old trees of about 15 cm diameter it was about 1.5 cm wide and, in contrast to the sapwood, free from starch, impermeable and containing tyloses; it was pale in color. A narrow impermeable zone was also observed surrounding the heartwood of *Manilkara multinervis* (Barnacle and Ampong 1974).

2.4 Transition Zone

Hartig (1894) described a pale, narrow zone, two growth rings wide, surrounding the heartwoods of a *Quercus* species and *Fagus sylvatica* and present only from mid-summer to winter before being transformed completely into heartwood. The first colored illustration of the "white" zone between sapwood and the brown heartwood of *Taxus baccata* (felled in late autumn) was provided by Craib (1923), who also showed that it had a lower moisture content than the surrounding tissues and that living ray cells were present. As "tannins" (polyphenols) appeared in that half of the zone adjacent to the heartwood, Craib suggested that the zone was a primary stage of the conversion of sapwood into heartwood. A narrow zone adjacent to the heartwood of *Nothofagus cunninghamii* was observed by Chattaway (1952), who found that it lacked starch but contained living ray and parenchyma cells. This zone was included in a description of wide "intermediate woods". Frey-Wyssling and Bosshard (1959) described a narrow "transition zone" surrounding heartwoods in which starch disappeared and phenols changed. Matsukuma et al. (1965), Sameshima et al. (1967) and Yazawa and Ishida (1965b) used the term "intermediate zone" and the latter researchers also used the term "white zone". Nobuchi et al. (1982) state that "the white-color zone" generally corresponds to intermediate wood. Although a white or pale-colored appearance is the most obvious feature (Fig. 2.3), "transition zone" is the most appropriate term. It has been used by Sandermann et al. (1967), who have provided colored illustrations of the zone in *Pinus* and *Picea* species, and by Harris (1954 a, b), who found living cells in the transition zone (about 60 mm thickness) of *Pinus radiata* in late winter to early summer.

2.4.1 Definition

The transition zone can be defined as a narrow, pale-colored zone surrounding some heartwoods and injured regions, often containing living cells, usually devoid of starch, often impermeable to liquids, with a moisture content lower than the sapwood and sometimes also than the heartwood (Fig. 2.4).

Fig. 2.3.a. Cross-section of a *Banksia* species with the light-colored transition zone (*TZ*) between the sapwood and a pink outer heartwood surrounding a dark inner heartwood. Part of another section (*left*) shows the transition zone at the periphery of a lesion (scale 10 cm).
b Dried cross-sections of *Callitris columellaris* showing transition zones (*TZ*) surrounding the heartwood of slow-grown (*left*) and fast-grown (*right*) specimens (diameter 18 cm). **c** Cross-section of *Eucalyptus camaldulensis* with a transition zone (*TZ*) surrounding a red-brown heartwood. The dark-colored band on the outer edge of the zone appeared on drying (scale 5 cm)

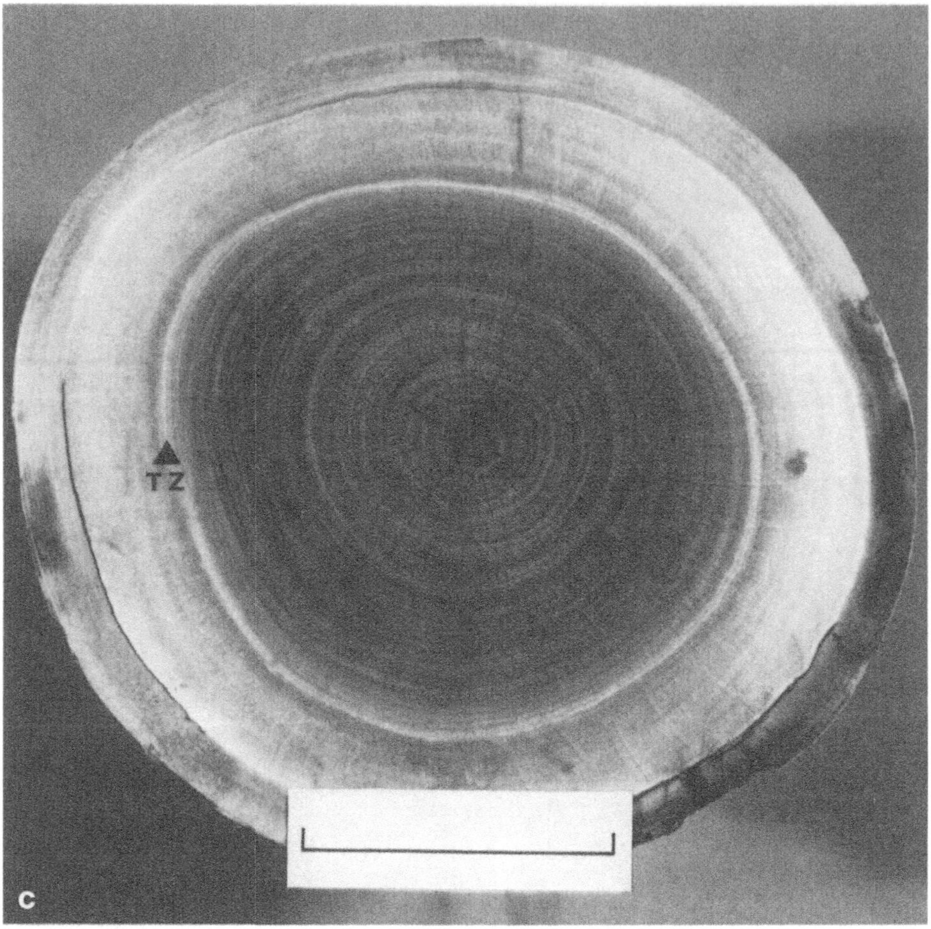

Fig. 2.3c. Legend see page 17

2.4.2 Description

The width of the transition zone, usually about 1–3 growth rings, can vary seasonally. It has a sharp boundary with both sapwood and heartwood. Samples taken at increasing heights in the tree reach a level where heartwood is no longer present, and only the transition zone and sapwood exist. When present, it is usually visible by eye, but in some species (e.g., *Juglans nigra*, Nelson 1976) magnification is necessary.

J.M. Harris (personal communication) observed an ephemeral region of transition zone up to about 60 mm thickness within the drywood zone of *Pinus radiata* which was present at all times. The transition zone of *P. radiata* grown in New Zealand first becomes evident in late winter at a point some distance up the tree,

Fig. 2.4. Part of a cross-section of *Quercus velutina* showing a blue-black stain (*arrow*) in the sapwood resulting from an injury about 2 cm from this section of the root. The products could not enter the impermeable transition zone (*TZ*) which is clearly delineated. (The stains in the heartwood were formed at its periphery. (Sachs et al. 1966)). (Photograph courtesy of J.C. Ward)

extends upward and downward in the tree during early spring and disappears in late spring and early summer, presumably with the formation of heartwood (Harris 1954a). With the same species grown in Australia, a dry transition zone was observed in all trees examined throughout the year; the differences between the two countries could be due to differences in site and climatic conditions (Shain and Mackay 1973a).

2.4.3 Occurrence

Although a transition zone has been observed in a range of coniferous species, it can be difficult or impossible to detect at times. It is less frequently observed in angiosperms; some with the most typical white transition zones belong to the family Rosaceae or to diffuse-porous wood species. It has been recorded in

Chamaecyparis obtusa, Larix leptolepis, Cryptomeria japonica, Prunus sargentii, P. maximowiczii, P. ssiori, Sorbus alnifolia, S. comixta, Syringa reticulata, Acer mono var. *glabrum, Cercidiphyllum japonicum, Taxus cuspidata, Quercus crispula, Ostrya japonica, Cornus controversa* (Yazawa and Ishida 1965b), *Magnolia obovata* (Fukazawa and Higuchi 1965), *Dacrydium cupressinum* (W.C. Ward and Reid 1949), *Larix laricina* (K.C. Yang et al. 1985) *L. leptolepis* (Hirai 1951), various *Pinus, Eucalyptus, Quercus, Callitris* species, *Sequoia sempervirens, Quercus robur*, and *Pseudotsuga menziesii.* In irrigated and nonirrigated specimens of various provenances of the latter species the transition zone had the same width of almost 1 growth ring (Polge 1982).

In *Taxus baccata* felled in late autumn the moisture content of the transition zone (50 percent) was less than that of the heartwood (100 percent) and of the sapwood (130 percent) (Craib 1923). Yazawa and Ishida (1965b) found a similar but less marked contrast with *Sorbus, Acer, Cornus, Magnolia, Acanthopanax, Octrya, Tilia*, and *Quercus* species. In *Pinus radiata* the moisture content of the transition zone is similar to that of the heartwood (about 40 percent on a dry weight basis) compared with 160 percent for the surrounding sapwood (Shain and Mackay 1973a).

Transition zones occur around mechanical injuries, as in *Sorbus alnifolia* (Yazawa and Ishida 1965b), or the wounds formed by *Sirex noctilio* in *Pinus radiata* and around the necrotic, phenol-enriched sapwood of *Picea abies* affected by *Fomes annosus.* Usually the transition zone is up to 3 mm wide in *P. radiata* and with a width of less than 1 growth ring, up to 7 mm wide in *Eucalyptus* species and a width of less than 2 growth rings and *Cryptomeria japonica* can have a width of 10–18 mm containing 4–10 growth rings (Nobuchi and Harada 1983).

2.4.3.1 Heartwood Boundary Stain

Three general types of dark discolorations have been described in the wood of living *Quercus* species (Sachs et al. 1966). One type is found in fully functional sapwood as a result of injury, a second is found in normally colored fully mature heartwood, and a third type of stain is formed during the transition of sapwood into heartwood. The specific gravity of the sapwood-heartwood boundary stain is greater than either the adjacent sapwood or heartwood and its ash content is lower than the adjacent sapwood, but there are more copious amounts of dark-colored substances than in normal heartwood. The distribution of these substances varies from completely surrounding the annual rings to others confined to one quadrant of the stem. Longitudinally, their length varies from one or two meters to the entire length of the stem connecting a large crown branch with a major root. Almost all the trees examined (Bulgrin and Ward 1968) which contained sapwood-heartwood boundary stain were found growing on moist soils which might even have been water-saturated for extended periods. Some stains at the heartwood boundary or transition zone were formed with neither bacteria nor fungi being detected in the discolored wood. It was considered that the general site characteristics, together with soil pH and soil texture, may be involved.

Similar stains have been observed in *Acer* and *Juglans* species (Good et al. 1955, Hart and McNabb 1963, Scheffer and Cowling 1966).

2.5 Heartwood

2.5.1 Definition

Most mature trees contain a central core of heartwood usually significantly darker in color than the surrounding sapwood or any intermediate wood, and often darkening considerably when the cut surface is exposed to air. The transformation from sapwood usually occurs abruptly over a few rows of cells. Heartwood is defined as "the inner layers of the wood, which, in the growing tree, have ceased to contain living cells, and in which the reserve materials (e.g., starch) have been removed or converted into heartwood substance" (Anonymous 1957). It is important to note that heartwood results from physiological death due to internal factors or the normal phenomena associated with the annual cycle of the life of the tree, and once the process has commenced it is spatially continuous. Similarly changed wood – but with different chemical and biological properties – can result from external factors (e.g., pathological death or injury) which occur independently of the season (Shigo and Hillis 1973).

Heartwood forms in the roots of *Picea abies*, *Larix leptolepis*, *Pinus sylvestris*, *Quercus robur*, *Fagus sylvatica*, *Populus canadensis*, and many other species. Only the root region near the stem wood is capable of forming heartwood and as a result of the same cytological and physiological factors as in the stem wood. The *Larix* species forms a dark heartwood, but several trees of *P. abies* up to 100 years of age contained only a ripewood in the roots. Some parenchyma cells retain their cytological and physiological activities in the heartwood (Böttcher and Liese 1975).

2.5.2 Description

2.5.2.1 Appearance

There are several groups of heartwood, but each has a wide variation in appearance and properties, so that a clear-cut subdivision of different species into different heartwood groups is not possible. For example, there are many groups of colored heartwood. When freshly cut, the color of various heartwood covers almost the entire spectrum, such as the jet-black (ebony) of a number of *Diospyros* species and *Dalbergia crassiflora* and *D. melanoxylon*, the violet-brown and blue of *Mansonia* species and some Amazonian species, purple of *Peltogyne pubescens*, red of *Dalbergia variabilis, Haematoxylon campechianum* and *Caesalpinia* species, yellow of *Chlorophora tinctoria, Rhus*, and *Artocarpus* species. Certainly the appearance of the heartwood is the simplest criterion to use, but there are difficulties with some species of the *Abies, Picea, Tsuga, Populus*, and other genera which can have an inner core of wood that is not always marked by a perceptible color change even though heartwood is present. The highly durable heartwood of *Dacrydium franklinii*, which contains colorless terpenoid compounds, is barely discernible from the sapwood when freshly cut. Accordingly, the differentiation of heartwoods based solely on color is inadequate for discussions on the physiology of formation.

Some heartwoods have a distinctive odor, such as sandalwood (*Santalum* species), camphor (*Cinnamomum* species), rosewood (*Dysoxylum fraseranum*), cypress (*Callitris* species), juniper (*Juniperus communis, J. virginiana*) cedar (*Heyderia decurrens*), raspberry jam wood (*Acacia acuminata*), or stinkwood (*Goupia tomentosa* and *Ocotea bullata*).

Although a terminology based on macroscopic features has its limitations, it is used here because of its convenience and common usage. At this stage there are uncertainties in the selection of the appropriate cytological characters which control the macroscopic changes so they are not used to define heartwoods.

Most trees possess a regular heartwood with a periphery which is similar in shape to that of the outside of the woody stem (Fig. 2.5). Heartwood which forms regularly (as in *Quercus robur*) is sometimes known as obligatory colored heartwood (Bosshard 1968). The heartwood usually has an undulatory outline, which does not correspond to the growth rings. Growth rings of sapwood do not change as entities into heartwood, and without apparent cause, the periphery of the heartwood may bulge radially or vertically into the sapwood (Fig. 2.6). The heartwood periphery may be irregular in shape at the base but regular in the stem, as in *Pinus sylvestris* (Werberg 1929; quoted by Dadswell and Hillis 1962). (Heart-

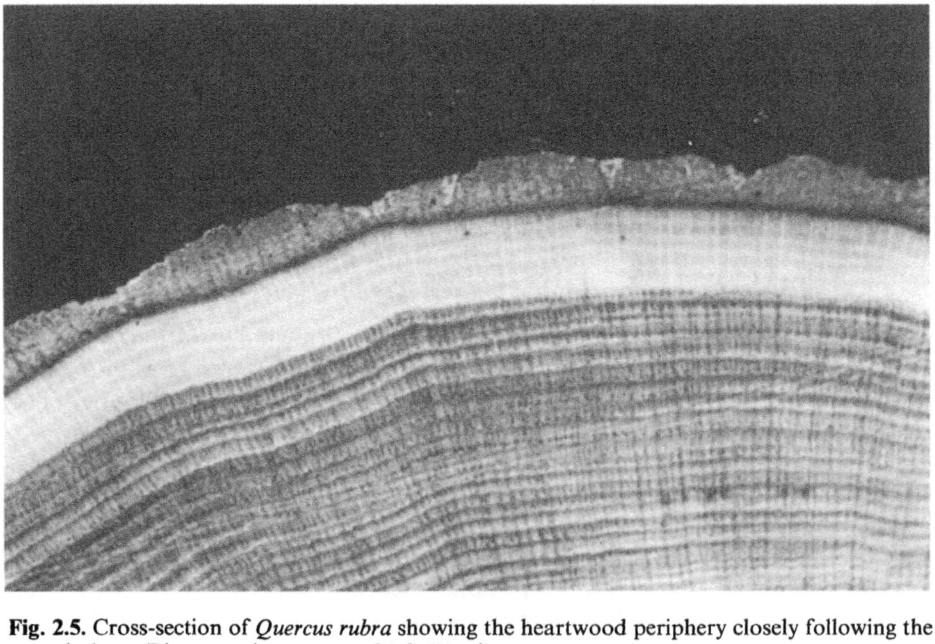

Fig. 2.5. Cross-section of *Quercus rubra* showing the heartwood periphery closely following the growth rings. (Photograph courtesy of J.C. Ward)

Fig. 2.6. Butt log of *Pinus pinaster* showing the undulating periphery of heartwood crossing several growth rings. The amount of extractives varies in the heartwood. (Length of ruler 46 cm)

wood which forms irregularly as in *Fraxinus excelsior* is sometimes known as facultatively colored heartwood (Bosshard 1968) (Section 2.5.2.4). Polyphenols infiltrate the cell wall of regular heartwoods, reducing the shrinkage and swelling capacity of the wood and increasing its durability and other properties.

The moisture content of the heartwoods of conifers is considerably lower than that of the sapwood, whereas the heartwood of some angiosperms contains more moisture than the sapwood, as in *Carya* species (W.R. Smith and Goebel 1952), *Fraxinus excelsior* (Bosshard 1955), and some *Eucalyptus* species (Stewart 1967). When dried, they are usually significantly darker in color and in those cases where the heartwood periphery cannot be distinguished, the application of stains (Chapter 5.8) can usually provide sufficient contrast. In *Pinus* species the resin which appears only on the cut surface of sapwood of freshly felled logs, such as those of *P. elliottii*, provides sufficient contrast at the heartwood periphery.

The heartwoods of the angiosperms are usually significantly darker than the sapwoods, and in some cases when the freshly felled wood dries the color becomes much darker. This is noticeable with the red heartwood *Eucalyptus* species and particularly with *Eusideroxylon zwageri*, which changes from bright yellow to light brown, reddish brown, and very dark brown. The outer heartwood (0.5–2 cm) of *Eucalyptus marginata* occasionally possesses a red color, in contrast to the red-brown of the inner heartwood (Hillis 1956). This band, containing extractives of much higher water solubility, darkens in sunlight to resemble the inner heartwood.

Whereas the color of gymnosperm heartwoods is usually relatively uniform, that of the angiosperms can have irregular bands of different color intensity with varying widths. These bands have been observed, for example, in *Betula alleghanienis*, *Olea europea*, *Nothofagus cunninghamii*, *Populus*, *Juglans*, *Diospyros*, *Pterocarpus*, *Berlinia*, and *Dracontomelon* species. Some bands have been traced or attributed to injury, resulting in abnormal wound parenchyma cells containing an excess of extractives which contrasts with normal tissues or those where cell differentiation has been arrested and small amounts of extractives formed (Phelps and McGinnes 1980a). [A detailed description of the measurements of color in heartwood has been reported by Phelps and McGinnes (1980b)]. Some eucalypts, such as *E. moluccana*, sometimes have a thin (2 mm) black band at the heartwood periphery.

2.5.2.2 Level of Maximum Area

Heartwood usually commences to form at about breast height, or within 1–3 m above ground level. Also, in mature trees the heartwood area, relative to disc area, at these levels can be greater than at ground level and examples have been reported with *Pinus* species (Pilz 1907, Harsch 1912, Paul 1930, 1932, 1952, MacKinney and Chaiken 1935, Lappi-Seppälä 1952, Trendelenburg and Mayer-Wegelin 1955, Hillis 1962b), *Pseudotsuga menziesii* (Smith et al. 1966, Wellwood 1955) *Thuja plicata* (Wellwood and Jurazs 1968), *Fagus orientalis*, *Fraxinus excelsior*, and *Larix leptolepis* (Hillis 1962b). However, the relative enlargement of heartwood at this level does not always occur, as has been found with specimens of *Pseudotsuga menziesii* (J.H.G. Smith et al. 1966) *Pinus taeda* (MacKinney and Chaiken 1935) and *Quercus* species (Trendelenburg and Mayer-Wegelin 1955).

In a study of *Pinus radiata*, the average of the heartwood area at zero and 2.4 m heights provided a more reliable basis for calculations than either area separately (Hillis and Ditchburne 1974). The heartwood tapers from the level of first initiation towards both the crown and the butt, but the extent of taper towards the apex differs with different trees.

2.5.2.3 Regular Heartwoods

Attention has been drawn to the variation of the width of sapwood within and between species (see Section 2.2.3). Other studies, notably on various *Pinus* species, have given attention to the converse, namely the age of initiation or volume of heartwood formation.

Heartwood first appeared in 50-67-year-old trees of *Pinus clausa*, *P. pungens*, *P. serotina*, and *P. virginiana* in 15 years or less, at 22 years with *P. rigida* (Koch 1972a), and in 13 years and 26 years respectively with *P. pentaphylla* and *P. densiflora* (Nobuchi et al. 1979). With *P. radiata*, heartwood often appears at 14 years of growth (Harris 1954a,b, Nicholls and Dadswell 1965); the most common transformation age is 14–18 years. However, in one group of 39-year-old *P. radiata* trees there were 29 growth rings of sapwood at ground level and 16 rings at the 20-m level (R.K. Bamber, personal communication). In one stand of 30-year-old trees, the transformation age of sapwood was between 14 and 28 years (Carrodus 1972) and up to 33 years of sapwood have been observed in suppressed trees. The age of first appearance of heartwood in a pine species can vary, as in *P. ponderosa*, from 36 years (Harris and Kripas 1959) to 150–200 years (Brown et al. 1949).

The age of the tree when sapwood tissues change to heartwood varies widely in different genera. In *Cryptomeria japonica* it is 6–8 years (Ihara 1972) and 12 years (Nobuchi et al. 1979), in *Robinia* species 3–4 years, *Nothofagus cunninghamii* about 50 years, 60–70 years in *Fraxinus excelsior*, and in *Alstonia scholaris* over 100 years. For a large proportion of the *Eucalyptus* species in their natural habitat, transformation to heartwood takes place when the sapwood width is 1.5–3.5 cm. Frequently, 5–7 growth rings are present in this width, although more rings may occur and up to 29 rings have been detected in *E. dalrympleana* growing at a high altitude. A regression analysis of heartwood appearance in samples of coppiced *E. viminalis* within the age limits of 5 and 20 years showed that it may be encountered between 3 and 4 years. For seed-propagated trees of this species it was at an age of less than 5 years (Nicholls and Phillips 1970). The influence of forestry practices on heartwood volume is discussed in Chapter 4.1.

2.5.2.3.1 Transformation and Age

It is often considered that there is a close association between the age of the sapwood tissues and the formation of heartwood in a species. Once the heartwood begins to form, there is a steady increase in its volume as the tree grows. In some cases, the heartwood increases at the same rate as the tree. Hirai (1951, 1952) found that the heartwood in young *Larix leptolepis* grown in Japan commenced to form in the 5th or 6th year; 35-year-old trees of the same species grown in Europe possessed 5 growth rings of sapwood (Trendelenburg and Mayer-Wegelin 1955). Whereas Hirai (1951) noted with this species that heartwood formation is not always periodic, annual, and uniform, Schober (1953) found that up to 60 years, the heartwood and new woody tissues formed at the same rate. Similar observations have been made with *Eucalyptus* species which form heartwood after 4–8 years of growth. Early studies of *Pinus sylvestris* growing in the Black Forest of Germany concluded that heartwood formation was a function of age of the tree and location, with the better locations yielding more heartwood (Harsch 1912). Later an investigation of over 2000 *P. sylvestris* growing in Alsace showed that the transition of sapwood to heartwood begins in trees of 20 years or more in age (in the Black Forest it is 30 years) and that there can be considerable variation in the number of sapwood rings in an age group (for example from 32 to 89 rings). However, once an average for the age group is established, it is constant within that age group on different sites (Pilz 1907). The transition age of heartwood formation at the butt is not representative of the whole bole. Transition begins in younger wood in the upper parts, and whereas heartwood can be 25 percent of the total rings at the base, it is 33 percent at 1 m height with maximum transition at about 6 m. Trendelenburg and Mayer-Wegelin (1955) found that data collected from nearby areas were similar and Werberg (1930; quoted by Dadswell and Hillis 1962) found that the pines from Latvia followed a different trend, although the general regression had the form: (number of rings in heartwood) = K log (tree age) + C.

Two regressions have been derived for the relationship between age and amount of heartwood expressed as a percentage of the disc diameter for *Thujopsis dolabrata* and *Chamaecyparis pisifera*, and as the tree age increases so does the heartwood percentage (Ito 1953). In *Pinus sylvestris*, *Picea abies*, *Pseudotsuga menziesii* and *Quercus* species (Dadswell and Hillis 1962), *Pinus taeda* (MacKinney and Chaiken 1935), and *Liquidambar styraciflua* (Hunter and Goggans 1968) the heartwood percentage often increased directly with the age of the tree. In some species, the amount of heartwood increased more slowly. The number of growth rings of sapwood in *Picea* species aged 95, 115, and 160 years were, respectively, 35, 40, and 48 (Trendelenburg and Mayer-Wegelin 1955). Also 20-year-old *Cryptomeria japonica* have about 9 growth rings of sapwood whereas 50-year-old trees have 18–22 growth rings.

A regression equation of the cross-sectional characteristics of *Pinus radiata* grown in different parts of Australia related the heartwood diameter at a given age of the tree to the diameter at breast height outside bark (d.b.h.o.b.) in cm and the product of the tree age(Age) and its diameter at 5 years (D5) in the following manner. Heartwood diameter = 0.1055 d.b.h.o.b. + 0.03114 Age x D5 – 2.83. A high degree of correspondence between observed and predicted heartwood diameter was found (Hillis and Ditchburne 1974).

The percentage heartwood by age in 34 mature trees of *Abies grandis* up to the age of 120 years was closely related to the age of the cross-sections when taken at the 3 and 7.6 m levels. The relationship was not as strong at base level but still significant (Puritch 1977). The relationship between percentage heartwood by area and disc area was less strong. When trees were infested by the balsam woolly aphid (*Adelges piceae*) the percentage heartwood was increased both in terms of age and area. The amount of heartwood formed was variable owing to different intensities of attack. Numerous investigations have all demonstrated that the balsam woolly aphid profoundly changes the anatomy and function not only of the outer bark and the phloem but also of the cambial zone and the ray parenchyma cells of the xylem. The entire literature has most recently been discussed and summarized by Timell (1986) in his monograph "Compression Wood in Gymnosperms," as well as the phenomenom of the formation of "premature heartwood," a well-known characteristic of *Abies* species infested by this aphid.

2.5.2.3.2 Transformation and Site

Contradictory conclusions have been drawn from various studies on the influence of site on heartwood formation. Early studies on *Pinus sylvestris* in the Black Forest (Germany) by Pilz (1907) and Harsch (1912) indicated that trees with broad canopies and growing in the warmer and drier regions delayed their heartwood formation. The site quality, the size of the crown, and the growth rate correlated negatively with the proportion of heartwood in the stem volume of *P. sylvestris* and *Picea abies* growing in Finland (Lappi-Seppälä 1952, Kärkkäinen 1972). The proportion of heartwood in *Liquidamber styraciflua* did not vary significantly with change in environment (Hunter and Goggans 1968). The maximum rate of increase of heartwood diameter in *Cryptomeria japonica* is attained with older trees on poorer sites (Ihara 1972). On the other hand, the factors conducive to heartwood formation in *Pinus radiata* are in general (a) soil conditions which favor the development of deep root systems; (b) a well-distributed and adequate rainfall, or other source of available ground water; (c) absence of drying winds or conditions of low humidity (Harris 1954b). In another study on the same species, the percentage heartwood area over disc area was higher on better than on poorer sites, but whereas the effect of site quality on heartwood diameter is noticeable at ground level it is negligible at 1–3 m (Hillis and Ditchburne 1974).

2.5.2.3.3 General Observations

The sapwood in a dominant or subdominant tree forms an outer band containing a relatively uniform number of growth rings which can be characteristic of the species although environmental conditions may modify the number. Even in trees of great age the number of sapwood rings is usually not different in the one species. The sapwood-heartwood boundary may undulate over a few growth rings, and occasionally small segments may change sharply in a radial direction. Heartwood is often widest at 2–10 m above ground level. Regular heartwood is relatively uniform in color, ranging from pale straw to dark red-brown, although the intensity often increases in the outer heartwood zones. The heartwood is not clearly differentiated in some species, and in others injury causes bands of different color intensity to be formed. Heartwood is drier than sapwood in conifers but not always in hardwoods.

2.5.2.4 Irregular Heartwoods

The irregular, sometimes called facultative, heartwoods found in trees such as in *Fraxinus excelsior*, *Olea europea*, *Fagus sylvatica*, and *Acer*, *Betula*, and *Carpinus* species can differ in many respects from regular heartwoods. During the life time of the tree the area of sapwood will change and usually decrease, the boundary with the heartwood usually moves sharply across several growth rings, the heartwood can be eccentrically placed and shaped, and it often possesses bands of varying color intensity, with dark streaks being common. The heartwoods of *Fa-*

gus sylvatica and *F. orientalis* usually commence to form at between 80 and 100 years [sometimes it has been 120 to 150 years (Nečesaný 1958a,b)], after which the diameter of the heartwood bears a linear relationship with the age of the tree (Trendelenburg and Mayer-Wegelin 1955). Although the heartwood of *Nothofagus cunninghamii* usually commences at about 50 years, it enlarges rapidly as the tree becomes older (Chattaway 1952). On the other hand, various cultivars of *Populus euramericana* and *P. maximowiczii* form dark-colored heartwoods and irregular peripheries at 3 to 5 growth rings from the cambium. The dark color of the heartwood is not due solely or partly to the action of fungi (Sachsse 1965) although a sharp increase in moisture content occurs on heartwood formation. Fungi can be present in the heartwood of *Fagus* species but are not responsible for its formation (Nečesaný 1958a,b), which is considered to be due to the entry of air which results first in the formation of tyloses in the vessels before the accumulation of extractives in the parenchyma and later in other cells. The extractives of irregular heartwoods line the cell walls of storage tissue as small drops or layers and do not penetrate them (Bosshard 1953, 1968), as they have been oxidized and polymerized so that their large molecular size prevents entry into the wall (see also Bauch 1980).

The discoloration and decay resulting from external injury to part of the sapwood can be readily detected. On the other hand, changes initiated internally (from branch stubs) may result in irregular heartwoods or may cause further changes in heartwood already present in the living tree or may result in color changes after felling. The wide range of appearance has resulted in names such as false heartwood, pathological heartwood, and frost heartwood, which have been used interchangeably.

2.5.2.4.1 False Heartwoods, Ripewood, and Black Heart

Büsgen and Münch (1929) stated that "false heartwood" formed extensively in ripewood species which do not normally form heartwood and especially in *Fagus* species and occasionally in *Betula*, *Populus*, and *Abies* species. Ripewood trees have been defined as those with light-colored heartwoods (Bosshard 1968). False heartwood can arise when branch wounds expose ripewood to the penetration of fungi which kill living cells and form colored components. Centrifugal development proceeds in stages, reputedly owing to the decrease in availability of oxygen, with the periphery appearing as a dark-brown border and containing tyloses filling the vessels to form irregular cloud-like figures. If radial cracks exist in the wood, the false heartwood follows these, so that stellate or star-shaped figures are formed. Most of the variations in color occur in concentric or crescent patterns around the pith.

The use of the term "black heart" hides a confusing situation in the terminology of various heartwoods. Black heart can be formed in a wide range of species such as *Ceratopetalum apetalum*, *Dracontomelon puberulum*, *Beilschmedia tawa*, *Gonystylus bancanus*, some *Diospyros* species, and J.M. Harris (personal communication) has observed it also in *Gymnelia*, *Alectryon*, and other genera. These black hearts form within uncolored heartwood and many of their physical prop-

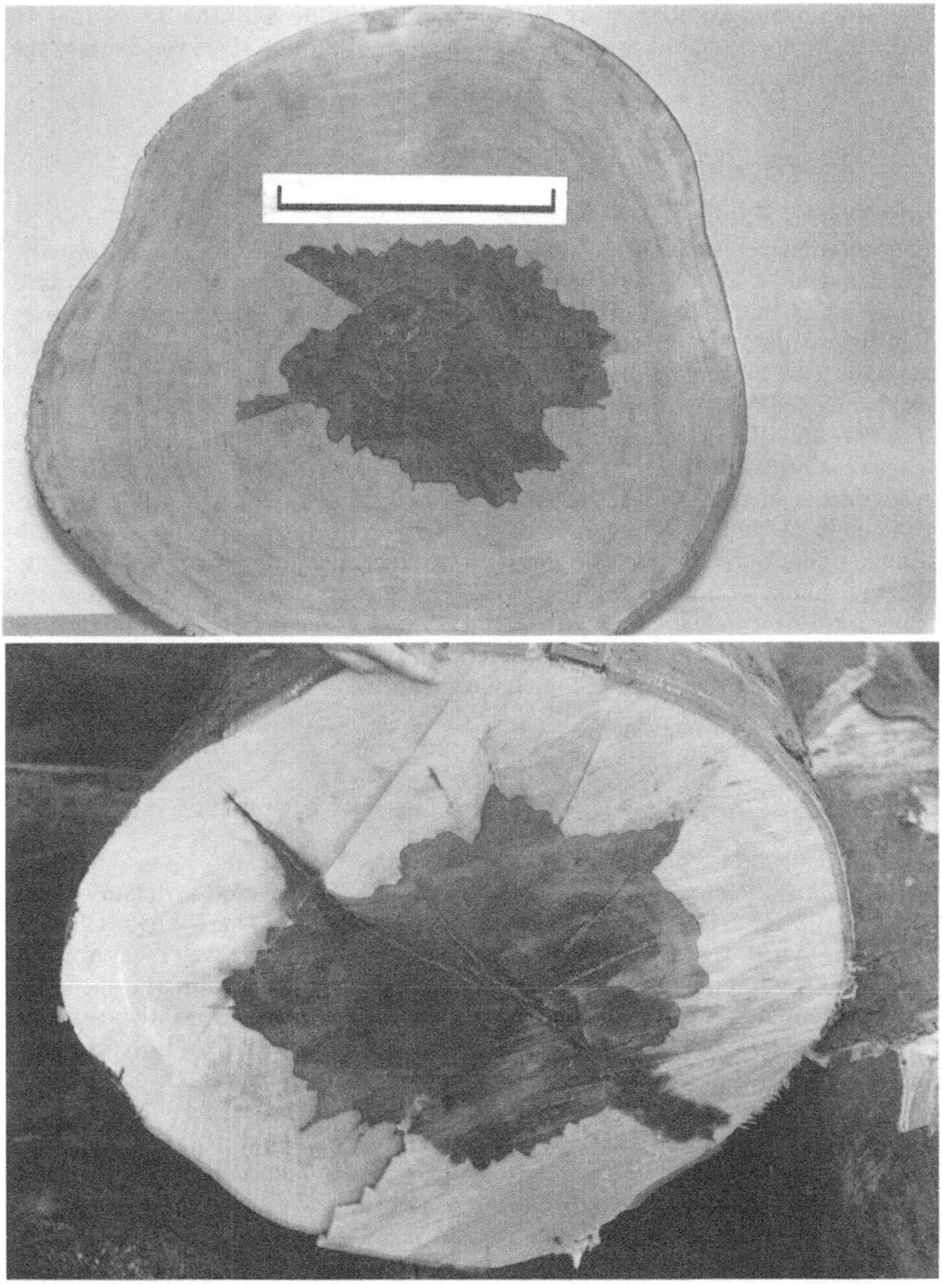

a

b

Fig. 2.7.a. Cross-section of *Ceratopetalum apetalum* with a black heart inside a lighter-colored heartwood and a narrow sapwood. Note intensity of darkness particularly at the periphery (scale 30 cm). **b** Cross-section of *Dracontomelon mangiferum* log with a black heart and cloudlike appearance of parts of the advancing area. The color is more intense at the periphery and some other regions with a favoring influence of a shake in the living tree

erties are unchanged as they are not associated with decay. Usually the margin of the black heart regions is darker than the rest, as though there were some type of self-limiting process in force (Fig. 2.7).

2.5.2.4.2 Red Heart

"Red heart is a discoloration of mature beech wood (*Fagus* species), caused by the formation of tyloses and by the enclosure of colored substances and is usually cloud shaped" (Erteld et al. 1964). Sound red heart does not affect the physical properties of the sawn timber (Erteld et al. 1964), its moisture content is only slightly less than that of the adjacent sapwood, and the term "false heartwood" has been used by some workers. A similar heartwood exists in *Nothofagus cunninghamii*. Red heartwood in *Fagus sylvatica* is characterized by the development of tyloses and of a brown coloration (Zycha 1948).

Nečesaný (1958a,b) also described a formation in beech as "composite hearts" (sometimes called mosaic hearts) or groups of small rounded heartwoods surrounded by a shell, which appears in cross-section through the trunk as a boundary line of darker color. He (Nečesaný 1968) also used biophysical techniques to distinguish regular and "false" heartwoods of *Quercus cerris*. The latter consisted of one or more irregular zones with distinct dark-colored border lines and the ray-like extremities had obvious connections with frost splits. He concluded that the dark borders were due to extra high concentrations of high-molecular-weight extractives in cell lumens.

2.5.2.4.3 Frost Heart

An abnormal and irregular heartwood formation, "frost heart," formed in *Fagus sylvatica, Acer* species, and *Fraxinus excelsior* after the severe European winters of 1928–1929 and 1941–1942, when the temperatures fell below −30° C (Liese 1930). Frost heart usually begins about 0.5 m above ground level; it can form in trees which do not normally form heartwood, and being concentric it can enclose any heartwood and within a few years change (in *Fagus* species) to the usual red heartwood (Erteld et al. 1964). In contrast to red heartwood, frost heart contains very few tyloses, but many appear at the border (see also Raunecker 1956, Nečesaný 1956, 1958a) of the zone and the bright colors of the freshly cut heartwood fade within a few hours. The moisture content is somewhat higher than that of red heartwood. Zycha (1948) and Nečesaný (1958a) consider that frost heartwood results from an abnormally strong penetration of air into the inner stem.

"Crescent" ring formation, which occurs more frequently, is also the result of severe frosts, but in this case, the area is not converted to heartwood.

2.5.2.4.4 Fungal and Pathological Heartwoods

Chattaway (1952) drew attention to heartwoods with irregular peripheries and coloration that also possessed fungal mycelia. In stem cross-sections from *Notho-*

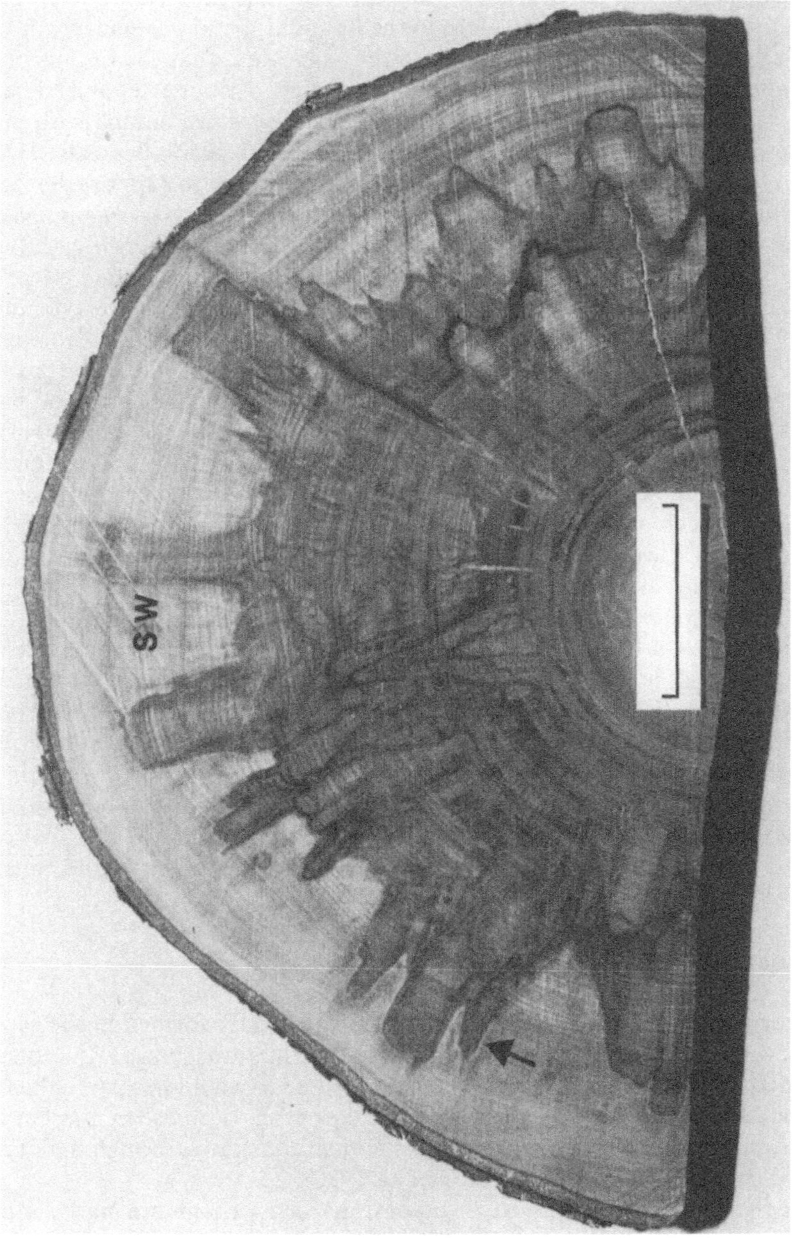

Fig. 2.8. Cross-section of *Nothofagus cunninghamii* with a red-brown irregular heartwood inside light-colored sapwood (*SW*). Traces of transition zones at the heartwood periphery can be seen (*arrow*) on the left (scale 5 cm)

fagus cunninghamii trees 120 and 350 years old, fungal hyphae were detected within the heartwood and very close to the boundary, and particularly in the thin black-brown boundary lines, leading to the suggestion that the fungus provided the stimulus for pathological heartwood formation. Also the "ebonized" areas in *Diospyros pentamera* contained fungal hyphae in the vessel lumina with the adjacent dead ray and parenchyma cells being filled with black deposits. However, Zycha (1948) and Nečesaný (1958a) found little evidence to support the view that fungi stimulated the formation of red and other heartwoods in *Fagus sylvatica*.

Heartwoods of *Fagus* species and *Nothofagus cunninghamii* (Fig. 2.8) can be similar in appearance when there are stellate hearts showing "flame-like" arms or sharp projections of dark lines (see also Erteld et al. 1964). This type of heartwood can contain a strong bacterial but limited fungal infection (von Aufsess et al. 1985). Nečesaný (1958a) subdivided the various heartwoods on the basis of the frequency of tyloses present. The irregular periphery of the heartwood in *Fagus* species follows a spiral-shaped course in a vertical direction (Zycha (1948).

Occasionally, trees which do not normally form heartwood contain areas with a dark color, high extractives content, and a most irregular shape, as in *Euroschinus falcatus*; such areas are frequently called pathological heartwood and sometimes protection wood (Jorgensen 1962). *Prunus domestica* (var. *Victoria*) occasionally contains an irregular heartwood which is associated with the fungus *Stereum* (now *Chondrostereum*) *purpureum*. Damage to *Rhus coriaria* can cause a similar heartwood formation. Narrow bands of dark-colored wood, which simulates heartwood, have been found frequently in injured sapwood, such as that of *Tectona grandis* (Büsgen and Münch 1929), and surround insect borer cavities in *Podocarpus* species. These tissues, which contain large amounts of extractives, are more properly classed as "reaction zones" such as that described by Shain (1971) in the response of *Picea abies* to infection by *Fomes annosus*. The formation of these zones depends on many factors including the genetics of the host tree, its age and vigor, the virulence of the pathogen, seasonal, climatic and nutritional factors.

2.5.2.4.5 Wound Wood

Wounds affect the anatomy of the tissues subsequently formed in the cambium, and several years can elapse before tissue formation returns to normal. In wounded *Acer* and *Betula* species the new tissues have a higher parenchyma content, particularly of rays. At the same time the vessels decrease in number and diameter, but the fiber proportion remains almost constant although the fiber walls become thinner (Bauch 1980, Bauch et al. 1980).

When branches are removed, or other transverse wounds are made, tyloses are formed in the vessels, if present, or resin is exuded and the wounds are blocked to hinder further development. Colored extractives can also be formed with the affected region extending for several centimeters. Wound-induced cores of discolored wood are found in a range of species, but the response of the tree to wounding varies with the species and with the tissue. Some investigators considered that because of the frequent similarity in color, wound-induced discolorations were an extension of normal heartwood into the sapwood (e.g., Büsgen and Münch 1929).

However, there are important differences between normal heartwood formed by internal stimuli and discoloration induced by wounding. The former continues to enlarge regularly and centrifugally but the latter does not; normal heartwood has a lower pH, moisture and mineral content than woundwood, which is also discontinuous and eccentrically placed.

Some species with symmetrical heartwoods occasionally contain isolated defects in the heartwood, such as "sinker stock" or wetwood, a slightly infected or partly decayed heartwood with an abnormally high water content. These wet pockets are often sharply defined and dry more slowly than the rest of the wood so that the moisture content in a nominally dry piece may vary within 1 or 2 mm from 15 to 80 per cent. The irregularly outlined "Nasskern" of *Abies alba* is a similar defect. Localized pink-colored zones are frequently in close proximity to "sinker stock" areas in large trees of *Araucaria hunsteinii* (syn. *A. klinkii*); they appear to be initial stages of bacterial decay in this species with the production of lignin degradation units (Hillis and Carle 1958a). A similar type of stain has been found in *Abies amabilis* (Barton and Smith 1971).

Two major types of "wetwood" have been proposed according to their origin of formation: (a) a sapwood-transition zone wetwood that is developed directly from sapwood, and (b) a wetwood that is derived from previously formed normal heartwood (J.C. Ward and Zeikus 1980). Wetwood in heartwood is often associated with discolored wood and is probably caused by wounds to the trunk or roots of conifers and hardwoods. Wetwood is often characterized by a darker color or a wetter appearance than normal heartwood. Some species (e.g., *Tsuga* and *Abies*) are easily susceptible to its formation. The important characteristic besides the increased water content is the presence of microorganisms. Some of the latter can produce a range of aliphatic acids and others, various gases (Bauch 1980, J.C. Ward and Zeikus 1980, Schink and Ward 1984, Passialis and Tsoumis 1984).

The formation and properties of discolored wood, wetwood, and related secondary changes entail primarily a consideration of microbial changes. The complex subject concerned with these aspects and those relating to changes in the xylem cells only occasionally involves extractives as a major aspect and is not considered in this treatise. It has received detailed attention from Shigo and Larson (1969), Schroeder and Kozlik (1972), Bauch et al. (1975), Shigo (1983, 1984), and other workers (Bauch and Baas 1984).

2.5.2.5 Variations in Appearance

2.5.2.5.1 Moon and Target Ring Patterns

"Moon rings are circular or sickle-shaped areas of heartwood that have not themselves been transformed into heartwood" (Erteld et al. 1964). In central Europe moon rings have been recorded in *Quercus petraea* and other oak species (Paclt 1953a, Trendelenburg and Mayer-Wegelin 1955, Dujesiefkern et al. 1984) and *Robinia* species (Erteld et al. 1964) and in North America in *Larix occidentalis* (Englerth and Scheffer 1955). There can be one or more concentric bands (of vari-

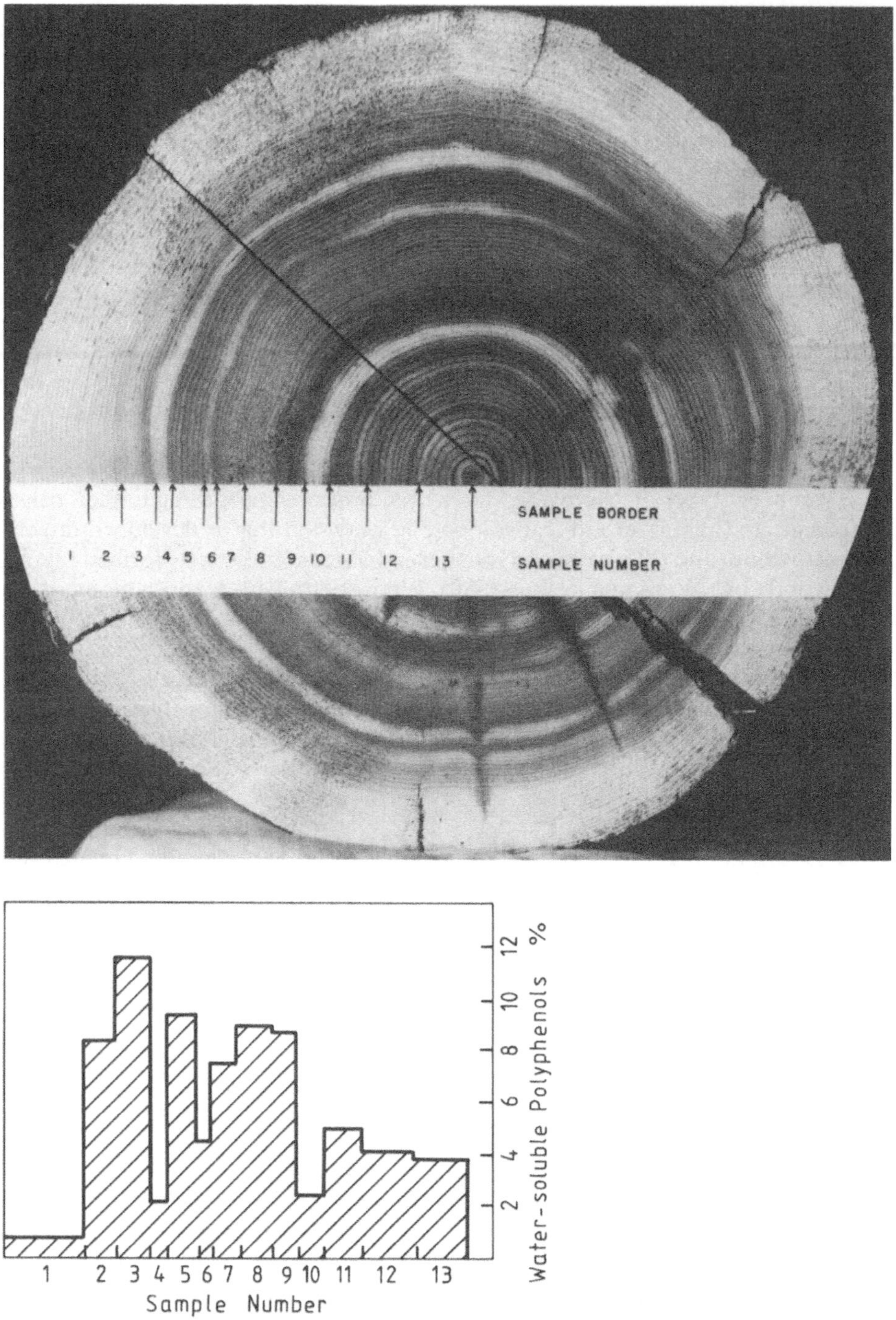

Fig. 2.9. Cross-section of a *Thuja plicata* log 30 cm diameter showing target ring pattern. The percentages of water-soluble polyphenols present in different rings are shown in the graph. (MacLean and Gardner 1958)

able width) of paler-colored wood in heartwood which for the most part follows the growth rings. This feature is known as "target ring" in *Thuja plicata* (MacLean and Gardner 1958) (Fig. 2.9) and *Pseudotsuga menziesii* (Kennedy and Wilson 1956) where the pale-colored rings contain extractives similar to the sapwood both in nature and amount. The cause of formation of moon rings of included sapwood has been attributed to severe frost when only those cells with a high sap concentration survive, and the remaining parenchyma cells are not subsequently converted to heartwood (Paclt 1953a, Trendelenburg and Mayer-Wegelin 1955, König 1957, Dadswell and Hillis 1962, Erteld et al. 1964). In moon rings caused by winter frosts, the rays extend through the damaged zone. However, in a specimen of *Acer pseudoplatanus* injured by heavy frost in early spring, and apparently at a time when the cambial cells were turgid but had not divided, the rays did not

Fig. 2.10. A concentric ring of included sapwood (*between arrowheads*) in *Eucalyptus marginata* (scale 9 inches, 23 cm)

run continuously through the injured zone, but gave rise to a number of parenchymatous cells with thick lignified walls (see also Glock 1951). As this species is a "sapwood" tree, the effect of the damage on heartwood formation cannot be seen.

The target ring pattern has been observed in *Eucalyptus marginata* (Fig. 2.10), *Tristania conferta*, *Syncarpia glomulifera*, and other species which do not grow in cold climates so that there are other, or more than one cause of this imperfection.

Crescent-shaped moon rings in oak have been found to be more resistant to fungal attack than sapwood (Lohwag 1968) and a similar behavior has been observed in *Thuja plicata*.

2.5.2.5.2 Color Differences

"Color-faults" are abnormal, usually darker, color tones in the wood. The color is of particular importance in good quality oak, maple, walnut, and ash (Erteld et al. 1964).

The heartwood of *Eucalyptus marginata* (and occasionally the heartwood of other eucalypts) contains a defect known as pencilling, which appears on the transverse face as thin (1 mm) black lines radiating outward to the sapwood-heartwood boundary. The color is due to deposits of black-colored material in the cells, and the formation appears to be due to *Fistulina hepatica* (Tamblyn 1937), which also causes the "brown oak" discoloration of English oak (Cartwright and Findlay 1936, von Pechmann and von Aufsess 1973). A similar but larger formation of flecks is found in the heartwood of *Entandrophragma utile* (Gottwald and Willeitner 1976).

"Black heart" is a light to dark brown discoloration in the central core of the trunk of trees with pale-colored heartwoods (Section 2.5.2.4.1). In maple, ash, and birch it is caused by a combination of heartwood deposits, oxidation, and the formation of tyloses (Erteld et al. 1964). In *Ceratopetalum apetalum*, *Beilschmiedia tawa*, *Atherosperma moschatum*, *Dracontomelon puberulum*, and in some *Diospyros* species this irregularly shaped zone of black heart is usually found in trees not less than 100 years old, and its origin is attributed to dead branch stubs and unsealed scars. The irregular coloration gives the appearance of overlapping black, green-black or grey-brown margins and slightly lighter-toned interior regions. Dried black heart is not markedly denser than normal heartwood or sapwood (although it can be wetter in the freshly cut wood) and the irregular outlines seldom extend into the sapwood region. When dried, the sawn wood is more prone to collapse and internal checking, and the veneers are more prone to buckle.

An association between the presence of wetwood and of brown or dark-colored streaks has been observed in some woods which are prized for their appearance, for example *Dacrydium cupressinum* (Reid 1956). The cause of the "mosaic-stained" heartwood in *Fagus sylvatica* has been attributed to cracks in the bark resulting from extreme temperature fluctuations during the winter when foliage was absent. These cracks penetrate to the cambium and initiate local coloring,

Fig. 2.11.a.

Fig. 2.11.a. Cross-section of a *Prunus* species with a brown inner heartwood, a red outer heartwood, a thin dark-brown band which appeared on drying, a transition zone which became less distinct on drying, and a sapwood zone with a low extractives content and pronounced shrinkage. **b** Cross-section of a log of *Eucalyptus marginata* showing sapwood (*SW*), an outer heartwood band (OHW) (red when freshly cut), a red-brown inner heartwood which became lighter in color towards the pith and contained several kino veins with included sapwood on the pith side (length of ruler 46 cm)

after which the surrounding tissues grow over the damaged cambium and leave a T-shaped discoloration in the cross-section (Bosshard 1965b). The oxygen tension present in the stem of species such as *Fagus sylvatica*, *Fraxinus excelsior*, and *Beilschmiedia tawa* mainly determines the depth of color of stained heartwood (Bosshard 1967).

The significantly darker color of the outer heartwood of some species is due to the larger amount of extractives present. The outer heartwood (0.5–2 cm wide) of mature *Eucalyptus marginata* occasionally possesses a red color in contrast to

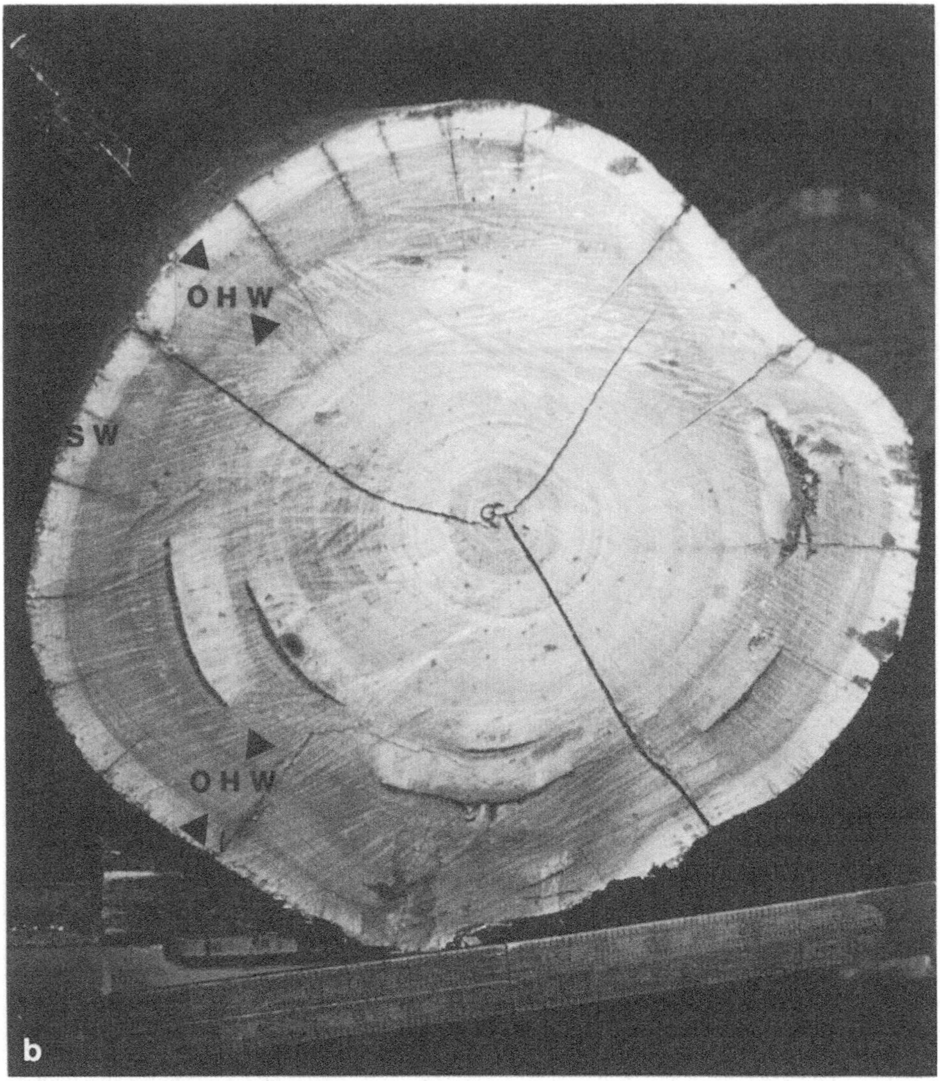

Fig. 2.11.b. Legend see page 37

the red-brown of the inner heartwood (Fig. 2.11). This feature occurs as distinct bands or lenticular shapes. Although the amounts of total extractives are similar in both regions, the proportion of water-soluble material in the outer heartwood zone is much higher (Hillis 1956).

2.6 Tree Exudates and Extracellular Materials

Exudates are formed by the stems and branches of a range of trees and shrubs. Although the basic biochemical processes for their formation are similar to those of heartwood extractives, there are significant differences in the anatomy of the tissues in which these processes are conducted. In some cases it is experimentally more convenient to study the formation of exudates than the formation of heartwood extractives. Also, although there are usually marked differences in the composition of exudates and extractives, conclusions concerning their biosynthesis are mutually supportive.

Due to a variety of reasons, such as the susceptibility of the tree to exudate formation on injury, the severity and season in which the injury was made, and the nature of the exudate, some exudates do not appear on the surface of the wood or bark. The exudates which remain within the tissues are more correctly defined as extracellular materials.

2.6.1 Definitions

2.6.1.1 Intercellular Secretory Spaces, Cavities and Shakes

Injury of the developing cambial cells, particularly in their most active periods of growth, of many softwoods and hardwoods results in the formation of traumatic intercellular secretory spaces. These spaces occasionally rupture and release their contents onto the bark where they harden into vitreous, and sometimes powdery, masses. When the intercellular cavities do not rupture they form "intercellular spaces of limited length, generally serving as a repository for resin, gum etc. and generally formed in response to injury to the living tree" (Anonymous 1957). When the injury is slight the resultant veins are traversed by parenchyma bridges and can be completely surrounded by heartwood when found in that zone. On the other hand, pockets can be large, are not traversed by parenchyma bridges, and on the pith side of the injury the sapwood is not transformed to heartwood. Whereas cambial damage results in sickle-shaped spaces along growth rings, the checks or "shakes" which also form in the living tree are usually in a radial direction.

Powdery and resinous extraneous substances are found in the shakes of a number of species. A shake has been defined as "a lengthwise separation of wood which usually occurs between and parallel to the growth layers." Shakes can be in two different configurations, the ring, cup or round shakes which are separations parallel to the growth layers and the heart or star shakes which are radial splits or cracks passing through the pith (Fig. 2.12). Shakes are formed in the living tree and can also be formed in the log due to felling damage or to growth stress. Extraneous substances are formed only in shakes of the living tree. Anatomical surveys have been made of the ring shakes in *Tsuga heterophylla, Abies* species, *Pseudotsuga menziesii, Juglans nigra* and *Quercus coccinea* (McGinnes 1968, Meyer and Leney 1968, Kandeel and McGinnes 1970, McGinnes et al. 1971).

Fig. 2.12. Portion of a log of *Intsia bijuga* showing a small radial shake (*s*) and yellow crystals of pure robinetin (*r*) in shakes (scale 5 cm)

Ring shakes formed in the standing tree occur as separations of latewood cells along the middle lamella. Little is known about the causes of the formation of shakes, and they may arise because of zones of tissue damage (Shigo 1972), traumatic resin ducts, wind action, or frost damage (Meyer and Leney 1968).

Some shakes are devoid of contents but there are a number of reports of extraneous substances present in shakes in *Tsuga heterophylla*, *Abies* species, *Pseudotsuga menziesii*, *Cardwellia sublimis*, *Dacrydium* and *Afzelia* species, *Intsia bijuga* (Fig. 2.12), *Eucalyptus maculata*, *Dryobalanops aromatica*, and *Thuja plicata*. Often these substances are concentrations, and even crystals, of one substance that in some cases may not be found elsewhere in the tree (see Section 2.6.2.9). The wood surrounding a ring shake in *Juglans nigra* contained larger than normal amounts of soluble extractives, and also calcium and other salts on the bark side of the shake (McGinnes et al. 1971).

2.6.1.2 Intercellular Canals

A second type of secretory intercellular space is the intercellular canal formed normally in the wood and defined as "a tubular intercellular space of intermediate length generally serving as a repository for resin, gum etc. secreted by the epithelium" (Anonymous 1957). The canal may be either axial (found in the latewood zones) or radial (within a ray) or both, and interconnected (e.g., Koch 1972c). When cut, these canals, which are often under pressure in the living tree, exude part of their contents onto the cut surface. This situation is often observed when the sapwood of living pine trees is cut and resin emerges onto the surface (the epithelium is dead in the heartwood). The resin has arisen from canals surrounded by epithelial cells which produce the resin and exude it into the canal.

Some tropical angiosperms, which contain large amounts of resin in their heartwoods, also exude resin onto the surface of boards and veneers and form flecks of different sizes. Some of these resins remain sticky for prolonged periods and cause problems in utilization. The species include *Prioria copaifera*, *Eperua falcata*, *Oxystigma oxyphyllum*, *Gossweilerodendron balsamiferum*, *Daniellia ogea*, *Tarrietia utilis*, *Guarea cedrata*, *Cedrela mexicana*, *Hopea odorata* (Ehlers and Sandermann 1961), and some species of *Sindora*, *Dipterocarpus*, *Shorea*, and *Dryobalanops*.

When secretory canals in the phloem are cut mechanically or punctured by insects they can yield latex of various types.

The various spaces mentioned above are formed within the tissues either schizogenously (when the space forms between cells), lysigenously (when the space results from the dissolution of cell walls and their contents) or schizolysigenously (when the duct is formed by both mechanisms, sometimes at different stages during its development). The ducts may anastomose to form extensive systems (see Chapter 6.5.2).

2.6.1.3 Pitch Tubules and Stones

Insects, such as bark beetles of the *Dendroctonus* genus, can bore beneath the bark onto the surface of the sapwood when they cause resin to exude from the living tree (such as *Pinus ponderosa*) and form "pitch tubules" on the surface of the bark (R.H. Smith 1966).

Sometimes "stone" or inorganic material is found in large lumps of tropical hardwoods such as *Chlorophora excelsa, Afzelia africana, Dalbergia latifolia, Millettia pendula*, and *Swietenia mahagani* (Sandermann et al. 1965, Jane 1970). These stones, largely calcium carbonate, can be large (about 13 x 17 cm, Jane 1970) and weigh up to 470 g; there have been reports of specimens larger than 50 kg (Sandermann et al. 1965, Sandermann 1966). On one occasion the stone in *Chlorophora excelsa* was accompanied by crystals of calcium malate (Farmer and Campbell 1951, Sandermann et al. 1965). Earlier workers and Campbell and McGowan (1939) have reported streaks of calcium hydrogen phosphate in insect bore-holes in *Tectona grandis* (see also Sandermann et al. 1965).

2.6.1.4 Pockets, Veins, and Streaks

Pockets and veins containing extracellular material form as a result of injury to the cambium either by fire, mechanical or by biological agents. Pockets are crescent-shaped cavities sometimes of large size and filled with liquid or solid extraneous materials in amounts of two liters or more. The usual size is about 0.5 cm in the radial direction, several cm in the tangential direction parallel to the growth increments and proportionally larger in the vertical direction. These defects are found in *Pseudotsuga menziesii, Picea, Larix*, and *Pinus* species where they are known as pitch or resin pockets, containing yellow resinous substances. Pitch streaks develop through the accumulation of resin in abnormal amounts in localized regions of *Pinus* species. Pockets and streaks can be extensive in *Pseudotsuga menziesii* after attack by the Douglas-fir beetle (Belluschi et al. 1965). In these regions the resin fills the lumina of various cells and permeates the cell wall so that the wood takes on a resin-soaked appearance.

Three distinct types of resin pocket defect have been recognized in *Pinus radiata* (Somerville 1980). The first type has been defined as the well-known, radially narrow discontinuity in the wood, oval in the tangential-longitudinal plane and filled with free resin and callus tissue. The second type originates, as does the first type, in the cambial region, but then expands radially when it fills with free resin under pressure. A rupture may then occur through the cambium, and external bleeding results. The whole defect, including any retarded cambium, is occluded by surrounding healthy cambium, and an occlusion scar is formed. The third type of defect, which is appreciably smaller than the other two, is a narrow, longitudinally oriented break in the continuity of the wood, filled with dry resin and parenchymatous tissue. The defect originates as a lesion in the cambial zone, and a rupture results in external resin bleeding. As with the second type, the surrounding healthy cambium occludes the injured region.

Fig. 2.13. Bulge containing a kino pocket in the stem of a *Eucalyptus calophylla* from which over 5 liters of kino flowed at the axe cut. Portion of a kino pocket in *E. goniocalyx* (*right*) shows the callused surface on the bark side (length of ruler 38 cm)

In *Eucalyptus* species gum, or more correctly kino, pockets are formed containing polyphenols in amounts of up to 10 liters or more (Fig. 2.13). The kino veins formed in eucalypts are crossed by parenchyma bridges and are rarely wider than 3 mm (Fig. 2.14). They are formed by the lysigenous breakdown of parenchyma bands produced by the cambium. Species belonging to three sections of the eucalypt sub-genus *Symphyomyrtus* form kino veins in the phloem (and are then shed in the bark). Species in the *Monocalyptus* and *Corymbia* sub-genera form veins in the xylem (Tippett 1986) (Fig. 2.15).

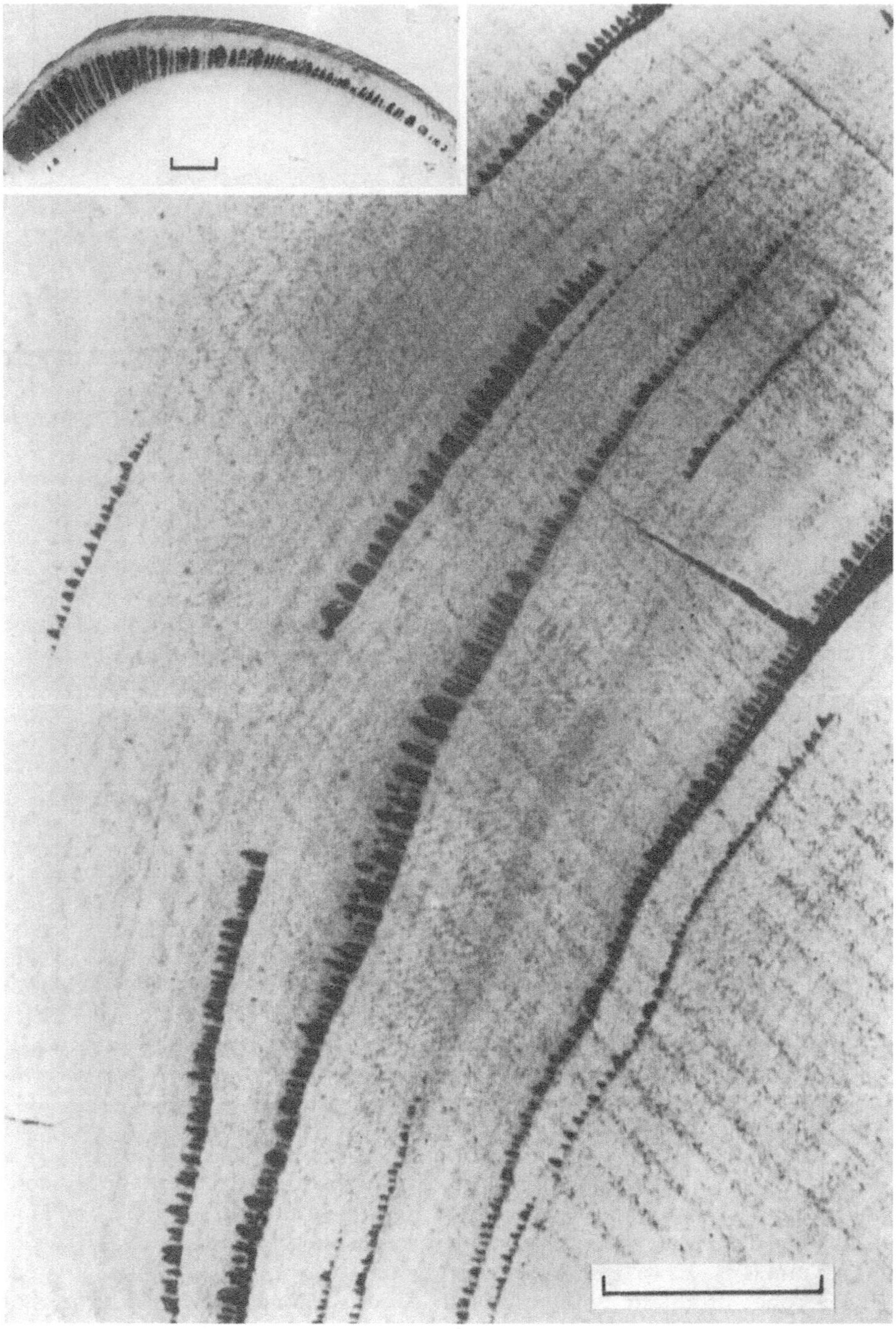

Fig. 2.14. Kino veins in an end-section of a log from a *Eucalyptus sieberi* injured in frequent fires (scale 5 cm). Inset shows in the same species a vein which enlarged to a pocket (scale 1 cm)

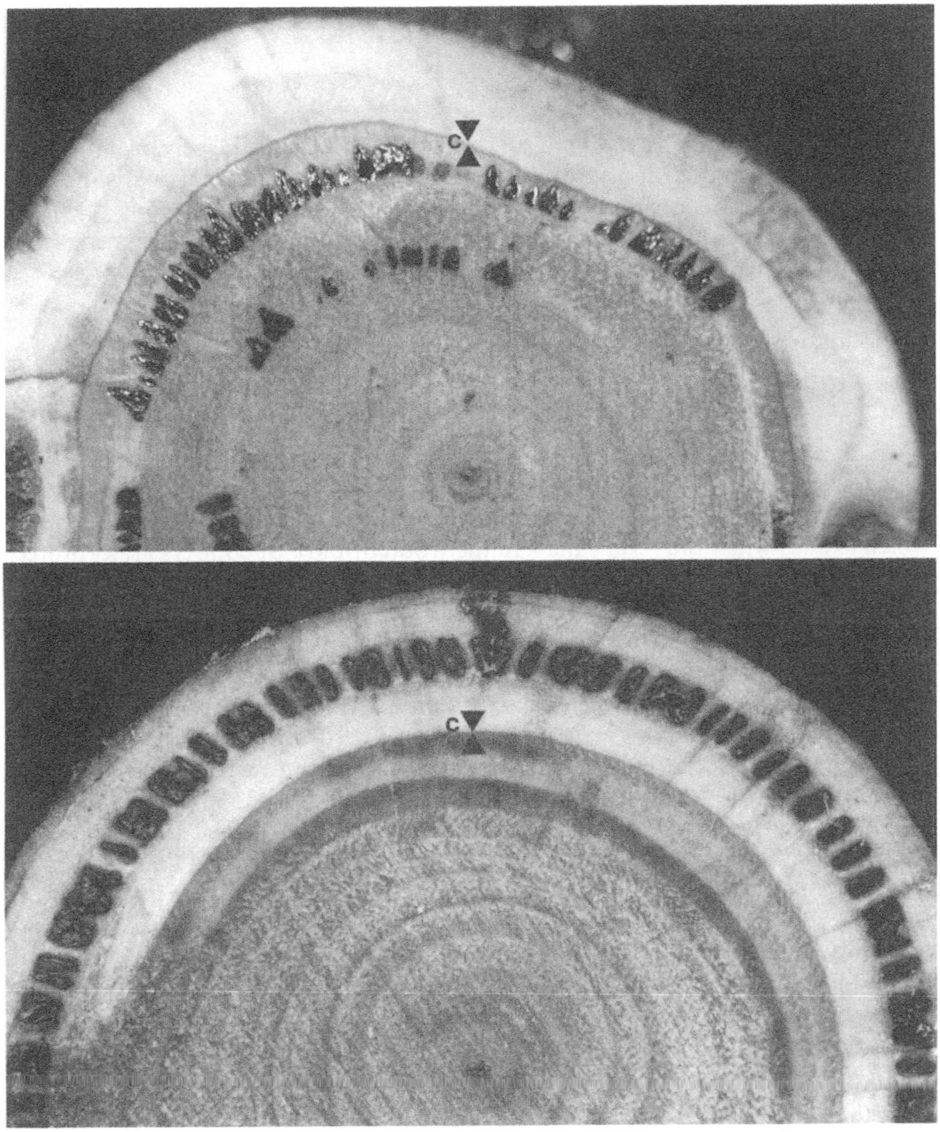

Fig. 2.15. Kino veins in the wood of a stem (diameter 10 cm) of *Eucalyptus camaldulensis* (*top*) and in the phloem of a stem (diameter 6 cm) of *E. wandoo* (bottom); c = cambial zone. (Photographs courtesy of J.T. Tippett and T. Hill)

2.6.2 Types of Exudate or Deposit

Some of the classifications of exudates were made before rigorous chemical tests were available or before satisfactory botanical stains were developed. A number of these descriptions still remain, such as the haphazard use of "resins" and "gums". Precise classification of exudates will enable greater understanding of their formation.

A notable feature with many exudates is that their composition differs markedly from the secondary components in uninjured cambium or from the wood and bark extractives. For example, all the eucalypt kinos are largely proanthocyanidin polymers in addition to minor amounts of flavanoids or lignans, none of which occur in the surrounding tissues. Injury of the cambial regions of *Acacia* and *Prunus* species also results in the formation of complex polysaccharide gums not found elsewhere in the tree. In another example, the resin from *Agathis* species differs in composition from the extractives of the wood.

2.6.2.1 Resin

There is no satisfactory classification for the various types of resinous exudates, which vary greatly among themselves. Broadly they are of two major types namely (i), oleoresin, which is essentially a solution of resin acids in turpentine, which exudes when the resin canals in the wood or bark of conifers are ruptured, and (ii), "natural resins" appearing on the bark or in cavities of various genera of the families Dipterocarpaceae and the Leguminosae/Caesalpinaceae.

Oleoresin may be defined as hydrophobic substances soluble in neutral, nonpolar organic solvents such as dry ethyl ether, hexane, and other petroleum solvents. It is a pale yellow, clear and sticky mass, which on evaporation becomes brittle. Oleoresin contains mainly terpenoids, hydrocarbons, and neutral compounds and on distillation produces gum rosin and gum turpentine in ratios between 4:1 and 6:1. Most of this type of resin is found in the intercellular canals of conifers native to the northern hemisphere. Because of their historic use for ship caulking, various products from pine oleoresin are termed "naval stores." They are obtained from three sources and the proportion of their production in the USA is about 2:25:73. (a) Gum naval stores. Various living *Pinus* species are tapped (wounded) and oleoresin collected, from which a high quality gum rosin is produced on steam distillation which removes the turpentine. Oleoresin is obtained mainly from *Pinus palustris* and *P.elliottii* (in the USA), *P.pinaster* and *P.halepensis* (France, Italy, Portugal, Spain, Greece), *P.roxburghii* (India, Pakistan), *P.massoniana* and *P.tabulaeformis* (China), *P.caribeae* var. *hondurensis* and *P.oocarpa* (Central America). The production of these stores has markedly declined in the USA in recent years but has increased considerably in China, Brazil, and Portugal. Gum naval stores are still the major source of the world's supply of rosin and turpentine. (b) Wood naval stores. Obtained by the solvent extraction of southern pine stumps from old trees which contain resin-soaked heartwood. The resin is less pure than that obtained by tapping. (c) Kraft naval stores. Turpentine and wood resin are obtained as byproducts from the kraft (sulfate) pulping of *Pinus* species. Turpentine is removed from the chip digester during the initial steaming and condensed from the relief gases. Sulfate turpentine often contains 60 to 80 percent α-pinene, 20 to 30 percent β-pinene, with the remainder being other terpenes. The proportion of β-pinene is much higher in *P.radiata* than in other pines. Crude tall oil obtained from the alkaline waste liquors is later fractionated into various products including tall oil rosin, tall oil, and fatty acids. Their production has increased considerably recently. The fractionating capacity in the USA is nearly three times that of the rest of the world and is about one million tons (Zinkel 1981, Stubbs et al. 1984) with production exceeding 90 000 liters of turpentine, 200 000 tons rosin and 187 000 tons of fatty acids.

Shakes radiating from the center of the tree, particularly with *Pinus elliottii*, can become filled with resin. Older trees appear to be more affected than younger ones and more so at the base of the tree. Trees exposed to high winds are more affected by resin accumulations than those that are protected, and those with low wood density are most affected (Clifton 1969, Cown 1973, Malan and de Villiers 1976, van der Sijde 1976). Stands that were thinned early in their rotation are more affected than those thinned later or not at all. In abnormal situations, the resin concentration in radiata pine may exceed 30 percent (Harris 1963). There is evidence that some form of wounding in the roots or root collar zone is the precursor of resin problems in older trees (Cown 1973, van der Sijde 1976).

The conifer *Abies balsamea* and other *Abies* species possess resin canals in the bark which sometimes form blisters. When the latter are punctured, a transparent and viscous oleoresin exudes, containing about 70 to 80 percent resin of sesquiterpenoid hydrocarbons and other terpenoids dissolved mainly in α- and β-pinene, β-phellandrene, camphene, limonene, and bornyl acetate (Zavarin and Snajberk 1965, Zavarin 1968, Smedman et al. 1969). It is known as Canada balsam. This is a misleading term as "balsam" (see Section 2.6.2.7) is characterized by the presence of benzoic and cinnamic acids, which are both absent from Canada balsam.

Oleoresin ("Oregon balsam") can also be found in the cavities or windshakes in *Pseudotsuga menziesii* with a pale green-yellow color and containing about one-third volatile components (mainly α- and β-pinenes) and resins. Although 2 gallons can be recovered normally from internal splits in one year, as much as 120 gallons have been obtained (Kurth 1952). The oleoresin obtained by slashing *Canarium luzonicum* of the Philippines is known as elemi (Howes 1949). Sandarac resins containing largely pimaric acid and pinenes exude from other conifers, namely *Tetraclinus articulata* (Howes 1949) and various *Callitris* species (Maiden 1894, Baker and Smith 1910).

"Natural" resins are more or less soluble in organic solvents and vegetable oils, totally insoluble in water, and chemically related to terpenes. They can be colorless, as with damars, or light brown glossy solids. They are stable, inert and amorphous, become sticky when heated and are fusible (above 100° C) with no sharp melting points. Hardness is one of the most noticeable as well as one of the most important properties, as with damars and copals, although some, such as the elemis, are soft. Their composition varies considerably but they consist essentially of organic acids and mainly resenes of low oxygen content (Tschirch and Stock 1933–1936). When heated, they melt with distillation of volatile oils as the temperature is increased, and unlike gums they do not carbonize or produce char when destructively distilled. By thermal processes the resins may be converted into materials soluble in hot vegetable oils to form varnishes. Formerly many of the resins were dug from the soil when they contained little essential or volatile oils and were harder than those obtained from the surface of the tree.

The amount of resin obtained on injury varies, but 900 g has been collected. It is not known whether resin exudates are products of normal metabolism or of a pathological condition, although usually exudation does not commence until the tree is injured. Wounding by beetles stimulates production from *Araucaria* species (Howes 1949), and wounding with acid causes *Pinus* species to form resin (A.B. Anderson 1955). A zinc deficiency in the soil is responsible for the resin exudation around the buds of *Pinus radiata* (Kessel and Stoate 1938).

Resins of one type or another have been formed in notable quantities by trees belonging to the following genera: *Abies, Agathis, Amphimas, Araucaria, Balanocarpus, Bursera, Callitris, Calophyllum, Canarium, Copaifera, Dacryodes, Daniellia, Dialium, Dipterocarpus, Gossweilerodendron, Hopea, Hymenaea, Larix, Olea, Picea, Pinus, Protium, Pterygopodium, Pterocarpus, Shorea, Styrax, Symphonia, Tetraclinis, and Trachylobium* (Howes 1949).

The damars (or dammar in commerce) from the Dipterocarpaceae are pale-colored and transparent, have low acid values and are soluble in alcohol, turpentine, coal tar hydrocarbons, and vegetable oils. The copals and manilas (from *Agathis* and *Copaifera* species) are harder, have higher acid numbers, and are alcohol-soluble. Kauri (from *Agathis australis*) is hard and oil-soluble, and a variety of other minor resins have different properties. The damars are largely obtained in the Indo-Malaysian region from a variety of species including *Hopea dryobalanoides* and other *Hopea*, *Shorea*, *Balanocarpus*, *Vatica*, and *Vateria* species (Hedley 1936, Browne 1955). Lumps of damar were once readily available from the soil in this region and were sometimes washed down the rivers and into the sea. The above species exude the resin freely although tapping is also practiced.

Copals are obtained in different regions from different genera. Congo copal is recovered from the marshy ground along the Congo river in mid-west Africa, and supplies are considered to be inexhaustible. The resin is thought to have originated from *Copaifera demeusei*, which still grows in the region, although other genera may have contributed small amounts. The existing *C.demeusei* vary in their facility to produce resin on wounding. West Africa copals probably originated as well from *C.copallifera, C.mopane, Daniellia oliveri, D.thurifera* and *D.similis*. Copal can also be obtained by removing sections of the bark at the end of the rainy season (Howes 1949). Holes (2 cm in diameter) drilled into the heartwood of *C. multijuga* and *C. langsdorfii* of the Amazon, Brazil, yield 40 liters a year of copaiba oil. The oil contains a number of C_{15} sesquiterpene compounds (Calvin 1983). *C. officinalis* of Venezuela is another important source of copaiba. *Trachylobium verrucosum* provides the copal of east Africa, which may be collected from the soil or from the tree in colorless to pale red lumps. South American copals are obtained from *Hymenaea courbaril* and other species. The pale yellow or reddish resin exudes from the bark and trickles to the ground where it hardens in lumps (Mantell et al. 1942, Howes 1949).

Agathis alba (*A. philippinensis*) is a large tree occurring in the region from Malaysia to the Philippines, producing high-quality timber and Manila copal. This light-colored resin was obtained as large lumps from the soil, or as lumps from forks in the branches of large trees (sometimes to the weight of 20 kg) or now by tapping the bark of the trunk of the tree (rootbark also exudes resin freely) (Howes 1949). Kauri "gum" or kauri copal was obtained from the soil as large lumps mainly in the fossilized or semi-fossilized form from the sites of present or former forests in New Zealand of the giant tree *Agathis australis*. Lumps up to about 25 kg in weight have been obtained from forks in the branches.

Mastic collected from *Pistacia lentiscus* var. *chios* growing on the island of Chio (in the Aegean Sea), and used since before 400 B.C., is a clear, pale yellow, brittle exudate used as a varnish and previously medicinally. It is obtained by incising the bark. A number of genera of the family Burseraceae yield resins collectively known as "elemi". After exudation, the high proportion of essential oils evaporate to give pale-yellow, semi-solid to harder solids. Manila elemi is obtained from *Canarium luzonicum*. Some trees yield large amounts of elemi freely except towards the end of the dry season and others require gashes in the trunk to produce up to 45 kg annually. Yucatan elemi is obtained from *Amyris* species, and Brazilian elemi from *Bursera* and *Protium* species. Frankincense, collected in the dry season after incising the bark of *Boswellia* species, has been in use from 1700 B.C. and is now mainly used in incense (Howes 1950). Myrrh is an aromatic exudate containing about 25-35 percent resin obtained from the trunks of several species of *Commiphora*, a large genus in the family Burseraceae. The bark contains schizogenous ducts which form large interconnecting cavities from which the resin flows on wounding (Howes 1949).

The "resinous" exudate which provides the basis of the lacquer ware of Japan and China is obtained by incising, during the summer, the bark of *Rhus vernicifera*. Exudation continues for a week, when another layer of bark is removed so that exudation recommences. After a few more removals, tapping is discontinued for a few years, and then the old wounds are reopened. The watery exudate, consisting of the "resin" urushiol (a phenolic acid with a long chain, unsaturated fatty acid substituent), gum and a small amount of protein, has at first a milky appearance and then darkens on exposure to air. Burmese lacquer is made from the exudate obtained in a similar manner from *Melanorrhoea usitata*, another species of the Anacardiaceae. Its main component is also urushiol (Howes 1949) (see Section 2.6.2.4).

The resin exudates from diseased or injured *Chamaecyparis obtusa* contain a range of terpenoids (Ohashi and Imamura 1973) different from those found in the heartwood (Erdtman 1963).

Other natural resins of variable composition and properties are formed by a variety of genera and have been observed with *Gossweilerodendron balsamiferum*, *Daniellia oliveri* (Jane 1970), *Dysoxylon fraseranum*, *Dipterocarpus*, *Alstonia*, and *Araucaria* species.

2.6.2.2 Gum

Gums are viscid secretions, exuding naturally or on incision or infection, generally colorless, nontoxic, odorless, and usually tasteless; and on desiccation, or exposure to air, they become hard, clear, glassy masses.

Commercial gums from trees are obtained mainly from *Acacia*, *Astragalus*, and *Sterculia* species (Urquidi 1978, Whistler 1982, D.M.W. Anderson et al. 1983, 1985). Various *Prunus* and *Acacia* species exude gums when injured by insects and fire or when in a moribund condition. While some gums (such as those from *Astragalus* species) may arise from normal physiological processes, most appear to be a pathological product (Howes 1949). The production and collection of gum is mainly associated with dry or relatively dry climates. Commercial gums are collected from natural exudates or following careful partial stripping of the bark or injury of the tree (for example with gum arabic and with karaya gum from *Sterculia urens*) or from cut roots (for example tragacanth). Gum karaya begins to exude immediately after tapping and continues for several days although the maximum amount occurs in the first 24 hours. The gum is in the form of large irregular tears and the amount obtained from a mature tree is 1–4.5 kg per season. A tree can be tapped five times during its life time (Howes 1949).

The "gum arabic" of commerce, from a small tree *Acacia senegal*, is probably the most important natural gum and is collected mainly in the Sudan. There are reports that it was used as early as 2000 B.C. by the Egyptians for making paints. Before 1940 about 20 000 tons were exported annually (Howes 1949), an amount which probably increased considerably after that time before it decreased as a result of drought (D.M.W. Anderson 1985). Large quantities of babul gum from *A.arabica* are also collected but it is of inferior quality.

In addition to the above examples, gums are exuded by trees belonging to the following genera: *Adansonia, Aegle, Afzelia, Albizia, Aleurites, Anacardium, Anogeissus, Bauhinia, Berlinia, Buchanania, Brachychiton, Caesalpinia, Cedrela, Ceiba, Ceratonia, Chloroxylon, Citrus, Combretum, Cordyla, Elaeodendron, Feronia, Flindersia, Garuga, Grevillea, Hakea, Khaya, Lagerstroemia, Lannea, Mangifera, Melia, Pithecolobium, Prosopis, Pseudocedrela, Puya, Sapota, Sclerocarya, Sloanea, Soymida, Spondias, Sterculia, Tarrieta*, and *Terminalia* (Howes 1949; F. Smith and Montgomery 1959).

2.6.2.3 Kino

The word "kino" (of Indian origin) was in use in 1757 for the reddish exudates, consisting essentially of proanthocyanidins (condensed tannins), which were used medicinally as astringents. Newly formed kino is usually of a thick consistency similar to that of honey, ranging from straw to blood-red in color; the appearance of the kino can be characteristic of the species. When kino exudes through the bark during spring and early summer, it becomes increasingly red and dries to a brittle, semi-transparent solid. Freshly exuded kinos are slightly acidic, usually soluble in water, often soluble in alcohol, and always soluble in alkali.

The kinos exude from *Butea frondosa* (Butea kino), *Pterocarpus marsupium* (Malabar kino), and from some of the *Eucalyptus* species. An extensive literature survey of the occurrence of kino was published by McGookin and Heilbron (1926). Kino-bearing species were reported to belong to the following genera: *Acacia, Adansonia, Angophora, Baloghia, Berlinia, Bombax, Brachystegia, Butea, Casuarina, Ceratopetalum, Coccoloba, Croton, Derris, Dipteryx, Eucalyptus, Eugenia, Jatropha, Lonchocarpus, Macaranga, Machaerium, Mezoneurum, Milletia, Myrica, Myristica, Parkia, Prosopis, Pterocarpus, Rhizophora, Schinopsis, Schizomeria, Sesbania, Spermolepis*, and *Xylia*. However, at that time the terms "gum" and "kino" were used indiscriminately in the literature so that the above list is probably misleading.

The extent to which kino veins or pockets form in eucalypts varies from species to species (Section 2.6.1.4). Extensive kino veins can be found in species such as *Eucalyptus obliqua* and *E.goniocalyx* and large pockets in some other species, such as *E. calophylla*, other bloodwoods and *Angophora* species. From a tree of *E.calophylla* as much as 45 liters has been obtained from a single reservoir (Phillips 1931). In most cases, kino veins occur only in the wood. Pockets containing more than 300 ml of kino have been found in the bark of *E. sideroxylon*.

It has been claimed that kino forms if an injury permits access of air to the undifferentiated cambium and newly formed sapwood of *E. calophylla* (Kessell 1921), but it is more likely due to other causes, such as liberation of ethylene by damaged tissues. Injuries resulting in kino formation can be caused by wood-boring larvae and other insects, fires which dry the bark so that it splits, damage by falling limbs, axe cuts, branch shed, etc. or by drought (Irvine 1936, Jacobs 1955, Day 1959, Bakshi and Singh 1964). Injury to young trees under the age of 5 years rarely results in kino veins, but kino can begin to exude within 21 days of injury to older trees.

Maiden (1889a,b,c, 1891) divided the eucalypt kinos into three main groups: (a) The *Ruby group*, the members of which are completely soluble in water and alcohol to give ruby-colored solutions. (b) The *Gummy group*, the members of which are partly insoluble in alcohol. (Maiden considered the insoluble portion to be polysaccharides). (c) The *Turbid group*, the members of which are soluble in hot water and alcohol, but whose aqueous solutions become turbid on cooling owing to the presence of either or both dihydrokaempferol (a flavanoid) and eudesmin (a lignan) (Hillis 1952, Hillis and Carle 1962). This group was further subdivided according to the presence or absence of one of these two components. Later, Smith (1913) found that the above classification was not satisfactory for a wider range (from over 100 species) of kinos. With the aid of color reactions and chemical tests he extended the classification. In turn, a chromatographic examination (Hillis, unpublished data) showed Smith's classification to be inadequate and that the composition of the kinos has limited use in assisting botanical classification of the genus.

2.6.2.4 Latex

Latex is a colloidal suspension or emulsion of water-insoluble substances, suspended in an aqueous phase, which may be released on cutting the bark. It is typically white (milky), but may be yellow to red, or colorless. It is produced in specialized internal secretory structures known as laticifers. The suspensions may be of terpenoids (e.g., rubber), proteins (e.g., the proteolytic enzyme ficin from *Ficus* species), essential oils, mucilage, and other components (Esau 1965, Fahn 1974, 1979).

Latex in various forms of rubber, chicle, and gutta percha is obtained from incisions in the bark of a number of species in the dicotyledonous angiosperms. Chicle, at one time the basis for chewing gum, was obtained during October to

March from zig-zag incisions made in the early morning in the bark of the central American tree *Achras zapota*. The flow ceases within a few hours and the tree is allowed to rest for a few years. Commercial chicle is obtained by boiling away most of the water in the exudate (Record and Hess 1943). Chicle is also obtained from *Dyera costulata* of Malaysia.

Gutta percha is another latex with a composition based on isoprene units. When a vertical herring-bone cut is made in the inner bark of *Palaquium gutta* in the early morning the white latex of gutta percha soon oozes from the wounds and hardens to a sticky mass. Yields of over 1 kg of dry latex are obtained from an average tree (about 20 m high), after which the tree must be allowed to rest for 18 months. A reputedly similar latex can be obtained from the leaves (Browne 1955). Other members of the Sapotaceae growing in Malaysia, namely *Ganua motleyana, Mimusops balata, Lucuma malaccensis*, and *Payena leeri*, also yield gutta percha and chicle in a similar manner (Browne 1955) as do *Manilkara zapota, M. bidentata*, and *Euonymus japonica* in other regions (Franke 1983). When heated up to 100 °C, gutta percha becomes plastic and can be molded into a desired shape which is retained on cooling, and has been so used from remote periods in Malaysia and elsewhere. Many industrial uses were developed on account of its resistance to acids, notably hydrofluoric acid, and excellent electrical insulating properties. The recent decline in exports to less than 50 tons annually by 1980 has been due to its replacement by synthetic plastics, the very low yields per hectare and the extremely labor-intensive procedures for collection (Franke 1983).

Rubber latex from *Hevea brasiliensis* is obtained by making V-shaped or sloping cuts at 30 °C to the horizontal into a panel in the bark of trees older than 6–8 years. When the vertically oriented latex ducts are cut, they contract and force out the latex which gradually coagulates on exposure to the air so that the flow stops after a few hours. The cuts are made in the early morning, and the latex runs into a cup from which it is collected at intervals. Rubber comprises 20 to 60 percent of the latex on a dry weight basis. The cambium is not damaged, so that the phloem and wood continue to form. In contrast to chicle and gutta percha, the formation of the latex responds to injury. Consequently, frequent cuts are made over the surface of the wounds every 1 to 3 days removing a few millimeters from the lower surface of the grove in the bark in the panel, resulting in continued production of the latex. One method to obtain rubber is to add dilute acetic acid to the latex which then coagulates with the rubber forming a sheet on the surface.

Another type of latex is that obtained after slicing, in different ways and with special tools, the bark of various lac or lacquer trees such as *Rhus vernicifera, R. succedanea* (in east Asia), and *Melanorrhoea usitata* (in Burma and Thailand). A yield of about 150 g of latex is obtained from a tree of 10 years age and 6–10 m height. After this tapping it is cut down (in Japan) or rested for 2–3 year intervals before the process is repeated (in China). The exudate is colorless at first but, when a film of it is dried at low temperatures and high humidity, it oxidises to a black highly durable lacquer; in China it is sometimes pigmented with red colored cinnabar. The fresh exudate from *Rhus* species contains urushiol (65–70 percent), hemicellulose and glycoprotein (5–7 percent), laccase and other enzymes (less then 1 percent), and water (20–25 percent). Urushiol is a mixture of 3-substituted catechol derivates each with a 15- or a 17-carbon olefinic side chain according to the region in which the tree grows. The active components in the exudate from *Melanorrhoea usitata* are isourushiol, a number of homologues of urushiol, and catechol and resorcinol derivatives with the long side chains possessing a phenylalkyl group (J. Kumanotani 1986, personal communication).

The latex from *Antiaria toxicaria* and *Artocarpus* species, which has a strong cardiovascular effect, is obtained by cutting the bark; it is used on arrow tips (Browne 1955). Cavities containing "latex" of unknown composition have been observed in *Alstonia* species.

2.6.2.5 Manna

Incision of the bark of some trees (such as *Fraxinus ornus*) results in the exudation of a liquid which dries to a soft, opaque, light-colored mass of manna (Flückiger 1894; Huber 1953). On the other hand, the cambium of *Myoporum platycarpum* must be injured in a particular manner before exudation commences. Mannas are soluble in water and largely soluble in alcohol, and some of them contain large amounts of mannitol.

Exudation of the liquor from *M. platycarpum* is most pronounced in spring, and as much as 5 kg of mannitol have been isolated from the liquor exuded by one tree during a 12-month period (Hatt and Hillis 1947). The cause of the exudation is unknown, but starch disappears completely, and gum and tannin accumulate in the rays and parenchyma of the wood near the injury (Fisher 1945).

Manna of high mannitol content has been obtained also from *Fraxinus ornus* (Flückiger 1894), *Platanus orientalis* (Jandrier 1893), *Olea europea* (Trabut 1901), and *Gardenia turgida* (Forster and Rao 1925).

2.6.2.6 Amber

Ambers are fossilized resins containing nonvolatile terpenoid materials which have been progressively oxidized and polymerized to a point where they can withstand chemical and microbiological attack (Langenheim 1969). They occur in a few isolated regions throughout the world and most commonly in the Cretaceous and Tertiary strata (70 million to 2 million years old), and no doubt from the huge prehistoric forests which then existed. The deposits in the Baltic region were probably from the forests of *Araucaria* species, and from ancient times these transparent light brown lumps were traded over great distances. The ambers found in other regions can have a different appearance, such as the pink, transparent masses from the Indo-China region, which probably arose from other genera.

2.6.2.7 Balsam

True balsams are resinous exudates whose fragrant aroma and pungent taste are due to the presence of 60 percent or more benzoic acid, or cinnamic acid, or both, and esters of them.

Storax (or styrax) and Peru balsam are obtained by bruising or damaging in early summer the bark of respectively *Liquidambar orientalis* (of Turkey), *L. styraciflua* (USA) (Rowe and Conner 1979) and *Myroxylon pereirae* (of central America) (Howes 1949). The outer bark is removed and covered with cloth to absorb the exudate, or the bruised bark is removed in autumn, when the cloth or bark are extracted with boiling water. The highest yields of storax from *L. styraciflua* growing in Louisiana, USA were obtained from horizontal cuts in the bark during the hottest months. If the cuts damaged the cambium, ducts were formed in the wood and these subsequently increased the yield of storax (Gerry 1921, Mahood and Gerry 1921). The purified products are semi-transparent, semi-fluid and yellow-brown in color. Tolu balsam is obtained by cutting V-shaped incisions in the bark of *Myroxylon balsamum* of South America. The freshly collected exudate is a soft, sticky, yellow-brown mass which eventually becomes brittle. Siam benzoin obtained through injury to the cambium of *Styrax benzoin* is very largely coniferyl benzoate (Schroeder 1968). Balsams were widely used for perfumery and medicinal purposes.

2.6.2.8 Maple Sugar

In late winter, the xylem sap of *Acer saccharum* or *A.nigrum* of eastern North America contains a relatively high concentration of sugars, mostly sucrose, and minute amounts of organic acids, nitrogenous compounds, and inorganic salts. The sugar content ranges from 0.5 to 7 percent with an average of about 3 percent. Maple sugar is obtained by boring small holes into the sapwood near to the base of the trunk of the tree and a channelled spout is driven into each hole so that the sap is directed into a bucket or plastic tubing. Much of the stored starch is converted to sugar in winter and the ready flow of sap occurs in early spring (Kozlowski and Keller 1966, Gregory et al. 1982).

2.6.2.9 Crystalline Compounds

Powdery or resinous deposits are occasionally found in the heart shakes or cavities in trees (Section 2.6.1). The deposit in *Podocarpus spicatus* contains components (lignans and flavanoids) either present in, or readily derived from, the heartwood components (Briggs and Cain 1959). The yellow powdery material in the fissure of an *Afzelia* species contained kaempferol-3-rhamnoside and traces of related polyphenols, whereas the heartwood contained the latter but not the former (F.E. King et al. 1955, Sandermann et al. 1965). The lignan gmelinol has been obtained from a white deposit in the shakes of *Gmelina leichhardtii* (H.G. Smith 1912, Birch et al. 1954), and isoolivil from the resinous deposits in the cracks of *Olea cunninghamii* (Briggs and Frieberg 1937). Heartshakes found in *Dacrydium nausoriensis*, *D. nidulum*, and *D. elatum* were completely filled with the triterpene podocarpic acid, which was present in adjacent tissues with much smaller amounts in regular heartwood of the same specimen, but absent in these and other *Dacrydium* species free from shakes (Hillis et al. 1976). Three different lignans have been identified side by side on the wall of a frost check in *Tsuga heterophylla* (Krahmer et al. 1970). One of these lignans, matairesinol, has been found as crystals in ring shakes in *T. mertensiana* (Barton and Gardner 1962) and western true fir (*Abies* species, Meyer and Leney 1968). In a specimen of *T. heterophylla*, the deposit in the ring shake was a mixture of matairesinol and small amounts of hydroxymatairesinol and in a ratio the reverse of that in the adjacent heartwood (G.M. Barton personal communication). The lignan α-conidendrin found in a shake in *T. heterophylla* was probably a contaminant (Schroeder and Kozlik 1972). Large deposits of aluminum succinate occur in the shakes in the heartwood of *Cardwellia sublimis* and *Orites excelsa* (H.G. Smith 1903, L.J. Webb 1954, Hillis and de Silva 1979); *Orites excelsa* can accumulate large

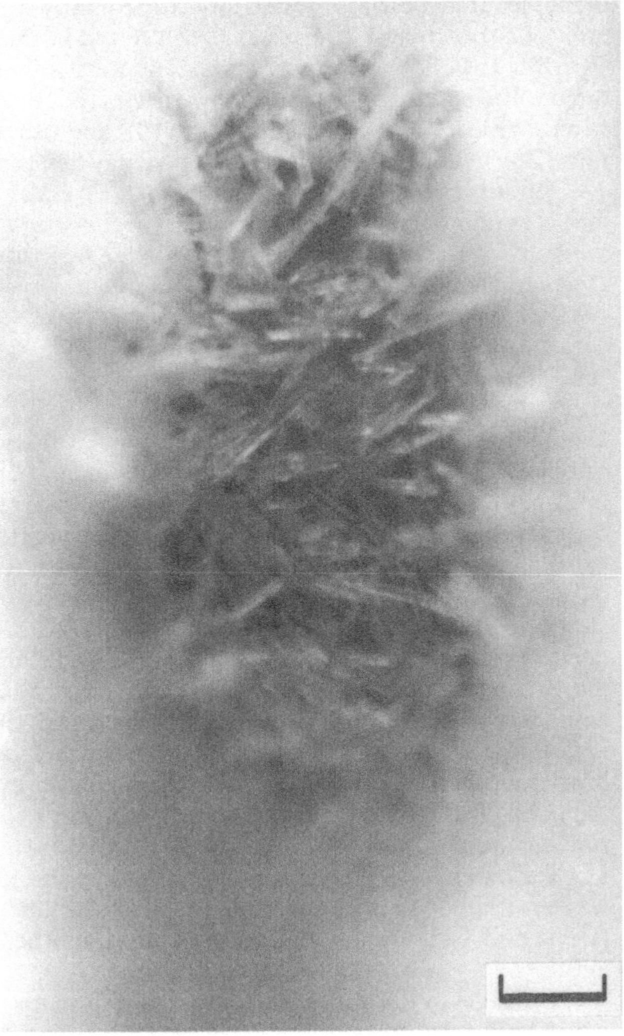

Fig. 2.16. Crystals of pure robinetin (a flavonol) in a vessel of *Intsia bijuga* (scale 100 μm). (Photograph courtesy of J. Ilic)

amounts of aluminum in its sapwood (6475 ppm Al) and heartwood (1425 ppm Al) (Lambert et al. 1983). The splits in a specimen of *Chlorophora excelsa* contained calcium malate (Sanderman et al. 1965). The yellow deposit in the shake of *Intsia bijuga* from New Caledonia was pure robinetin (a flavonol) which was found in crystalline condition in the vessels (Fig. 2.16) (Hillis and Yazaki 1973a, W.E. Hillis, J. Ilic, Y. Yazaki 1977 unpublished results, Hillis 1977), but that in a shake from an *Intsia* species grown in Borneo was a mixture of components including a new type of flavonol (Imamura et al. 1972, 1974). A pocket in the bark of *Haplormosia monophylla* contained mainly a di-C-glucosyl apigenin (a flavone), whereas the heartwood contained traces of this compound, other flavonoids, and stilbenes; the white specks in the wood were due to the flavonone sakuranetin (Hayashi et al. 1974). Crystalline camphor occurs within heartshakes of *Dryobalanops aromatica* and *D. lanceolata* (Chu Fei-Tan 1975 personal commununication, Desch 1941) and large (up to 9 x 18 x 20 mm) crystals of the methyl ester of dehydroperillic acid have been found in the heartwood of *Thuja plicata* (Kurth 1950). Some components readily form sublimates on the surface of undried wood e.g., guaiol from *Callitris* species (Baker and Smith 1910), β-eudesmol from *Pterocarpus indicus* (Gonzales 1980), and occidentalol from *Thuja occidentalis* (Roy et al. 1984). Such compounds may be found in shakes in the future.

Deposits in the vessels of a number of species can be seen when the radial surface is examined, such as black deposits in *Sindora, Khaya, Swietenia, Toona, Entandrophragma,* and *Samanea* species, white deposits in *Anisoptera, Dillenia* and *Dialium* species, *Lophira alata*, and *Chlorophora excelsa*.

2.6.3 Rate of Formation

Determination of the rate of appearance or formation of heartwood extractives or of exudates would indicate the site and mode of their formation. There is little information on the rate of formation of extractives at the periphery of heartwood owing to the experimental difficulties in studying undamaged heartwood. Labeled glucose (or sucrose) moves from the phloem to the sapwood and within 19 days in early summer reaches the sapwood-heartwood boundary of *Eucalyptus sieberi* (Hillis and Hasegawa 1963). Although this was consistent with other evidence, it was not proven that the normal translocation path was used. Furthermore, any lack of conversion of sugars to extractives at the heartwood periphery could be due to a period of inactivity in the transition zone at the time of administration. The rate of conversion of primary metabolites to polyphenols at the heartwood periphery is likely to be rapid. Because there was no phenylalanine ammonia lyase activity in the transition zone of *E. polyanthemos* and *Pinus radiata*, mature leaves of *E. sideroxylon* and *E. sieberi* with a low activity of this key enzyme were studied. Radioactivity was found in the polyphenols of the leaves within 24 hours of administration of the labeled phenylalanine or glucose (Hillis and Ishikura 1970), indicating that another biosynthetic pathway may exist.

A parallel to the abrupt increase in the formation of extractives at the heartwood boundary exists in the localized formation of different types of plant exudate. Although different species exude different types of exudate, certain similarities are evident in the mode of formation of the exudates. The emergence of exudates provides an easier means to study the rate of formation of extractives in woody tissues. The tapping of *Acacia senegal* in southern Sudan is practised by cutting the thin fibrous bark and tearing strips about 4 cm wide from both sides of the cut. Gum forms along parts of the wound within 3 weeks under favorable conditions when the tapping is conducted in the dry season from October to May, as during the wet season the gum does not form. Starch begins to change to gum in the ray cells within 6 hours of injury of the sapwood of *Prunus* species and hydrolyzed starch also passes into the vessels where gum accumulates (Brooks and Moore 1925, quoted by Hillis 1962b). Gum formation takes place from the beginning of spring to the end of summer, and the infection of twigs of *Prunus* species with *Stereum* (now *Chondrostereum*) *purpureum* proceeds rapidly, so that within 6 weeks of infection the gum formation may extend 10–18 cm. On the other hand, gum tragacanth from *Astragalus* species forms within a shorter period (mostly within 24 hours) when the roots (and sometimes the trunk) are incised (Howes 1949).

When eucalypt cambium is injured in the spring season and at a time of greatest activity, kino will begin to exude from the wound in 14–17 days (Hillis and Hasegawa 1963, Hillis 1975). When labeled glucose is administered to a newly formed kino vein, labeled kino can be collected within 5 days of injection (Hillis and Hasegawa 1963).

An extensive examination of the causes of kino vein formation in various *Eucalyptus* species has been made by Jacobs (1937). Veins form in most species during the warmer months when the temperatures exceed 10 °C–16 °C. A mid-winter injury may not develop until the warm weather commences, but in some cases these injuries heal without the formation of a vein. In warm weather, the veins form very quickly and may be complete within 6 weeks of injury; the spread of the vein is greater above small injuries than below, and the over-all length after that period is between 5 and 15 cm.

Chapter 3

Historical Aspects

The absence of technological facilities for extraction and concentration of the extracts until recent times restricted the use of the nonstructural components of wood mainly to the exudates of trees. Different extractives have been used nevertheless for specific purposes. These may be exemplified by the volatilization of the perfumes in sandalwood by burning for ritual purposes, by soaking hides in crushed tissues rich in tannins, or by using woods naturally containing toxic components as arrowheads or in spears.

The following examples of the historical use of extractives and exudates have been arranged in the order of increasing sophistication of knowledge of the active components required by the user. Although convenient for the present description, such an arrangement cannot provide precisely defined categories.

3.1 The Use of Durable Woods

There is little doubt the properties resulting from the presence of extractives in wood have been recognized for a long time. The export (in the 10th century B.C.) of the cedar wood from Lebanon (*Cedrus libani*) to Jerusalem for building the temple of Solomon was doubtless due not only to the size and medium density of the cedar logs, but also partly due to their durability. Teak (*Tectona grandis*) was exported 4000 years B.C., from India to Babylon and Yemen where apparently its durability was recognized (Sandermann and Simatupang 1966). The Romans removed the nondurable sapwood from *Quercus robur*, pine, and cypress when those woods were used for buildings (Hosli 1982). The Chinese appreciated the durability of wood for many centuries, and the wood used in the Imperial Palaces of Beijing was largely of *Persea nan-mu*. Nan-mu was also frequently used for the buildings attached to the Ming tombs erected in the early 15th century. To satisfy the high demand, the authorities found it necessary to transport the wood for long distances and in the year 1558, 15 712 logs (of 27 m in length and of large diameter) were transported from the Yunnan province, for over 3500 km, over mountain ranges and then by river and canal transport to Beijing (Shaw 1914; Chinese official notes). The present unchanged condition of those timbers not destroyed by war or fire indicates that careful observations must have been made previously to lead to the selection of this durable wood in a distant location.

Other durable timbers, notably those of the family Cupressaceae, have been used for buildings in many countries (such as Japan) and for other purposes such as boats and coffins, and have survived microbiological and insect attack for centuries. Remnants of a coffin built with *Cedrus deodara* and *Dalbergia latifolia* have been found in an Indian site dated at 2300 B.C. (R.K. Rao 1983). Teak (*Tec-*

tona grandis) has long been valued for its very high resistance to fungal decay and to insect and borer attack and its dimensional stability after drying. It was widely used for house building in south-east Asia and Malaysia. Teak is still used for boat building and is the highest-priced timber in international trade.

3.2 Exudates

The most notable exudate is amber, prized by man since its discovery in Neolithic times (7000–8000 B.C.). Its reputed mystical healing and protective powers, as well as its esthetically appealing appearance – such as its use in modern jewellery – and its development into varnishes has resulted in its continued use to the present day. This is partly due to the large deposits in eastern Baltic locations, which probably arose from *Araucaria* species. The discussions of Pliny in his "Historia Naturalis" (A.D. 77) show that the possible origins of amber had been discussed in the literature before that time. Pliny was the first to propose that amber was a plant exudate (rather than of mineral origin) although he erroneously considered pine to be its origin (Langenheim 1969). Ambers are fossilized resins containing nonvolatile terpenoid materials which have been progressively oxidized and polymerized to a point where they can withstand chemical and microbiological attack and this was the basis of their use in varnishes.

Records of the earliest use of other exudates are limited because of the difficulties of dating the material. An examination of a sarcophagus of the 19th Egyptian dynasty (1400–1300 B.C.) showed that an oleoresin or balsam had been used as a coating. The preservation of human bodies as mummies in the period 1000 years B.C. involved the use of storax (from *Liquidambar orientalis*), mastic, Aleppo pine resin, and probably sandarac and amber (Howes 1949). The exudate benzoin from *Styrax* species of south-east Asia has been found in a sarcophagus dating from the second century A.D. (Lucas 1962). The South America Incas used balsams (possibly from *Myroxylon balsamum*) for embalming (Howes 1949).

3.3 Varnishes and Lacquers

The principles of varnish making were first recorded in the eleventh century; they consisted of heating the resin, probably amber, in linseed oil. Rembrandt and Leonardo da Vinci used amber varnishes as vehicles for their paints (Mantell et al. 1942). The discovery by Europeans in the 17th century of damar resins in Malaya and neighboring countries resulted in an extension of their uses. Originally they were used in resin candles for illumination, the caulking of boats, and for preparing batik, then they became important for the preparation of varnishes, enamels, polishes, and printing inks. The damars originate from several genera of the Dipterocarpaceae and particularly *Shorea, Hopea*, and *Vatica* species, and large quantities have been removed or washed from the soils formerly covered by these forests.

Kauri "gum" or the resinous exudate from *Agathis australis* was first recorded by Capt. Cook in November 1769. Lumps of the resin, some of large size (30 × 15 × 15 cm), have been dug from the soil in many parts of the North Island of New Zealand (Reed 1954). The peak annual production was 11 116 tons in 1899, and altogether over 442 000 tons were exported between 1847–1939 (Harrison-Smith 1940–41). In 1924 surveys conducted by the New Zealand government indicated that there were sufficient resources to meet global demands for well over 100 years. However, its replacement by synthetic resins resulted in only 228 kg being exported in 1982. Kauri provided the basis for high-quality varnishes, while the lower-quality grades were used for the manufacture of linoleum together with pine rosin. Manila resins have been obtained by tapping the bark of *Agathis alba* growing in Indonesia, the Philippines, and New Guinea and have been used for varnishes in recent times.

Mastic collected (from *Pistacia lentiscus*) on the island of Chios in the Aegean was once the monopoly of Greek emperors. Its use to sweeten the breath and for dental purposes was first recorded in 400 B.C. (Howes 1949). Mastic and sandarac (from *Tetraclinis articulata*) dissolved in alcohol were used as varnishes for the protection of oil and water color paintings or for coating metals.

The sap exuding from the incised bark of *Rhus vernicifera* in China and Japan and of *Melanorrhoea usitata* of Burma contains urushiol. This and other components oxidize during drying at 20 °C–30 °C and at a humidity of 75–85 percent to produce the basis of the black or red lacquerware of the Orient. Lacquerware has been discovered with its original color in Chinese burial mounds from the late Zhou and Han Dynasties (ca 770 B.C. to 220 A.D.). Lacquerware imported over 1200 years ago into Japan is preserved in the "Sho-So-In" in Nara, and in that region the craft was further developed.

The high durability of the lacquer in many situations and its stability have been recognized for thousands of years in the Orient and it has been used to protect various materials. Interest in the chemistry, production and use of lacquer has recently increased (J. Kumanotani and Y. Du 1986, personal communication).

3.4 Gums

Gums have satisfied a number of needs with their property to dissolve in water to form viscid solutions or to absorb water and form jellies or gelatinous pastes. Gum arabic, probably from the North African *Acacia senegal*, was used by the ancient Egyptians in the preparation of their colors for painting (Howes 1949). The trade in gums for this purpose continued through the Middle Ages and gums are still so used. The Australian aborigines and others used the gums of *Acacia* species (Maiden 1901, Maiden and Smith 1895) as a source of food, and gums are still used commercially in confectionery, lozenges etc.

3.5 Tannins

Leather production is one of the world's oldest crafts, and tannin-bearing pods of *Acacia nilotica* with part-processed and finished leather have been found on a 5000-year-old tan-yard site in Upper Egypt. By about 600 B.C. tannage with vegetable tissues was common throughout the Mediterranean area – if not elsewhere (White 1957, Knowles and White 1958). Tanning was carried out by placing alternate layers of hide and crushed vegetable tissues – generally of oak origin – adding water and allowing the hides to remain until a satisfactory leather was formed. In Australia, before contact with Europeans, it was the custom of some aboriginal people to partly fill animal skins with water which contained pieces of eucalypt kino (largely polymerized prodelphinidins). The skins were carried on the aboriginals' backs during their nomadic existence, and the agitation facilitated tanning.

It was not recognized until 1796 that there was an active principle in oak bark which combined with protein in the hide to form leather (Sequin 1796) and the word tannin was introduced to define a material which produces leather from hide. Pliny (Knowles and White 1958) mentioned the blue–black iron tannate reaction as a method for identifying iron and that this reaction was the basis of the hair dye used by ancient Egyptians. The same reaction has been the major source of ink until recent years. The ability of tannins to combine with iron has resulted in the preservation of iron objects that were in association with leather in archeological sites (Knowles and White 1958).

Mangrove tannin extract or "cutch" obtained from the barks of *Rhizophora* and *Bruguiera* species was used for centuries in tropical countries to preserve cotton or linen clothing, sails, and fishing nets. Recently the use of nylon for nets has obviated the use of cutch.

Tannin extracts from various sources are still very largely used for the conversion of hides and skins to leather. In 1959, the major sources were the bark of *Acacia mearnsii*, the heartwoods of *Schinopsis balansae*, *Castanea sativa*, and *Eucalyptus wandoo*, and the cups and unripe acorns of *Quercus aegilops*. In 1959, a total of 434 500 tons of these extracts were used globally but in 1982 this had fallen to 267 400 tons.

3.6 Dyes

The employment of natural dyestuffs dates back to antiquity, and many of these probably arose from wood or bark. Most gave a dull brown or a yellow color but the discovery of mordanting in ancient Hindustan and in the Orient resulted in brighter permanent colors. The first mordants were aluminum sulfate and iron sulfate. Knowledge of these spread westwards to Egypt, Greece, Rome, and Europe. With the subsequent importation of logwood, brazil wood, and other materials from the "New World", the natural dyes became increasingly important until they were replaced by synthetic dyes (Perkin and Everest 1918).

The most notable wood sources of dyes have been a group of bright red woods under the general name of Brazil or Brasil. About 1290 Marco Polo mentioned the use of *Caesalpinia sappan* in the Orient for this purpose. The wood subsequently became an important article of commerce in Europe during the Middle Ages and eventually became known as Sappan. Later the discovery of a similar bright red wood in eastern South America in large quantities as well as its use as a dye by the native people lead to the naming of the region "Brazil" after the type of wood which had the color of glowing embers ("brasa" in Portuguese). Confusion exists concerning the scientific names of the exported wood but it appears to have been *Caesalpinia echinata* (Record and Mell 1924, Record and Hess 1943).

The first exploitable wealth found in Brazil was indeed brazil wood. Originally large quantities were present and the demand for it led to the first three European settlements at Recife and elsewhere in 1502 (Anonymous 1976). Uncontrolled felling resulted in the rapid destruction of the coastal forests and by the middle of the 16th century exports of the wood had fallen considerably although the trade continued for another three centuries. At times it was the prize of piracy and its sale was controlled by royal monopoly. With various mordants, bright reds to violets were obtained, but the fugitive character of the colors reduced their use.

Logwood (*Haematoxylon campechianum*) of tropical America was exported from about the middle of the 16th century. Subsequently the wood extracts were also exported. In the latter half of the 19th century the annual exports of the wood exceeded 100 000 tons. The major component of the extractives, namely hematoxylin, when oxidized forms the dye hematin, which produces blue to black colors depending on the mordant of iron or copper salts or potassium dichromate.

Old fustic from the heartwood of *Chlorophora tinctoria* of central America and young fustic from *Rhus cotinus* were used in the middle of the 17th century to produce a range of light stable olive–yellow to gold shades depending on the mordants. Similar colors were obtained with quercitron bark (*Quercus discolor*) of the USA.

Other woods were used as sources of dyes but they were much less important than the above. These other heartwoods included jakwood (*Artocarpus* species) of India, Burma, Sri Lanka, and Indonesia, that provided a dye which in conjunction with alum gave the yellow color for the robes of Buddhist priests. Sanderswood (*Pterocarpus santalinus*) from tropical Asia, camwood or barwood (*Baphia nitida*) of the west coast of Africa, and narrawood (*Pterocarpus* species) of the Philippines were formerly exported to impart red to violet colors with different mordants.

3.7 Perfumes

Just as at the present time, ancient people prized various types of perfumes. Historical records show that some of these were obtained from wood or from tree exudates. Frankincense (from *Boswellia* species) and myrrh (from *Commiphora* species) were obtained from incisions in the bark of trees growing in eastern Mediterranean countries, and the resins were traded by the Phoenicians and others more than 2000 years ago to provide perfumes, incense, and unguents (Howes 1949). There are suggestions that frankincense was traded in the 17th century B.C. (Howes 1950), and its use continues today (Abercrombie 1985).

It is recorded that cinnamon barks were prescribed by Moses for the ritual of the Hebrew Tabernacle, and at the time of the Hebrew exodus from Egypt the barks apparently originated from India. The true cinnamon bark of *Cinnamomum zeylanicum* from southern India and Sri Lanka entered the Mediterranean trade about 300 B.C. The Portuguese captured and suppressed Sri Lanka in 1500 in order to control the best source of the bark, and in turn they were followed by the Dutch and then the English (Burkhill 1935). The bark is still used for flavoring foods but now less for perfumery purposes.

A wide range of woods have been used for incense in religious ceremonies. For example, the dark resinous pathological heartwood of *Aquilaria malaccensis* and the wood of *Amyris balsamifera* have been used in tropical regions (Burgess 1966). The use of the fragrant wood and oil of the Indian sandalwood species *Santalum album* is much better known and is one of great antiquity. The wood was traded up through the Red Sea from India to Mediterranean countries in the 6th century A.D. and at the same time was imported by China (Burkhill 1935). In 1511 the Portuguese occupied Malacca to establish a trading base to monopolize the spice and sandalwood trade which was carried out between China and other countries. Early in the 19th century the British had established plantations in Mysore, India and used Singapore as their chief market.

Another *Santalum* species was discovered in Hawaii about 1790, and its export to China reached a peak between 1810 and 1825. This trade brought great wealth to the people of the islands (once known as the Sandalwood Islands) until the resource was almost completely destroyed by 1840 and with it the economy of the islands (Record and Hess 1943). In 1843 reports reached Perth, Western Australia, of the high prices being paid in Singapore for sandalwood. At that stage Western Australia had a very large imbalance of trade but in clearing land for agriculture it was removing sandalwood trees (*Santalum spicatum*) burning or using them for buildings or posts. Export was begun in 1846 and by 1848, 1335 tons earned more than one-half of the export income of the colony as it was at that time. Sometime about 1870, restrictions were placed on the export of sandalwood from India owing to the failure of the Mysore plantations, and from 1872 to 1882 the export of sandalwood from Western Australia rose from 3942 to 9605 tons. An all time peak was reached during the year 1920 with 13 945 tons (free from sapwood) being exported, earning about US $467 000 – with the present day equivalent being considerably higher. The amount cut has declined considerably in recent years, and the remaining material is found only in remote desert regions (L. Talbot 1983). The wood is used for incense and joss sticks in Buddhist rituals; sandalwood oil was formerly used as a medicine and is now used in perfumes and cosmetics.

3.8 Rubber

The natives of tropical America were preparing articles from the rubbery exudates of trees of *Hevea* and *Castilla* species well before their discovery by Europeans in the early 17th century. Following the developments of vulcanization and other treatments, an enormous exploitation of rubber trees began, particularly of *Hevea braziliensis* growing in the delta of the Amazon River and in its upper reaches west of Manaus. This took place from 1827 onwards. Large fortunes were made by a few European entrepreneurs at the expense of the lives of many local and imported laborers. After some difficulties, rubber trees were cultivated in Malaysia in 1880, and subsequently in neighboring countries, from which most of the natural rubber of today is derived. Following severe competition from synthetic rubber, improved plantation techniques have ensured their continuation as a source of rubber. In 1980, over 3.8 billion tonnes of natural rubber were exported; of this 67 percent originated in Malaysia and Indonesia and 7 percent in Brazil. Almost 6.5 billion tonnes of synthetic rubber were produced at the same time.

3.9 Medicines

Many plants have been used in the past for medicinal purposes, and the search continues today for superior sources. Various extracts of trees and of exudates have been used for different complaints.

The use of the deep red exudate of "Dragon's blood" from *Dracaena cinnabri* (or more recently *Daemonorops* species) as an astringent for the treatment of dysentery and diarrhoea was recorded by Pliny and early Greek writers. It has been found in tombs of that period (Howes 1949). Its incorporation in violin varnishes as a dye was probably a more effective use.

"Cutch", the aqueous extract of the heartwood of *Acacia catechu* grown in India, was imported by the Chinese in the late 16th century, or earlier, and it had been used as a medicine (and as a dye) in India earlier in that century. Other polyphenols and tannins have been used as astringents for the control of diarrhoea. For example, malabar kino (an exudate from *Pterocarpus marsupium*) and Bengal kino (from *Butea frondosa*) were exported to Europe in the late 18th century. Their use and that of kinos from some eucalypt species (e.g., *Eucalyptus camaldulensis*) have been prescribed by some pharmacopoeias for this purpose until recently.

Mannitol is a major component of the exudates ("manna") of a range of trees including *Fraxinus ornus*, *Olea europaea*, *Alhagi camelorum*, and *Myoporum platycarpum* (Flückiger 1894, Trabut 1901, Hatt and Hillis 1947), and was formerly used as a mild laxative. The manna which sustained the migration of the Israelites to Palestine in biblical times is reputed to have contained mainly raffinose.

There are a few cases of effective medicines obtained from wood or bark for the cure of specific illnesses. Until recent times the barks of *Cinchona* species have provided quinine and thence control or cure of malaria. Its first use by Europeans was recorded in 1636 in Peru with bark collected in the Andes. The subsequent exploitation of the trees almost lead to their extinction until plantations in other tropical countries were established. *C. calisaya* var. *ledgeriana* and *C. officinalis* provided the richest sources of quinine, and high-yielding hybrids were developed. This situation was achieved with difficulty by smuggling the seeds, over the period 1854 to 1859, from South America to the East, followed by the lack of germination of some of the seeds (Burkhill 1935). Cinchona has lost much of its importance because of the synthesis of anti-malarials with greater effectiveness and fewer side-effects.

Cavities in the trunks of *Dryobalanops aromatica* of Malaysia and Indonesia can contain crystals of camphor used for medicinal purposes. These crystals were sold in the Mediterranean region 2000 years ago and were the first type of camphor to reach Europe. The Chinese were importing camphor from Malaysia before the 8th century but subsequently replaced it with camphor from *Cinnamomum camphora*. Camphor was manufactured by steam distilling the wood from old trees by the Chinese and Japanese before the 9th century. The introduction of water-cooled condensers early in the 19th century increased the yield of camphor and also produced camphor oil and the industry prospered. Natural camphor, camphor wood, and its oil are still used, but camphor is now synthesized from α-pinene in increasing amounts.

From the converse aspect, a number of woods can cause sickness, and severe illness in the course of processing, and the occupational risks from processing were defined in 1700 (Bolza 1980). Some woods cause dermatitis and general irritation (Hausen 1981), and some users are particularly sensitive to them. Constituents responsible for most of the observed cases of allergic contact dermatitis belong to the benzo- and naphtho-quinones. These sensitizing quinones have been found in particular in *Dalbergia*, *Cordia*, and *Manchaerium* species, *Tectona grandis*, *Paratecoma peroba*, *Mansonia altissima*, *Diospyros celebica*, and *Thuja plicata*. Some woods cause severe stomach complaints (*Erythrophleum le-testui*), others congestion of the lungs and bleeding nose (*Dysoxylum muelleri*), nausea, cramps, and irritation of mucous membranes (*Aspidosperma peroba*). *Cryptocarya pleurosperma* contains an alkaloid which causes several severe complaints, and other woods cause different problems.

3.10 Lessons from History

From prehistoric times to the present day, extractives and exudates have provided for a number of mankind's needs. Originally they were supplied by local sources and a large number of them were used for relatively simple purposes. Eventually as international trade increased, and the sophistication of requirements rose, some materials were supplied in enlarged quantities when their superior properties became known. They became the objects of royal monopolies, trading nations were established by their exploitation, wars were fought over their supply, they enabled some countries or regions to become temporarily wealthy or to avoid bankruptcy, and they gave their names to countries. Eventually many exudates and extractives were replaced by other resources, or by synthetic or alternative substances with superior properties, which could be reliably supplied in uniformly high qualities and at acceptable costs. In some cases, their ready availability, high qualities, and easy profit led to the destruction of the resources. Very few of the materials used 50 years ago are now used commercially. Previously, the user sought the best natural product available and adapted his requirements to them. Today his precise needs can often be supplied from other sources.

In a world of diminishing resources per capita there is a growing need for the development of renewable resources, such as extractives, exudates, and forest products. What conclusions regarding their use can be drawn from history? Unless the product can be supplied in the highest appropriate quality, with assurances of the maintenance of that quality and continued availability at acceptable costs, the product will be supplanted. Many of the woody plants producing these materials are slow-growing and hence costly to maintain. Furthermore, the collection and preparation of extractives and exudates is labor-intensive. Such activities, even in regions of increasing unemployment and low wages, will be fruitless unless the consumers' requirements are provided.

The provision and use of these renewable resources can be increased by research and understanding. Vulcanization and other developments of rubber greatly increased its value to modern man. Biological research and strict merchandizing control have increased the supplies and quality of rubber from tropical rubber plantations so that it can compete with synthetics from nonrenewable resources. Improved tapping methods and strict quality control has enabled China with its labor resources to increase its exports of rosin and other naval store products. The use of stimulants for resin production in the wood of certain pines will assist the maintenance of production of naval stores as a byproduct of chemical pulping. Plantations of tannin-bearing trees, improvements in the production and purification of the extracts, and development of uses such as adhesives will ensure the continued use of these extracts. With the decreasing availability of low-cost petrochemical intermediates and the increasing costs to control hazardous effluents of the industry, the production of minor forest products in low-cost countries may become viable in the future. However, unless adequate quantities of a material, such as a damar, can be provided at low costs, in uniform quality, and with the supporting technical expertise for a particular need, developments are unlikely.

The history of the vicissitudes of the commercial use of extractives and exudates points to short- and long-term needs for their increased understanding. A greater use of unfamiliar wood species to provide man's increasing needs of building materials, pulp and paper, and other products will encounter difficulties in the way of efficient utilization. Understanding the extractives responsible for certain adverse properties has in the past led to the control of resultant problems. On the other hand, examination of favorable properties of wood due to extractives could provide understanding of the mechanism of that property and lead to its enhancement or the development of alternatives of higher quality.

An exciting aspect of the study of extractives and exudates is the recent developments in enzymology. These will help in the provision of answers to the questions frequently asked in the past of why and how the products extraneous to the lignocellulosic cell wall are formed. The different chemistry of the extractives and exudates from the same tree points to specific biological mechanisms. The different composition of the groups of compounds found in the oleoresin in the resin anals and in the adjacent parenchyma of conifers exemplifies the selectivity of these mechanisms. Even more dramatic examples of the detailed specificity of formation of individual compounds are shown in *Intsia bijuga* (and other species), where the vessels may be filled with a pure crystalline compound but the adjacent parenchyma will contain a mixture of several compounds from different classes (Hillis 1977). When these mechanisms, which the tree readily controls, are understood we will be closer to the formation by biological methods of the complex chemicals required by the consumer in a pure form without byproducts. This will require knowledge of the complex array of cell reactions in woody plants and the products they produce.

Chapter 4

Influence of Forestry Aspects

4.1 Variation in Heartwood Volume

The proposals put forward for the varying proportion of heartwood in trees have included its service as a respository for "toxic" products of metabolism (Stewart 1966a,b) and to keep the sapwood volume at an optimum level for the functioning of the tree. Another view is that adequate volumes of sapwood are needed for storage materials such as starch for metabolism and crown growth (Bamber 1976b) and fruit development (Kile and Wade 1975) and for water. This view implies that heartwood formation is a growth-regulating process related to development and to maintaining an optimum sapwood volume. In this way the nutritional balance in the living part of the tree is conserved, as it has been established that the essential elements are resorbed by the sapwood on heartwood formation. Consequently death of the parenchyma is seen as the result of heartwood formation rather than the cause (Bamber 1976b). Extensive data on a *Eucalyptus obliqua* forest supports this view (Attiwill 1979, 1980).

The regularity of initiation of heartwood formation and of the number of growth rings in the sapwood (Chapter 2.5.2.3.1) indicates a genetic control of the transformation. However, the age of transformation is influenced by environmental features and forestry practices with a number of species. The effect of these influences varies with different species. An examination has been made of the extent of spatial and of temporal control of heartwood formation in *Juglans nigra* and *P. serotina* (Nelson 1975, 1976). Temporal control refers to the death of the xylem parenchyma cells solely in response to the age of the cells, and evidence indicates that this form of control has the strongest influence in *J.nigra.* In spatial control the death of the xylem parenchyma is solely in response to an increase in the distance between the cells and the outer portions of the tree stem, and cell age has no effect. Data on the influence of growth rate on sapwood width and number of rings in the sapwood of *P. serotina* indicate that both temporal and spatial types of control are important in that species.

The presence of heartwood influences the utilization of wood in different ways and affects its uniformity. It is a contributor to the uncertainty between forest inventory and the quality of wood available for utilization. If the controlling factors for its formation could be predicted and a relationship found, then the selection of the most suitable group of trees for a particular purpose would be assisted, as well as the choice of the most suitable silvicultural practices.

4.1.1 Heritability

Heritability, or the ratio of the genetic variance to the total variance (genetic plus environmental variance), has been examined in relation to heartwood formation in *Pinus radiata*. In a preliminary study of 13 clones, Nicholls (1965) estimated the gross heritability to be low and at 0.37 at a 5 percent probability level. In another study of a group of 19 clones in another locality it was found to be 0.49 (J.W.P. Nicholls personal communication). This figure indicates that the variation due to genetics and that due to environmental factors are about equal. It was concluded also that individual rapidly grown trees will tend to have heartwood proportions that are lower than the mean. Twelve years later, the results of the preliminary trials were confirmed (Nicholls and Brown 1974) with values close to those of the second group of clonal material. In a study of groups of *P. pinaster* with the same seed source, age, and climatic conditions, the group which did not receive fertilizer and was grown under severe moisture stress during summer had the greater proportion of heartwood (J.W.P. Nicholls personal communication). There were significant differences in heartwood area between 7-year-old trees of *Eucalyptus obliqua* from different provenances and those showing better height growth having a greater proportion of heartwood (Nicholls and Matheson 1980).

4.1.2 Effect of Growth Rate and Crown Size

Büsgen and Münch (1929) refer to earlier work in which width of sapwood was found to be correlated with the size and weight of the crown in *Picea abies* and *Abies alba*, and in *Quercus sessiliflora* with the size of the stem. The proportion of heartwood to merchantable volume varies (0 to 15 percent) directly with age in 20- to 90-year-old *Pinus taeda* and inversely with rate of growth (MacKinney and Chaiken 1935). Various attempts have been made to relate growth rate to the proportion of heartwood or sapwood in a tree by means of the size of the crown.

In an even-aged stand, those trees of the same height (5 m) with a small crown (3 m width) or from closely stocked areas, had proportionately more heartwood (28 percent) than larger-crowned (10 m width) trees (with 8 percent heartwood) (Paul 1932). *Pinus palustris* showed a similar behavior in relation to growth rate and heartwood content. In a detailed study on *P.glabra* a regression was determined predicting heartwood volume (percent of total woody stem) in terms of live crown length (in feet) as :–

$$\text{Heartwood volume} = 9.515 - 0.174\ (\text{crown length}).$$

This equation accounted for 20.4 percent of the variation in heartwood content (Manwiller, quoted by Koch 1972a,b).

Crown class is a relative assessment for a particular stand and it would directly influence growth rate, bole size, and bark thickness. Crown size, or class, is directly related to sapwood thickness (Chapter 2.2.3.5) but the characteristics of the bole can be more important. This has been shown further by the highly significant correlation found between heartwood diameter (d_h) and stem diameter under bark (d) for *Cryptomeria japonica* in the regression $d_h = 5.6607 + 0.9018d$ (Ihara 1972). A strong correlation has been found between the surface area of sapwood and tree diameter of *Juglans nigra* (Nelson 1976). In one stand of *Pinus radiata* of about 30 years of age, the relationship between the width of sapwood (y) and diameter of stem (x) was found to be $y = 0.694x - 1.67$ (Carrodus 1972). The observed and predicted heartwood diameters in cm of the same species have been highly related (Hillis and Ditchburne 1974) to the diameter outside bark at breast height (D) and the diameter under bark at 5 years of age (D5) with the regression equation :–

$$\text{Heartwood diameter} = 0.1055\ D + 0.03114\ \text{Age} \times D5\ - 2.83$$

That is, those trees with the most rapid growth in the first 5 years had the greatest diameter of heartwood. The regression applied for samples obtained from a wide variety of environmental and soil conditions and also with the inclusion of samples with erratic heartwood boundaries. A lower degree of accuracy of prediction was obtained when the diameter at 10 years was used instead of at 5 years. There may be more sensitive ages than the latter. For a given age *Pinus taeda* trees having the fastest growth rate during the last 10 years have a relatively smaller proportion of heartwood, and this relationship is maintained irrespective of the nature of the site (MacKinney and Chaiken 1935). In a study of *Pseudotsuga menziesii* the radial growth in the last 10 years was the most significant growth variable, and the relationship between sapwood thickness (T) and the radial growth for the last 10 years in inches (G) and diameter outside bark (D) was found to be $T = 0.21 + 0.070D + 0.73G$ (J.H.G. Smith et al. 1966). Part of the value of this type of regression is that the necessary information can be obtained without felling or damaging the tree if suitable records have been kept.

4.1.3 Influence of Environment

There are contradictory conclusions on the influence of the environment. Whereas it is claimed that the extent of heartwood formation in European pines increases with the quality of locality (Büsgen and Münch 1929), MacKinney and Chaiken (1935) showed that the percentage heartwood content of *P.taeda* varies indirectly with rate of growth, and that this relationship is maintained irrespective of the nature of the site. Similar conclusions to the latter were drawn from the behavior of *P.radiata* grown on a number of good sites (Hillis and Ditchburne 1974). Paul (1952), however, found that the age of heartwood initiation for pines is not relatively constant at a given location. He contended that the age of development is influenced by site, and such other factors as crown size and growing space, so that heartwood may begin to form in suppressed trees slightly earlier than in dominant trees. A decrease in the site quality results in a decrease in heartwood volume of *P.sylvestris* of the same age (Harsch 1912), an increase of transition age of sapwood of *Cryptomeria japonica* (Ihara 1972), but it has no effect on *Pseudotsuga menziesii* (Wellwood 1955).

Studies of the effects of the level of humidity and availability of water have yielded contradictory results. High humidity and availability of water appear to slow down the formation of heartwood in *Pinus sylvestris* (Pilz 1907). On the other hand, Harris (1953, 1954a) found that the factors conducive to heartwood formation in *P. radiata* were a well-distributed adequate supply of water, absence of drying winds or of conditions of low humidity, and particularly the development of a deep-root system. These conditions are in contrast to those reputedly favoring a greater formation of sapwood (see Chapter 2.2.3).

4.1.4 Influence of Injury and Health

The "admission of air to the interior of the stem" through branch stubs or injury has long been considered as an accelerator of heartwood formation (Büsgen and Münch 1929). Pruning of *Cryptomeria japonica* increases the heartwood ratio and volume (Ihara 1972) in agreement with work on *Populus* species (Sachsse 1965, Poluboyarinov and Nekrasova 1977) and *Pseudotsuga menziesii* (J.H.G. Smith et al. 1966). Turpentining (by incising the bark) appreciably increases the volume of heartwood in *Pinus palustris* but, on the other hand, wounds retarded the formation of heartwood in *Quercus alba* (McGinnes and Shigo 1975). Infestation of *Abies grandis* bark with balsam woolly aphids results in significantly greater amounts of heartwood (Puritch 1977).

It has been concluded that the increase in heartwood proportion with age of *Pinus sylvestris* is due to the diminution of the vitality of the tree (Lappi-Seppälä 1952). A marked decrease in the number of sapwood rings occurs in *Nothofagus cunninghamii* after about 50 years of growth.

4.2 Formation of Exudates

4.2.1 From Bark and Wood

Commercial oleoresin is obtained from a small number of pine species (see Chapter 2.6.2.1) in an annual yield of about 3 kg per tree. The yield and quality of the oleoresin can vary according to the species and its varieties, ambient temperature, rainfall, size of tree, and method of tapping (Coppen et al. 1984). Oleoresin is collected from trees of diameter exceeding 25 cm after repeatedly removing strips of bark from a panel on the stem; usually an incision of a few millimeters is made into the sapwood. Tapping may be continued upwards for 4 years (using a wide face) to 20 years (on a narrow face) on the same group of trees with the same overall production. The oleoresin is collected during spring to mid-autumn.

The pattern of tapping varies in different countries with the system in Portugal being more restricted than elsewhere. In general, the rough outer bark is removed to a height of about 2.5 m, an apron or gutter system and collecting cup is placed at the bottom, and to cause the oleoresin to flow, a horizontal strip of bark 2–2.5 cm high is removed across the width of the tree and an incision made in the wood. In France, Portugal, and Spain, the operating face may not be more than 9 cm wide or more than 45 cm in height for each year of working, although two or more separate faces can be worked at the same time depending on the size of the tree. In other countries the faces are three or more times wider. In another method, after removal of the bark, a fine spray of 50 percent sulfuric acid or a paste of 60 percent acid, is applied along the top edge of exposed tissue so that it is not necessary to cut into the wood to open the resin canals. The acid increases the yield of oleoresin and maintains the flow so that the removal of bark above the first streak is not necessary until about 2 weeks later. About 16 visits to the tree are made in the course of an 8-month tapping season, and the top surface of the cut is moved upwards (Howes 1949, Koch 1972d, Coppen et al. 1984). Good yields of oleoresin have been obtained also from *Picea jezoensis* after stimulation with sulfuric acid (Abe et al. 1953).

In addition to the fossilized copals, fresh resin can be obtained by removing sections of the bark of *Copaifera demeusei, C.copallifera*, and *Daniellia oliveri* at the end of the rainy season and collecting the hardened nodules of resin during the dry season. *C.demeusei* growing in different parts of the Congo vary considerably in their ability to produce resin on wounding (Howes 1949). Similarly, Manila copal (or almaciga resin) is obtained by making a horizontal cut in the bark of *Agathis alba* (*A. philippinensis*) without damaging the cambium. The wound is enlarged in an upward direction at weekly intervals, and the resin flows over the surface, from which the bark has been removed, and hardens. An annual yield of 12 kg per tree can be obtained. Tapped resin from some areas differs from the normal in that it never hardens (Howes 1949, Anonymous 1978, Gonzales and Abejo 1978). Addition of strong sulfuric acid to the tapping wound increases the yield of copal.

Damar is obtained from mature trees of *Balanocarpus heimii* by making a number of incisions about 10 x 10 cm and 1–2 cm into the wood. The damar is sticky when first exuded and hardens in about 3 months. The surface of the holes is recut to encourage further exudation, which eventually ceases. Frankincense is obtained by scraping away portions of the bark of *Boswellia* species in early spring until the autumn or early winter. About 14 days later the resin is removed and the wound rescraped (Howes 1949).

4.2.2 From Wood

In softwoods, resin acids, fatty acids, and their esters are formed in the ray parenchyma cells. The oleoresin, a solution of resin acids dissolved in volatile terpenes, is synthesized in specialized epithelial cells lining the resin ducts, which arise as schizogenous intercellular spaces. These ducts are a normal feature of *Pinus, Picea, Larix* species, and *Pseudotsuga menziesii*, although they reach their greatest development in certain *Pinus* species. The mechanism of oleoresin secretion across cell membranes and water-saturated cell walls into the duct lumen is obscure (Birchem and Brown 1979). The secretion process occurs centripetally against a pressure gradient often exceeding 10–12 atmospheres (Bourdeau and Schopmeyer 1958). Often the dominant trees have the highest oleoresin pressure, which enables them to seal off injured areas and branch stubs, thereby inhibiting the entry of fungi. When a part of the three-dimensional anastomosing resin canal system is ruptured, oleoresin flows to that location, exuding onto the surface.

Owing to the increasing costs of labor required for tapping (Section 4.2.1) and the decreasing availability of resin-soaked pine stumps, the products of oleoresin are now obtained in the USA mainly during pulping. The gum, wood, and sulfate (Kraft) sources contribute to the current USA oleoresin production of nearly 450 million kg (Stubbs et al. 1984). The USA, China, the Soviet Union, and Portugal together account for about 80 percent of the world rosin production of approximately one million tonnes. The USA and Soviet Union produce rosin, mostly for local consumption. The estimated annual production of Portugal and China of 90 000 tonnes and 200 000–250 000 tonnes respectively is almost entirely of gum rosin and these countries dominate international trade. Turpentine production is estimated at about 250 000 tonnes per annum, one half of which is sulfate turpentine (Coppen et al. 1984). Production of rosin and turpentine is decreasing in the USA owing to declining resources. However, if the present stands of southern pines in the USA were treated with paraquat (Chapter 6.4.2) and the oleoresin recovered by the sulfate process, an estimated 43.4 million tonnes could be obtained (Stubbs et al. 1984).

4.2.3 Addition of Stimulants

A range of chemicals increases the formation of exudates above that caused by mechanical damage. These chemicals have been studied most intensively to stimulate the latex flow and rubber yield from *Hevea braziliensis*. From about 1950 onwards, 2,4-D and 2,4,5-T were applied to the bark to increase the yield, but they could contaminate the product. Of the wide range of compounds subsequently tested those reportedly most effective included 2,4 dichloro-5 fluoro-phenoxyacetic acid, synthetic growth substances similar in action to auxin and indoleacetic acid, paraquat, picrolam, triazine, and acetamide herbicides, ethylene oxide, chloroform, and acetylene. Later (Abraham et al. 1968) it was found that the substances which enhance latex flow in *Hevea* species have been reported to produce ethylene in plant tissues (see also Abeles 1973). 2-Chloroethylphosphonic acid (CEPA, "Ethephon" in the proprietory formulation "Ethrel") which hydrolyzes to ethylene in plant tissue, is more effective than 2,4-D or 2,4,5-T, which promote cellular production of ethylene over a period of time. Whereas ethrel reduces the dry rubber content of the latex, the period of flow and the volume of latex harvested is increased. Naphthalene acetic acid (NAA) increased the latex yield from young *Hevea* trees without affecting the bark cells, but the application of ethrel in palm oil gave better results. It was found important to devise a tapping system which would allow regeneration of latex within the latex vessels (Gener and Du Plessix 1975); stimulation of yield with ethrel is now a routine practice in most commercial plantations with rubber trees of 15 years of age and older

(Tan and Leong 1975). Response was best when ethrel was applied to bark that had been renewed after two previous tappings, less effective with bark that had been renewed the first time, and only a poor response was obtained with bark that had not been tapped previously. Ethrel is applied in different ways, either to the lightly scraped bark below the tapping cuts every 2 months or to the cut bark monthly, with some clones of the species responding better to form more latex than others. A minimum interval of 12 h after application was necessary for the best response from one clone (Sivakumaran 1982). In some aspects 2,4,5-T gave better yields than ethrel, although the latter resulted in about 60 percent higher yield than the control (Abraham et al. 1975), but there is a tendency for a decreasing response with successive applications.

To reduce the costs of application with ethrel and the possible side effects from the residual acids after the release of ethylene, other ethylene-producing formulations have been tried. These include a mixture of granulated zinc, ethylene bromide, and ethanol applied directly to bored holes (Pakianathan 1970), derivatives of 2-chloroethyltrialkoxysilane and also gaseous ethylene adsorbed on special powdered material. The adsorbents are molecular sieves involving inorganic compounds with required porosity, and which displace the ethylene when exposed to a moist atmosphere. This latter release is controlled by mixing the adsorbent with suitable viscous carriers of hydrocarbon greases and oils (Dickenson et al. 1975). The adsorbent containing the ethylene is applied as a grease to grooves in the bark at monthly intervals when the ethylene is gradually released. The direct process of ethylene release to the tissues resulted in a high initial reponse of latex formation which then declined; but even so the yield of dried rubber obtained was higher than with ethrel treatments.

Following a 3- to 9-year period of use of ethrel, the initial high flow rates of latex were reduced in stimulated trees when compared with unstimulated trees. The severity of reduction of flow rates varied in different clones and with the frequency of tapping (Pakianathan et al. 1982).

Independently of the studies to increase latex flow in *Hevea braziliensis*, attention was being given to the stimulation of kino formation in eucalypts (Hillis 1964, 1975). Injection into the active cambial tissues of *Eucalyptus regnans* and *E.sieberi* of sodium arsenite resulted in a copious exudation of kino. It is not known whether the arsenite resulted in the formation of ethylene or if it acted to block the utilization of pyruvate in the tricarboxylic acid cycle and its transference to the synthesis of the acetate-engendered portions of the proanthocyanidins. Kino exudation resulted to a lesser extent from the injection of potassium cyanide, sodium arsenate, ferrous sulfate, and iodoacetic acid (Hillis 1964). Copious exudation of kino from a number of eucalypts has been observed following injection of the weedicides "Tordons" which contain picrolam alone or together with 2,4-D and 2,4,5-T; these weedicides are known to form ethylene in living tissues (Abeles 1973). Injection of ethrel into the cambial region of *E.sieberi, E.calophylla, E.astringens*, and *E.sideroxylon* resulted in copious formation of kino (Hillis 1975). To a lesser extent the application of lanolin pastes containing indoleacetic acid (IAA) and naphthalene acetic acid (NAA) resulted in kino formation. Dowden and Foster (1973) found that whereas eucalypts under 5 years of age rarely form kino, it may be induced by applying NAA on the bark of seedlings at younger ages (compare behavior with young trees of *Hevea braziliensis*). NAA can stimulate ethylene formation when applied to plant tissues (Abeles 1973). Injection of ethrel into the sapwood of a *Rhus* species resulted in the formation of heartwood polyphenols. In addition, ethrel resulted in the formation of copious amounts of gum in *Prunus* species (Hillis 1975) (see also Chapter 6.5).

Chapter 5

Chemical Features

The analyses of chemical features relevant to the formation of heartwood and exudates have often been conducted on samples of uncertain anatomical origin. Consequently the sections in this chapter draw attention to specific chemical features rather than to the location of the feature. The anatomical and biological features will be considered in more detail in Chapter 6.

5.1 Water and Gases

Moisture content in tree stems varies with the species, site, and environmental conditions as well as with the age of the stem. Moisture content saturation (calculated as the percent moisture content per maximum moisture content) provides the most meaningful comparison (Chalk and Bigg 1956) but these calculations are rarely made. Wood attains its maximum possible water content when all the spaces in the walls and lumens are filled with water. Because the volume of these voids is negatively correlated with the wood specific gravity it follows that the maximum moisture content is found in wood of lowest density in the stem and hence is affected by the proportion of juvenile wood. Data have been obtained which suggest that the average moisture contents of the sapwoods of ring-pored and diffuse-pored hardwoods and of nonpored species are, respectively, <75, about 100, and >130 percent (on the basis of dry wood) (Peck 1953). Other work indicates minimum moisture contents of 35–60, 60–75, and 75–100 percent respectively (R.D. Gibbs 1958). In addition, the moisture content in undamaged stems can be affected by the presence of heartwood, and these differences are much more marked in softwoods than in hardwoods.

Because of several causes of variation in moisture content, only a few species are given in Table 5.1 to indicate the range that can exist. The moisture content can vary considerably seasonally (Burmester 1983, R.D. Gibbs 1958) in the sapwood [which probably serves as a reservoir (B.S. Meyer 1956)] as well as showing sometimes a sharp drop over one or two rings at the sapwood-heartwood boundary. Examples of these changes are the decrease from 160 percent in the outer sapwood of *Pinus contorta* to 30 percent in the heartwood (Reid 1961) and 220 percent to 30 percent in the earlywood tracheids of the sapwood and heartwood respectively of *P. sylvestris* (Vintila 1939) (see also Koch 1972a) (Fig. 5.1).

The extensive survey of the seasonal changes in moisture content in different woods made by R.D. Gibbs (1958) shows that the maximum is present in spring and early summer. In the dry season the amount falls, sometimes considerably, as with *Tectona grandis*, *Picea abies*, *P. sitchensis*, and *Pseudotsuga menziesii* when the loss is greater in the inner than in the outer sapwood (R.D. Gibbs 1958,

Table 5.1. Variations in moisture contents of sapwood, transition zone and heartwood in various species

Species	Moisture content (% dry weight)			Reference
	Sapwood	Transition zone	Heartwood	
Abies balsamea	138–186		57– 75	Clark and Gibbs (1957)
Betula maximowicziana	61– 79		72–116	Yazawa et al. (1965)
Chamaecyparis pisifera	250		50	Craib (1923)
Cryptomeria japonica	150–200	80–95	110–220	Nobuchi et al. (1984a, 1985); Sameshima et al. (1967)
Dacrydium franklinii	148–170		32– 45	Bauch (1980)
Fraxinus mandshurica	40– 56		83– 90	Yazawa et al. (1965)
Picea glauca	136–162		47– 48	Clark and Gibbs (1957)
Picea rubens	93–142		33– 42	Clark and Gibbs (1957)
Pinus contorta	85–165		30– 40	Reid (1961)
Pinus ponderosa	148		40	Dobie and Wright (1978)
Pinus radiata	135		40	Cown and McConchie (1980)
Pinus radiata	170	40		Harris (1954a)
Populus maximowiczii	84		165	Yazawa et al. (1965)
Populus deltoides	146		162	Dobie and Wright (1978)
Pseudotsuga menziesii	102–148		37– 39	Scott (1964)
	100–200		48	Harris and Orman (1958)
Quercus robur	38– 46	37.5	38– 40	Ebermann and Stich (1985)
Taxus baccata	130	50	100	Craib (1923)
Thuja occidentalis	236–262		31– 38	Clark and Gibbs (1957)
Thuja plicata	249		58	Dobie and Wright (1978)
Ulmus serotina	92		95	Dobie and Wright (1978)

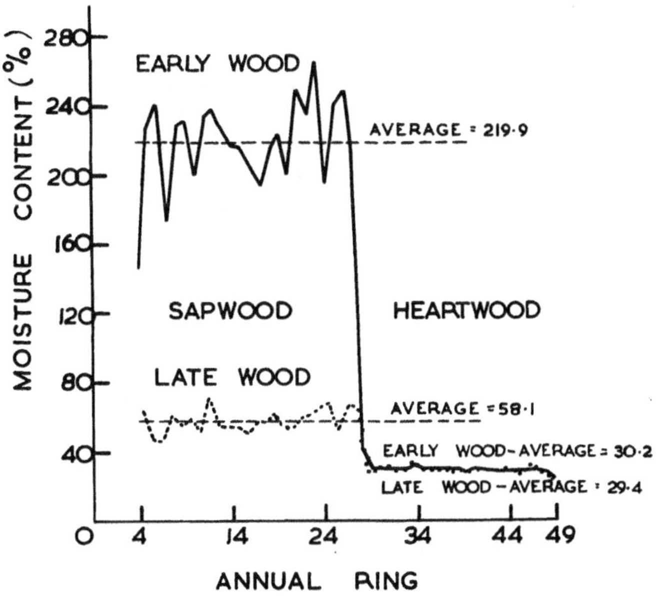

Fig. 5.1. Variations in moisture content across a stem of *Pinus sylvestris*. (Vintila 1939)

Chalk and Bigg 1956). The decrease in the percentage of water saturation in dominant trees is not as great as in suppressed trees (Dadswell and Hillis 1962). The moisture content can show a gradual increase so that it reaches a maximum in the innermost sapwood rings (cf. Stewart 1967), where also the hydroxyl group accessibility reaches a maximum in *Pseudotsuga menziesii* (Chow 1972).

The moisture content across cross-sections of angiosperms is less well-documented than with softwoods. In many species a similar, but less marked, dehydration occurs during the transformation of sapwood to heartwood. However, larger amounts of water can exist in some heartwoods than in the inner layers of sapwood. Although in some cases wetwood, resulting from microbiological infection, may be present, it is unlikely to be the cause of a higher moisture content in all cases. Genera known to have heartwoods containing more moisture than sapwood are *Carya, Fraxinus, Ulmus, Quercus, Nyssa, Juglans, Populus, Abies, Betula, Morus*, and *Eucalyptus* (Peck 1953, Yazawa and Ishida 1965a, Sauter 1966a, Stewart 1967, Hillis 1968a). All species of these genera do not possess heartwoods containing more moisture than sapwood. The water content in *Populus* "wet heartwood" shows maxima in spring and autumn and can decrease under certain conditions such as drought (Sauter 1966a, Sachsse 1965). The moisture content of the sapwood of *Fagus crenata* is higher than that of the heartwood in the winter but can be lower in the summer season (Yazawa 1960). On the other hand, Zycha (1948) found that the moisture content of *Fagus sylvatica* decreased steadily from the outermost sapwood to the pith, and that where the content had fallen to 60 percent (based on air-dried wood) the boundary of the red heartwood could be seen. It is interesting in this context that on the average the free water/gas ratio in sapwood pores near the heartwood boundary of several trees of a *Populus* cultivar was about 61 to 39 and at all heights in the stem (Sacchse 1965). In some species, such as *Fraxinus mandshurica*, *Populus maximowiczii* and *Ulmus davidiana*, the heartwood is wetter than the sapwood in all seasons of the year in both the trunk and root wood (Yazawa and Ishida 1965a, Yazawa et al. 1965). The moisture content increased sharply in the heartwood at the sapwood-heartwood boundary, for example an average of 40.5 to 83.6 percent, in a tree of *Fraxinus mandshurica*, and it increased at all levels in the tree (Yazawa et al. 1965). The ratio of the moisture content in the two zones usually decreased with height; for example, at 0.3 m above ground level the moisture content of the sapwood of one tree of *F. mandshurica* was 69 percent and of the heartwood 119 percent; at 17.3 m it was 41 and 86 percent respectively (Fig. 5.2). In *Ulmus laciniata* at 0.3 m it was 70 and 109 percent respectively and at 13.3 m 85 and 95 percent. Even with a high moisture content in an *Ulmus* species heartwood a pale-colored transition zone surrounding it has been detected (Yazawa et al. 1965). These high distributions of moisture in wet heartwood species are normal for a particular species and have been recorded also in *Carya laciniosa* (W.R. Smith and Goebel 1952) and *Fraxinus excelsior* (Bosshard 1955). In some other wet heartwood species, such as *Betula*, the differences in moisture content in sapwood and heartwood were not so great and they had an irregular distribution in their trunks (Yazawa et al. 1965). The pale-colored eucalypts have heartwoods with moisture contents about 20 percent higher than the sapwoods (Hillis 1978). *Fraxinus excelsior* produces colored heartwood only if the moisture content of the inner portions exceeds 55

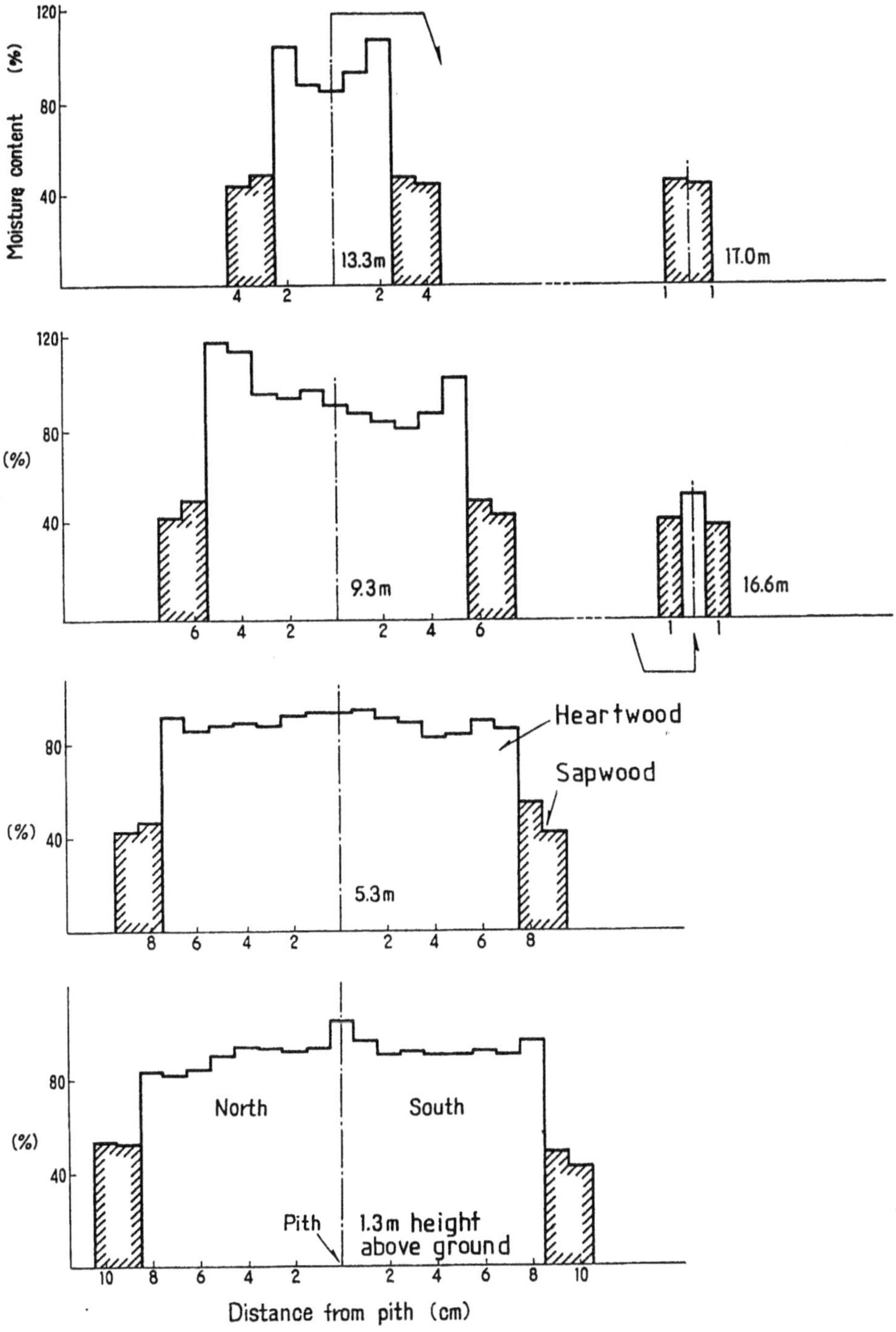

Fig. 5.2. Lateral distribution of moisture at different heights above ground level in a trunk of *Fraxinus mandshurica* felled in mid-summer. (Yazawa and Ishida 1965a)

Table 5.2. Percentage volume of water and gas in the sapwood and heartwood of living trees (Trendelenburg and Mayer-Wegelin 1955, Ziegler 1968)

Species	Water		Gas		Water and gas	
	Sap-wood	Heart-wood	Sap-wood	Heart-wood	Sap wood	Heart-wood
Picea spp.	60	20	16	57	76	77
Abies spp.	67	23	9	54	76	77
Pinus strobus	68	28	11	52	79	80
Pinus sylvestris	60	13	13	61	73	74
Larix spp.	60	21	13	48	73	69
Quercus spp.	46	45	21	18	67	63
Fagus sylvatica	50	40	14	23	74	63

percent and if oxygen is present; otherwise colorless heartwood is produced and the extractives remain in the ray and vertical parenchyma and do not diffuse into neighboring cell walls (Frey-Wyssling and Bosshard 1959).

Distinctly demarcated transition zones do not exist in all species at all times of the year. When the moisture content of these zones has been determined, it is lower than the adjacent sapwood and sometimes than the heartwood (see Table 5.1). Yazawa and Ishida (1965a) found less marked differences with *Sorbus, Acer, Cornus, Magnolia, Acanthopanax, Octrya, Tilia*, and *Quercus* species than with *Taxus baccata*.

Concomitant with the decrease in the volume of water in the cells is the increase of gas (Vintila 1939). The sum of the volumes of water and gas in the sapwood and heartwood of a number of species remains approximately the same (Table 5.2).

An association exists between the proportion of water and gas in cell lumens on the one hand and on the other the metabolic activity of the parenchyma and the blockage of the tissues with deposits of extractives, aspiration of pits, and formation of tyloses (Zycha 1948, Trendelenburg and Mayer-Wegelin 1955, Dadswell and Hillis 1962, Dietrichs 1964a,b). There are indications that when the water/gas ratio in the cell lumens falls below a specific threshold value, the changes from sapwood to heartwood or woundwood are initiated. In a study to determine the existence of any relationship between the moisture content in wood and the initiation of heartwood formation, Sachsse (1967) determined the volumes of free water and gas in the cell lumens across stems of *Fagus sylvatica* and a *Populus* hybrid. Within each species a characteristic similarity of the water/gas ratio was found irrespective of the level in the stem and the felling dates. However, there were typical differences between the two species due to the high moisture content in the *Populus* heartwood and the low content in the red heartwood of *F. sylvatica*. It was noteworthy that the ratio of free water to gas at the sapwood-heartwood boundary was similar for both species although the minimum water content was lower with the latter species. Sachsse (1967) concluded it was not possible to develop a simple relationship between the moisture content in the wood and the initiation of heartwood formation. Nevertheless, because of the constancy of the water/gas ratio in the lumens in the sapwood near the heartwood periphery, there is likely to be an indirect relationship between this ratio and the vitality of the parenchyma in these species.

Hartig (1888 as quoted by Yazawa et al. 1967) considered that oxidation of wood with air plays an important role in the formation of heartwood. Later, other workers (e.g., Frey-Wyssling and Bosshard 1959) suggested that an oxidative polymerization is involved in the formation of regular heartwood formation; to examine this proposal knowledge of the gases in wood is required.

Analysis of gas from the interior of woody stems in living trees is not easy owing to difficulties in eliminating leakage of air when samples are removed under vacuum. Moreover it is not possible to take samples from particular places across the stem. Examples of the analyses of gas from the interior (probably mainly the heartwood and nonrespiring) of a living stem of *Quercus macrocarpa* showed that the oxygen content fluctuated between 8 and 20 percent throughout the year and was lower than that in the air (21 percent) (Chase 1934). It was lowest in late spring to mid-summer, and the carbon dioxide content was less than 4 percent at all times. On the other hand, the composition of samples of gas taken from the interior of a young *Populus deltoides virginiana* (in which heartwood is presumed to be absent) changed during the year. The oxygen content was close to 20 percent in the winter months but less than 5 percent at other times; the carbon dioxide showed the reverse trend with the maximum of 15–20 percent occurring from mid-summer to late autumn (Chase 1934). Other work has shown the composition of the gases to differ markedly from those found in air, with the concentration of carbon dioxide being as high as 26 percent of the gases removed and oxygen as low as 3 percent (MacDougal et al. 1929, MacDougal and Working 1933, MacDougal 1936). The lack of oxygen in the transition zone of *Fraxinus excelsior* resulted in a respiratory coefficient considerably larger than 1 so that an almost anaerobic environment resulted (Ziegler 1957).

By analyzing the gases emitted from small blocks of wood obtained immediately from cross-sections of freshly fallen trees (about 40 cm in diameter) of *Fagus sylvatica, Quercus robur, Fraxinus excelsior*, and *Picea abies*, Möller and Müller (1938) found that most of the *Fagus sylvatica* blocks behaved as sapwood, whereas the other three species contained mainly dead heartwood cells. The amount of carbon dioxide emitted (in a few hours) increased in the central portions of the cross-sections, and with *P. abies* the increase was very marked between 30 and 40 growth rings from the cambium of a 42-year-old tree. More recently and with refined techniques, Jensen (1967, 1969) found with *Quercus rubra* that within 15 seconds of collecting the samples the oxygen content was 0.9 percent and the carbon dioxide 4.2 percent. The latter rose to 15.2 percent in 60 seconds and the oxygen fell to 0.4 percent. He ascribed the increase in carbon dioxide to its release from solution in water in the tree. With this and other *Quercus* species he found that the oxygen concentrations were less than 2 percent, and that there was no relationship between gas concentration and season, height, or depth in the trees. It can be concluded that the amount of oxygen inside a tree trunk is very small and the amount of carbon dioxide relatively large.

The presence of carbon dioxide in stems of *Acacia mearnsii* (a "dry" heartwood species) was carefully monitored by Carrodus (1971) and Carrodus and

Triffett (1975) throughout a year. They found with two trees of *A. mearnsii* sampled in early winter that the percentage of carbon dioxide and oxygen in the inner sapwood was 41–65 percent and 1–1.5 percent respectively. In the outer heartwood of one tree the carbon dioxide and oxygen percentages were 41 and 2 percent respectively. Attention was drawn to the work of others that high concentrations of carbon dioxide can inactivate respiratory and succinate dehydrogenase enzymes. When water is lost from cell lumens it is replaced by gas which probably is rich in carbon dioxide. If this inactivates the normal respiratory enzymes, then energy for metabolic activity must be derived from other routes. If oxidative polymerization of the initial extractives on heartwood formation occurs, only small amounts of oxygen are available for this purpose.

Evidence of a gas canal system in *Pinus sylvestris*, *Picea abies* and *Fagus sylvatica* has been found (Sifton 1957, Ziegler 1964, Back 1969). The system is formed from the spaces at the borders or corners of adjacent parenchyma cells in uniseriate or multiseriate rays. The volume of the system is less than 0.5 percent of the wood and presumably serves the metabolism of the parenchyma and exchange of gas via the pits; it would be expected that the network would be blocked on the formation of regular heartwood.

Carrodus (1971) suggested that very high concentrations of carbon dioxide were responsible for the production of polyphenols. However, cell-free preparations from cross-sections of *Robinia pseudoacacia, Tilia cordata*, and *Fraxinus excelsior* showed a decreasing ability to transfer carbon dioxide into organic compounds from the stem periphery inwards (Höll 1974, Höll and Meyer 1977), and so far there is no evidence of the dark fixation of carbon dioxide in inner sapwood. It is more likely instead that ethylene, which is physiologically active at very low concentrations, is responsible for initiating the reactions leading to polyphenol synthesis. Gaseous carbon dioxide stopped the formation of polyphenols in injured *Pinus radiata* sapwood (M.P. Coutts 1967 personal communication). Moreover, ethylene (5 ppm in air), but not carbon dioxide (27.5 percent in air), stimulated the production of pinosylvin in sapwood blocks of the same species (Shain and Hillis 1973).

The examination of samples taken at one point in time does not provide the full picture of the importance of the role of the transpiration stream in supplying oxygen to, and removing carbon dioxide from the living, internal tissues of the tree. The velocity of the transpiration stream through the bole of the tree varies significantly within and among groups of woody plants. With some ring-porous angiosperms the velocity is 20–40 m h^{-1}, with diffuse-porous ones it is 2–6 m h^{-1} and with conifers 1–2 m h^{-1} (Hook et al. 1972). Obviously, as these authors point out, the transpiration stream plays a vital role in the flow or movement of gases within the sapwood, and all living cells obtain their oxygen and release their carbon dioxide chiefly in solution by the long-range transport of the transpiration stream. Nevertheless, only the outer part of the sapwood of *Quercus phellos* has a transport function (Ziegler 1968).

5.2 Inorganic Compounds

The inorganic components are low in those woods from temperate zones which contain 0.1–1.0 percent and those from tropical and subtropical zones with up to 5 percent. The amount of individual components varies from species to species but the main components are calcium and then potassium and magnesium salts followed by those of manganese, sodium, phosphorus, and of other elements (Young and Guinn 1966; Cutter et al. 1980).

It has been known since the investigations of Daube (1883, Ellis 1965, Ziegler 1968) that considerable resorption of inorganic elements (mainly phosphorus, potassium, and sulfur) can occur in many species when sapwood is transformed into heartwood. Heartwood has the lowest concentration of inorganic nutrients in the tree, and sometimes the concentration is considerably lower than elsewhere. An increased concentration can occur with the discoloration and decay of wood in the living tree (Shigo and Hillis 1973). A selective and contradirectional flow of materials occurs with nutrients from the heartwood region and with primary metabolites to it. The inorganic components accumulate in the rays and, when present, in the resin ducts. The nature and extent of redistribution and storage vary with the particular element involved, the species, and the site on which it grows. The distribution between sapwood and heartwood may be influenced by the soil and those trees with a poor nutrient status may have greater nutrient redistribution levels.

Crystals, usually of calcium oxalate or calcium carbonate and sometimes of large size, are found much more frequently in heartwood than in sapwood of a wide range of species (Chattaway 1955a, 1956). Most of the massive crystals in *Populus maximowiczii* are located in the vessels in the heartwood, especially near the vessel-ray pit region and rarely in other tissues (Lee et al. 1985). Crystals and silica usually occur in different genera and/or species although there are cases where they occur in the same species and sometimes in the same cell. Silica conglomerates are found in a number of species and are formed in the ray parenchyma in the middle sapwood of *Syncarpia glomulifera* and long before the formation of heartwood extractives (Hillis and de Silva 1979). In one sample of this species 0.50 percent silica was present in mid-sapwood and 0.44 percent in the heartwood. The deposits in the vessels of *Tectona grandis* are mainly silica (Hillis and de Silva 1979) rather than a calcium phosphate complex previously reported. Silica is found in numerous species (Amos 1952, Hirata et al. 1972, Scurfield et al. 1974, ter Welle 1976 a,b) as an inclusion but not as skeletal material. It is most commonly found in the ray parenchyma in a pure form of characteristic berry shape and medium size of 6–15 μm (e.g., as in *Syncarpia glomulifera* and *Tristania suaveolens*) or infrequently as a deposit in vessels, as with *Tectona grandis*. The mechanism of formation of the two types may be different. The silica "berry" found in the sapwood of *S. glomulifera* is attached to the parenchyma cell wall by a nonsiliceous stalk (Hillis and de Silva 1979) (Fig. 5.3).

The distribution of nutrients in some subtropical rain forest species is less than that of non-rainforest species (Attiwill 1980, Lambert 1981, Lambert et al. 1983). For example, *Eucalyptus grandis* growing on a former rainforest site typically has about 100 ppm phosphorus in the sapwood and only 5 ppm in the heartwood,

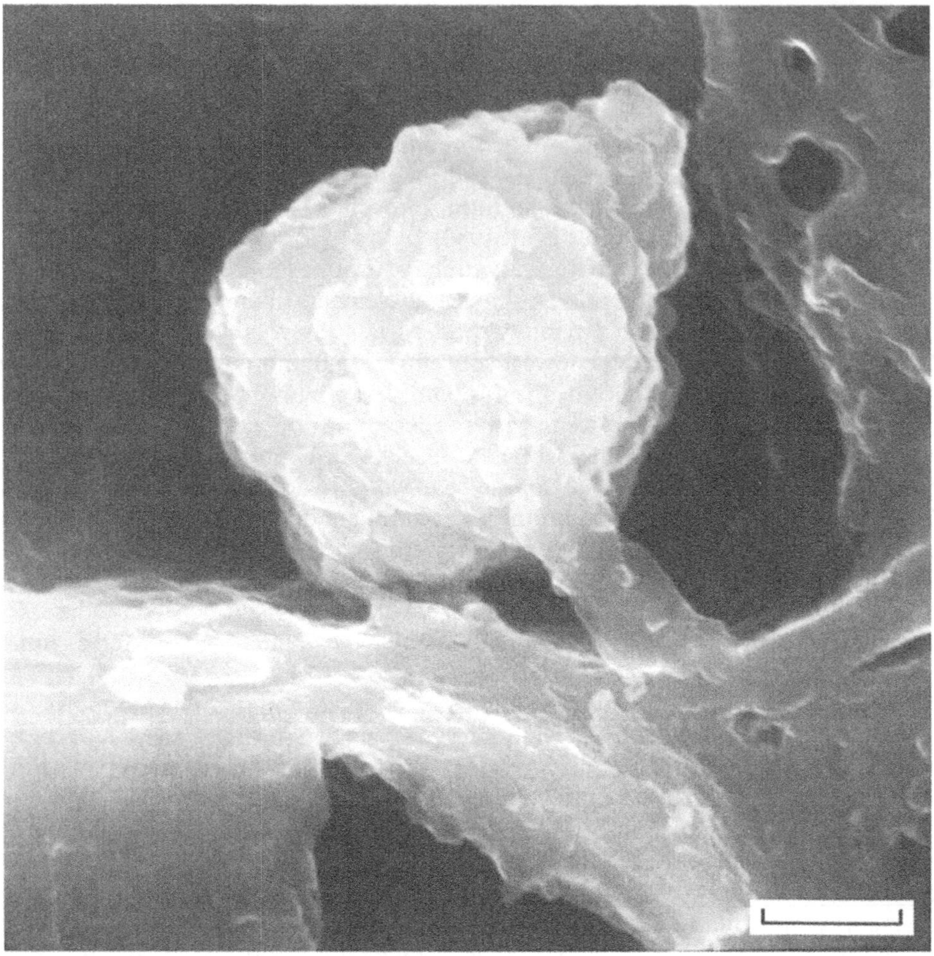

Fig. 5.3. Silica berry attached to a nonsiliceous stalk in a ray cell of sapwood of *Syncarpia glomulifera* (scale 2 μm; SEM by D. de Silva)

whereas one of the rainforest species, *Orites excelsa*, had 63 and 25 ppm phosphorus respectively (Lambert et al. 1983). Various eucalypts growing on different sites had one-tenth or less phosphorus in the heartwood than in the sapwood and one-fifth or less of potassium (Lambert 1981). The mean concentrations of phosphorus in the sapwood and heartwood of *E.obliqua* were 0.2 and 0.016 g kg^{-1} and of potassium 0.60 and 0.075 g kg^{-1}, respectively (Attiwill 1980). Some species can accumulate the minor nutrients, for example, *Orites excelsa*, the shakes of which contain aluminum succinate (see Chapter 2.6.2.8), can contain 6475 ppm aluminum in the sapwood and 1425 ppm in the heartwood. Another species, *Caldluvia paniculosa*, can accumulate 2345 ppm magnesium in the sapwood and 1670 ppm in the heartwood (Lambert et al. 1983).

A common pattern in the redistribution of nutrient elements occurs on heartwood formation in many species, and this may reveal a role for specific elements in the formation of various heartwood features.

Elemental analysis of cross-sections of trees of *Picea excelsa, Pinus sylvestris, Betula verrucosa, Populus tremula*, and *Quercus robur* samples in late summer showed two different patterns. The amounts of phosphorus and potassium steadily decreased in the sapwoods and were low in the heartwoods. Radial gradients for the multivalent cations magnesium, manganese, calcium, zinc, copper, and iron were quite similar to each other but different from those of potassium and phosphorus. With the first four species, the concentration of the latter group of elements generally decreased from the cambium to the third or fourth growth ring and then a trend of increasing concentration was observed with increasing age of the tissue, irrespective of the presence of heartwood. With *Quercus robur* the minimum concentration of the multivalent cations was observed at the sapwood-heartwood transition zone (Basham and Cowling 1975). The patterns for *Q. robur* (in Sweden) were essentially the same as *Q. alba* sampled in North America in summer. During the transformation of sapwood of *Robinia pseudoacacia* and *Maclura pomifera* to heartwood, the ash content fell by about 30 percent, but the amount of phosphorus dropped about 95 percent, while the calcium levels fell 24 and 41 percent, respectively; the other elements remained constant (Hart 1968).

With *Pinus radiata, P. nigra* var. *poiretiana*, and *P. sylvestris* the amounts of potassium and phosphorus decreased and calcium increased from the outer heartwood into the sapwood except for the growth rings in the transition zone immediately surrounding the heartwood, where the reverse occurred, in that the amounts of potassium and phosphorus increased and the calcium decreased (Wright and Will 1958, Orman and Will 1960). The most marked changes in the transition zone were the increases in the amounts of potassium, nitrogen, and phosphorus particularly in the upper levels of the tree. In view of the involvement of calcium in enzyme systems for the translocation of carbohydrates and for starch-sugar changes and also the involvement of potassium and phosphorus in other enzyme systems, this situation indicates considerable metabolic activity in, and transportation of carbohydrate to, the transition zone. In a limited number of *Quercus alba* growing on the one site and collected in summer, it was found that potassium decreased sharply, and calcium, magnesium, and manganese decreased in the sapwood adjacent to the heartwood boundary. Phosphorus decreased across the sapwood and remained at a low concentration in the heartwood (Wardell and Hart 1973). Potassium was the only element to increase across the sapwood and then showed the sharpest decrease in concentration at the transition zones. In *Populus robusta* trees, phosphorus decreased considerably from the fifth sapwood ring to the heartwood whereas calcium, magnesium, and potassium were high in the heartwood and low in the sapwood (Janin and Clément 1972).

The ash contents of discolored and wound woods are higher than those of the sapwoods and particularly the heartwoods; the increases are largely due to potassium, calcium, and magnesium salts (Hart 1968, Shain 1971, Shigo and Hillis 1973). Although potassium, phosphorus, and calcium probably have roles in the formation of heartwood, their importance is not clear.

An intensive study of the distribution of inorganic constituents in *Picea mariana* revealed concentration of elements in various regions (Saka and Goring 1983). Notably the concentration of copper, sodium, potassium, aluminum, sulfur and chlorine compounds was much higher in the transition zone between sapwood and heartwood than elsewhere. In addition, almost all elements were concentrated and localized in the tori and half-bordered pit membranes to a much greater extent than elsewhere. This was attributed to the high concentration of pectic substances in the pit membrane regions. In pines, inorganic substances accumulate in resin canals and ray cells (Fossum et al. 1972).

5.3 Storage Substances and Intermediates

Reference has been made to the general nature of the storage materials in the sapwood, namely starch or fat or both (Chapter 2.2.4). These substances on dissolution, together with the primary photosynthates (mainly sucrose) in the translocation stream, pass inwards through the ray parenchyma to provide the basis for the formation of secondary components in wood. More is known of the distribution, amounts, and variation of starch in sapwood due to its easy recognition (a blue color with iodine stains) (Fig. 5.4). The presence of starch is a useful characteristic of sapwood but its absence cannot be taken as indicative of the transition zone or heartwood, even when starch is the normal storage compound. Starch can be very local in its distribution, being present on one side of the tree and sparse or absent on the other. The amount of starch across the sapwood varies and decreases towards the heartwood boundary, where it is absent. It usually decreases in amount with increasing height in the tree (e.g., Hillis et al. 1962, Higuchi et al. 1964). The amount of starch present varies with the species and it can be as much as 7 percent of the wood. The amount of starch stored in sapwood is seasonal, it can fall to low levels early in the growing season; with bud-swelling and unfolding of the leaves and at times of need such as with profuse flowering, disease (Kile et al. 1981), defoliation (Bamber and Humphreys 1965), or with injury (Sauter 1966b). Changes in starch content can be induced by temperature changes but the extent varies with species and conditions (Sauter 1966b, 1967) and usually in greatest amounts during the cold periods (Höll 1975). Abundant conversion of starch to sucrose occurred in the ray tissues of a *Salix* hybrid at low temperatures, and hexose accumulation resulting from hydrolysis of the sucrose was much higher at 2° compared with 21° C (Sauter 1983). On a few occasions small groups of cells containing starch have been observed in the heartwood of some tropical species, as with *Shorea* species and in *Fagus sylvatica* (Dietrichs 1964b).

Lipids are present in a range of genera (Table 2.2) and also in small amounts in the sapwood of *Chamaecyparis obtusa, Cryptomeria japonica, Thujopsis dolabrata, Magnolia obovata, Pterocarya rhoifolia*, and *Betula maximowieziana* (Higuchi et al. 1964). The seasonal fluctuations of the amount of fats in the wood and the bark of trees has been reviewed by Kozlowski and Keller (1966). Fats contain more than double the energy per unit weight than do carbohydrates. The amounts

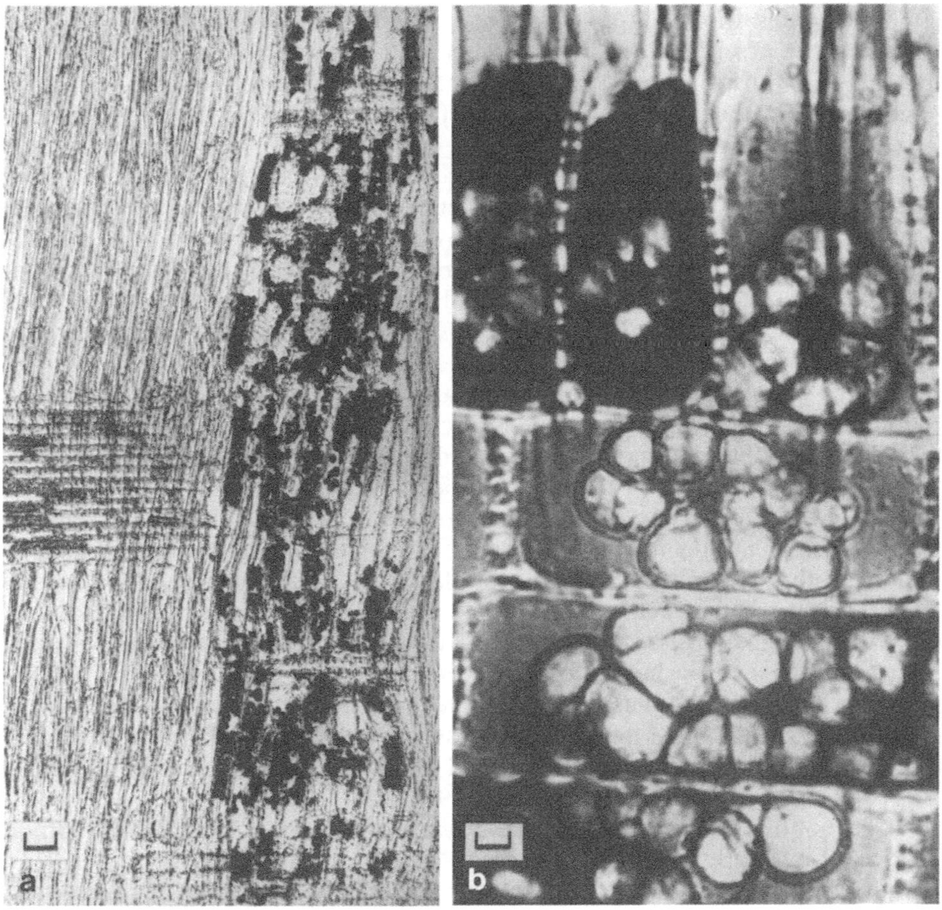

Fig. 5.4.a. Starch grains (stained with iodine) in vertical parenchyma, and polyphenols in ray parenchyma of the inner sapwood of *Acacia implexa* (photomicrograph by H.E. Dadswell) (scale 100 μm). **b** Starch grains clustered around nuclei in the ray cells of sapwood of *Nothofagus cunninghamii* (photomicrograph by M.M. Chattaway) (scale 10 μm)

decrease less than does starch with increase in sample height in the tree and show slight increases with increasing distance from the cambium. Lipid production is favored by low temperatures (Ziegler 1964). Examination of the triglycerides and free fatty acid concentrations in the trunk of a 29-year-old *Tilia cordata* between early winter and late spring showed that marked changes occurred before budbreak and unfolding of the leaves (Höll and Priebe 1985). Most of the triglycerides were hydrolyzed in the trunkwood in early spring, and their degradation products were translocated to sites where they would be needed or to metabolic sinks such as the roots. The amount of free fatty acids was highest at this time. The increase in triglycerides from early to mid-spring probably arose from stored carbohydrate. The triglycerides isolated from the peripheral trunkwood were made up of saturated fatty acids, whereas those of the inner parts contained larger amounts of unsaturated fatty acids. The lipid contents of the sapwoods of *Pinus banksiana*, *P. densiflora*, and *P. strobus* showed seasonal and internal variations but less than those of starch. The lipid contents were least in the spring and summer periods in the sapwood, but absent from the heartwood at all times (Fukazawa et al. 1980).

A clearer association between storage materials and nuclear activity was found in *Robinia pseudoacacia* (Nobuchi et al. 1984b). Starch grains disappeared from the innermost annual ring of the sapwood in late summer and this was immediately followed by a considerable increase in size of the lipid droplets. Just after this the formation of the heartwood became conspicuous and it enlarged rapidly. In the winter season starch grains and nuclei were again present in the ray parenchyma.

Starch grains can be stored in the ray and axial parenchyma cells. Numerous observations have been made of their abundant appearance in the axial parenchyma cells of the outermost sapwood and gradual decrease towards the transition zone. In a number of tropical trees from Venezuela the starch was stored mainly in the interfibrous axial parenchyma cells and in the living fibers, but the ray parenchyma were mostly without starch (Fink 1982). Starch grains are abundant in both axial and ray parenchyma cells of the outer sapwood of *Bridelia retusa*. Their gradual decrease towards the transition zone resulted in the procumbent ray cells having less starch grains than the erect cells. This decrease was accompanied by a strong acid phosphatase activity in the middle sapwood and increase in lipid and polyphenols in the inner sapwood and transition zone (Nair et al. 1981). On the other hand, although starch was detected from large to small amounts from the outer to inner sapwood of *Melia azedarach*, neither zone possessed phosphatase activity (Baqui et al. 1979).

The mobilization of stored substances depends on the activity of appropriate enzymes, so that the rate of movement to meet particular situations depends on various circumstances. Movement from the cambial region can be relatively fast, as shown by the translocation of labeled glucose in early summer from the phloem of *Eucalyptus sieberi* through 1.5 cm width of sapwood (3 growth rings) to the heartwood periphery in 19 days (Hillis and Hasegawa 1963). During this period some of the substrate had been converted to extractives and incorporated in the cell wall. Labeled kino was formed within 5 days of injection of the glucose.

In a study of the distribution of simple sugars, species which contained regular heartwood (*Quercus* species), irregular heartwood (*Fagus sylvatica* and *Betula alba*), ripewood (*Picea abies*), and a partly colored heartwood with a transition zone (*Terminalia superba*) were sampled over a late autumn and winter period (Dietrichs 1964a). There was a general decrease across the sapwood of glucose, fructose, sucrose, and mannose. [There is a similar decrease in *Robinia pseudoacacia* (Höll 1972, 1973)]. The sugar content decreased from the cambium to the heartwood boundary in *Angophora costata* trees sampled in early winter and in early spring, but the reverse distribution was found in early autumn when the amount was more than double at the heartwood periphery than at the cambium (Hillis et al. 1962). The heartwood of *Picea abies* contained mannose, that of a *Quercus* species glucose, arabinose and xylose, whereas heartwood of the other species did not contain simple sugars. It is interesting that when cross-sections were dried slowly, the amount of glucose and fructose increased in the sapwood close to the heartwood periphery in *Picea abies*, a *Quercus* species, and *Fagus sylvatica*. Administration of labeled sucrose to the latter species and to *Prunus yedoensis* resulted in the formation of shikimic acid and probably quinic acid within 4 and 12 days, respectively (Dietrichs 1964a, Hasegawa and Shiroya 1965, 1966). The formation of shikimic acid is important because it is not only a key intermediate in the biosynthesis of polyphenols but it has been detected in the cytoplasm of the ray cells of *Illicium religiosum* (Higuchi et al. 1967a) and found to increase sharply to over 2 percent in the transition zone of *Quercus iberica*, the heartwood of which contains ellagitannins (Nutsubidze and Abashidze 1980). Shikimic acid has been detected in the axial and ray parenchyma of the inner sapwood and the transition zones of *Melia azedarach*, concomitant with the degradation of starch grains (Baqui et al. 1979), and also in the transition zone of *Acacia nilotica* (Shah et al. 1981a) but not elsewhere in these species.

Although the seasonal variation in amounts of starch and fatty substances in sapwood is well documented, there is little information available on changes in the transition zone. Microscopic examination showed small seasonal differences of the starch content of the ray parenchyma in the outer and middle parts of the transition zone of *Cryptomeria japonica*, but there was no change in the inner parts. The lipids responding to Sudan IV were found in ray parenchyma cells from the cambium to the heartwood, whereas those staining with Nile blue suddenly increased at the boundary of sapwood and transition zone. No seasonal changes were found (Nobuchi and Harada 1983). The lipid droplets changed in appearance and increased in size at the boundary between sapwood and the transition zone in *C. japonica* (Nobuchi and Harada 1983) and *Robinia pseudoacacia* (Höll and Poschenreider 1975, Nobuchi et al. 1984b) and across the sapwoods of *Abies alba, Larix decidua, Pinus sylvestris, Fraxinus excelsior*, and *Fagus sylvatica* in which "fatty acids" appear to predominate in the heartwood (Frey- Wyssling and Bosshard 1959). At the boundary between transition zone and heartwood, and in the heartwood, lipid droplets were seen also in the tracheids of *C.japonica*. As they occurred at the pit pairs between the tracheids and parenchyma, it was suggested they were transferred to the tracheids (Nobuchi and Harada 1983, 1985). The colorless droplets in the transition zone became yellow or brown in the heartwood. The highest acidity of the lipids in *Robinia pseudoacacia* was found in the innermost sapwood growth ring. When the heartwood boundary was crossed, there was a decrease in lipid content, a rapid drop in acid value and a parallel increase in ester groups (Höll and Poschenrieder 1975). On the other hand, in *Pinus radiata* there was a pronounced decrease in fatty acid esters and an increase in free fatty acids of different composition when the heartwood boundary was crossed (Hemingway and Hillis 1971). The larger amounts of fatty acids, which are longer in length than those in the sapwood, apparently arose at the heartwood periphery of *R.pseudoacacia*. On the other hand the fatty acid pattern of *Tilia cordata* and *Betula verrucosa*, which did not contain heartwood, did not change towards the center of the trunk (Höll and Poschenrieder 1975).

The amount of starch and number of starch grains decrease in the inner sapwood, for example with the hardwood *Angophora costata* (Hillis et al. 1962). Also, it has been observed that starch disappears abruptly in the transition zone or adjacent to the heartwood periphery (Frey-Wyssling and Bosshard 1959, Hillis 1962b, Shah et al. 1981a). In *Acacia nilotica* starch, when present, occurs largely in the axial parenchyma cells rather than in the ray cells (Shah et al. 1981a). The procumbent ray cells in the middle sapwood of *Bridelia retusa* have a smaller number of starch grains than the erect cells (Nair et al. 1981). The starch content in the middle sapwood of *Pinus densiflora, P. strobus*, and *P. banksiana* is maintained at the highest levels through most of the year. Yamamoto (1982) has suggested that the increase in starch levels in the inner sapwood of *P. densiflora* during early autumn is connected with the parenchyma wall development in the transition zone during that period. There is a decrease in the inner sapwood of the other two pines during this period. Starch grains are often found in the living ray parenchyma of trees using starch as a reserve material, and after administration of labeled, glucose, the starch grains in the transition zone can be strongly labeled, thus indicating the vitality of parenchyma cells in that region (Yamamoto 1982). Parenchyma lacking nuclei in the transition zone do not contain starch grains (Nobuchi et al. 1982).

Lipid materials, when present in sapwood, can be used as reserve metabolic materials but the difference in composition from the end-products of metabolism found in heartwood is not easy to detect by microscopic examination. However, their shape and size alters in the innermost two or more sapwood rings of *Cryptomeria japonica* and nuclei, starch grains and other cell organelles are also missing from these cells (Nobuchi et al. 1982). A marked decrease in lipid content occurred in the transition zone of *Pinus densiflora* and *P.banksiana* from spring to autumn and in *P. strobus* from summer to autumnm. In *Acacia nilotica*, the lipid and nucleic acids contents decrease towards the transition zone, but shikimic acid was detected in the ray and axial parenchyma cells in that zone only (Shah et al. 1981a).

5.4 Nitrogenous Compounds

Mature xylem tissues contain very little nitrogen – only about 0.03–0.10 percent – whereas the cambial tissues and newly formed sapwood may contain ten times that amount. Nitrogen is a major constituent of amino acids, peptides, enzymes, and other proteins and nucleic acids. Cowling and Merrill (1966, Merrill and Cowling 1966) made several conclusions from an extensive study of nitrogen distribution in the tree stems of *Populus grandidentata* (sampled in summer), *Fraxinus americana* (sampled in autumn), *Quercus rubra* (sampled in autumn), *Picea abies* (sampled late winter) and *Pinus strobus* (sampled in winter, autumn, and summer). Their conclusions were that (i) during differentiation and growth, nitrogen in the cytoplasm and primary walls of xylary cells is diluted by the formation of cell wall substances (see also Grozdits and Ifju 1969), (ii) nitrogen in the cytoplasm of wood-fiber cells is eluted into the transpiration stream as these cells become functional in conduction, (iii) death of cells during aging of sapwood and formation of heartwood is accompanied by retrieval of nitrogen from those cells, (iv) heartwood is physiologically nonfunctional and its nitrogen content is stable.

In contrast to wood-fiber cells, xylem-parenchyma cells remain alive and retain their nitrogen-rich cytoplasmic constituents for many years. Sapwood cells eventually die and lose their cytoplasm (Frey-Wyssling and Bosshard 1959). With species which do not form regular heartwoods (e.g., *Populus* species) these changes occur very gradually across the sapwood; with regular heartwoods (as in *Quercus* species) the changes are abrupt. Further conclusions by Merrill and Cowling (1966) concerning the amounts and distribution of nitrogen in tree stems were (v) the distribution of nitrogen among various tissues can vary substantially and in the sapwood can be 1.2 – 2.5 times higher than that in the heartwood, (vi) when expressed as a percentage by weight, the nitrogen content of wood decreases progressively from the cambium, reaches a minimum, and increases again at or near the pith. With all five species tested the decrease in nitrogen in the sapwood from the cambium inwards extended only to the sapwood-heartwood boundary. With the *Populus, Fraxinus* and *Pinus* species the decrease was relatively continuous, with *Quercus* and *Picea* species a more gradual decrease was followed by an abrupt and rapid decrease in the transition zone. It can be observed from their data that the *Pinus strobus* sample felled in winter showed an abrupt decrease in nitrogen content in the transition zone, whereas other trees of this species felled at other times did not. In those species without heartwood the nitrogen decreased steadily to the pith, and (vii) the nitrogen content of living parenchyma cells is much higher than that of dead parenchyma cells or of fiber cells. Nitrogen is stored in the parenchyma cells, and the higher nitrogen content in the sapwood of angiosperms is due to the larger amount of parenchyma present than in the gymnosperms examined. In *Juglans nigra*, the yearly pattern of nitrogen content is bimodal with one low during dormancy and another during the early growing season. In *Pinus serotina*, the nitrogen contents of the middle and inner sapwoods and transition zone were low throughout the year (Nelson 1978).

Only a minor portion of the nitrogenous substances in both the sapwood and heartwood of *Liquidambar styraciflua, Populus grandidentata*, and *Pinus strobus* could be extracted with neutral solvents. However, free amino acids, including common protein amino acids, were present in abundance in sapwood but only in trace amounts in heartwood (see also Fukuda 1963, Laidlaw and Smith 1965). On the other hand the amount of water-soluble nitrogen increased considerably in early summer from the periphery inwards in stems of *Tilia cordata* and *Fraxinus excelsior* (Höll and Meyer 1977). Merrill and Cowling (1966) concluded that at least a portion of the nonextractable nitrogen consisted of protein, peptides, or amino acids.

The retrieval of nitrogen from dead ray parenchyma cells in the inner sapwood indicates a selective and contradirectional flow of materials, with nitrogenous and inorganic compounds (see Section 5.2) moving centrifugally and primary metabolites moving centripetally. There is no information that these selective flows may occur at different periods of time. For example, an initial flow inwards of primary metabolites and their subsequent conversion to extractives might be followed later by an outward flow of the small amount of nitrogenous compounds from disintegrated cytoplasms.

In the above studies, and the similar findings of Nickel (1960) and Becker (1962), the total nitrogen content of the individual zones has been converted to protein but there was no proof that the ratio of nonprotein nitrogen to protein nitrogen remains the same in all the annual rings.

When precautions are taken to separate the protein nitrogen in the individual annual rings across log cross-sections of *Pinus strobus, Fagus sylvatica*, and *Robinia pseudoacacia* collected in early winter, appreciable concentrations of protein were found in the sapwood rings adjacent to the heartwood; the peak was less pronounced in *Larix decidua* (Ziegler 1968). In this connection it is noteworthy that after a decrease in concentration across the sapwood there was an appreciable increase in the vitamin content in the transition zone before it fell to very low values just inside the heartwood (Zeigler 1968).

5.5 Ethylene

Two features associated with the formation of polyphenols in plants are the activity of phenol-oxidizing enzymes (Chapter 6.2.7) (Conn 1964, Harkin and Obst 1973) and the production of the phytohormone ethylene (Burg 1962, Chalutz et al. 1969, Hyodo and Yang 1971, Abeles 1973, Craker and Wetherbee 1973, Rhodes and Wooltorton 1973, Archer and Hislop 1975, Hillis 1975). All plant tissues are capable of producing ethylene, although the production rate is normally low. A marked ethylene production is induced during certain stages of growth such as ripening of fruit, senescence of flowers and abscission of leaves. Production can also be induced by mechanical wounding and environmental stresses such as drought, infection, chilling, and flooding (McMichael et al. 1972, S.F. Yang and Hoffman 1984). Auxin promotes ethylene production (Fuchs and Lieberman 1968), which in turn is autocatalyzed. Ethylene can in some cases, but not others, have a direct effect on phenylalanine ammonia lyase, a key enzyme of the shikimic acid pathway, and some other enzymes (e.g., Hyodo and Yang 1971, Rhodes and Wooltorton 1971, Sarkar and Phan 1974). It increases the activity of α-amylase (Jones 1968), acid phosphatase, succinate dehydrogenase, lipase, and ATPase (Shah et al. 1981), peroxidase and polyphenoloxidase (see Hillis 1977).

Polyphenols are found in increased quantities in heartwood, and ethylene is thought to be closely associated with heartwood formation in both gymnosperm and angiosperm species (Shain and Hillis 1973, Hillis 1975, Nelson et al. 1981). The application of 2-chloroethylphosphonic acid ("ethrel"), an ethylene-releasing compound, to the sapwood of a small tree of *Rhus succedanea* by Hillis (1975) resulted in the formation of typical heartwood polyphenols. When this compound was injected into the sapwood of young *Azadirachta indica* trees, heartwood was produced at the site of application. Histochemical studies revealed the disappearance of starch but the presence of increased amounts of lipids and polyphenols in the cells at the site of the treatment. The axial and ray parenchyma cells at the site, and a short distance from it, showed increased activity of enzymes such as lipase, acid phosphatase, succinate dehydrogenase, and ATPase (Shah et al. 1981b). Injection of 2-chloroethylphosphonic acid into the sapwood of *Samanea saman* resulted in high peroxidase and succinate dehydrogenase activities, high lipid, protein, and extractives contents, and reduced amounts of starch (Patel and Bhat 1984).

One difficulty with investigations of the liberation of ethylene by woody tissues in vitro is the very small amounts (about 1 x 10^{-9} liter/g dry wood per hour) of ethylene emanating from active tissues. Also the wounding caused by removal of samples, and their storage under stress, affects many of the same metabolic processes thought to be involved in producing heartwood. It is recognized that ethylene analysis involves the production of ethylene by an excised tissue over a relatively long period of time in a gaseous atmosphere very different from that within the tree with possibly a changed direction of metabolism. Nevertheless there is accumulating evidence of the importance of the involvement of ethylene in the formation of heartwood, protective tissues, and exudates.

Ethylene emanation decreased from the high levels in the outer sapwood (including cambial tissues) to the inner sapwood of *Pinus radiata*, *Juglans nigra* and *Prunus serotina* (Shain and Hillis 1973, Nelson 1978), and still lower in the transition zone of *Chamaecyparis obtusa* examined in winter (Yamanaka 1985). On the other hand, ethylene production by the transition zone of *Eucalyptus tereticornus* was highest in late autumn and early winter (Hillis 1975). Significantly greater quantities of ethylene were produced during winter by the transition zones of *Pinus radiata* than those collected during other times of the year (Shain and Hillis 1973). Furthermore, the amounts were greater than those formed during the entire year by middle and inner sapwood, and in addition these amounts did not change significantly throughout the year. It is noteworthy that the very small amount of parenchymatous tissue in *P.radiata* sapwood is capable of producing considerable quantities of ethylene. The production of ethylene by the heartwood was negligible. Freshly collected sapwood, when exposed to ethylene, produced characteristic heartwood stilbenes (Shain and Hillis 1973), and the formation of stilbenes when similar blocks were kept in high humidity (Coutts 1969) was probably initiated by wound ethylene. With *Juglans nigra* (which does not have a distinct transition zone), the ethylene production of the sapwood adjacent to the heartwood reached a maximum early in the dormant period of tree growth and was low when the cambium was active and at other times (Nelson 1978). The middle sapwood showed a major peak in the dormant season and a smaller one in the spring season. However, *Prunus serotina* did not show the pattern of heartwood formation in the dormant period that is evident in a number of other species. Ethylene production in the sapwood–heartwood transition zone of *P.serotina* is greater than in the middle and the outer sapwood during early dormancy but it peaks in this zone during the late growing season. The inner sapwood (2–13 mm width) adjacent to the transition zone (which is always clearly visible) showed a prominent peak of ethylene formation during the late dormancy period and a smaller peak during the mid–spring season, as was also shown by the middle and outer sapwood samples (Nelson 1978).

Large differences in mean ethylene production have been found between two plantations of the same five families of *Juglans nigra* for each outer, middle, and inner sapwood region (Nelson et al. 1981). These differences are probably largely due to different soil types and other environmental factors rather than to genetic differences.

If the increase in production of ethylene in the sapwood adjacent to the heartwood and the apparent formation of heartwood asociated with this increase are valid, then ethylene biosynthesis in the sapwood could be related to the genetic variation in the amount of heartwood formed. Nelson et al. (1981) examined the possibility that ethylene levels and heartwood/sapwood ratios are genetically correlated so that ethylene assays could be incorporated in techniques for early selection of genotypes that would have high heartwood/sapwood ratios at maturity. On the basis of earlier work, they examined ethylene production in the sapwood of genotypes of open–pollinated families of *Juglans nigra* during the early dormant period of the trees and in the early stages of heartwood formation. Their meticulous study of five families showed that half–sib family differences in the production of in vitro ethylene by sapwood were not significant, so that genetic variation could not be demonstrated, although more genetically diverse genotypes may have done so (Nelson et al. 1981).

Samples of *Juglans nigra* sapwood which had low ethylene production rates had slightly more parenchyma than high ethylene–producing samples. However, when expressed as a function of oven–dry weight of total wood substance, total parenchyma volume or individual parenchyma cells, ethylene production was approximately five times greater in high ethylene–producing samples than in low ethylene–producing ones. This indicates that ethylene production in sapwood tissue is controlled not by the amount of parenchyma, but by the rate of ethylene production within each sapwood parenchyma cell. Thus it appears that biochemical control of ethylene biosynthesis is of primary importance in determining ethylene production rates per unit weight (Phelps et al. 1983). A similar conclusion had been drawn previously for the biosynthesis of phenolics within the inner sap wood of *J.nigra* during heartwood formation (Nelson 1975). That is, the quantity of phenolic extractives produced per unit weight of wood by axial and ray parenchyma cells during heartwood formation was not influenced by the amount of parenchyma cells present.

In addition to its involvement in polyphenol formation at the heartwood boundary, increased ethylene levels have been associated with reaction wood formation in the cambium in *Pinus strobus* (K.M. Brown and Leopold 1973), *P.radiata* (Barker 1979), and *Eucalyptus gomphocephala* (Nelson and Hillis 1978c). Injection of ethylene–forming compounds into the sapwood of *Pinus* species also results in the increased formation of resinous compounds and with *Prunus* species of carbohydrates (see Chapter 6.4.3).

It is not yet proven that ethylene production has a direct biochemical effect on the rate of heartwood formation or whether ethylene–releasing regulator chemicals are involved in formation of heartwood.

The inner phloem tissues of *Chamaecyparis obtusa* produce higher levels of ethylene more rapidly in response to wounding or chemical treatments than the xylem tissues. The production of stress ethylene by these tissues was considerably greater during the growing period than in the dormant period. Ethylene is apparently a trigger for the formation of wound–induced, resin canals in the phloem, for the subsequent resin synthesis, and for resistance to the sugi bark–borer (Yamanaka 1985). During oviposition, the wood wasp *Sirex noctilio* inoculates the outer sapwood of *Pinus radiata* with a mucus secretion and arthrospores of the decay fungus *Amylostereum areolatum*. A dry, metabolically active transition zone usually surrounds the reaction zone when the host is responding to the attack; the transition zone is the origin of the ethylene. The production of ethylene clearly precedes polyphenol synthesis (Shain and Hillis 1972). The wound reaction appears to be most rapid and effective as a defense mechanism during the growing season (see Shain 1967). Furthermore, the quantity of ethylene produced by the transition zones surrounding *Sirex* lesions was considerably greater than that produced by the transition zones surrounding heartwood.

Dominant trees of *Pinus radiata* produced greater quantities (about 10 times) of ethylene in response to *Sirex noctilio* attack, and to mechanical injury, than did suppressed trees; it was detected one day after attack (Shain and Hillis 1972). Three weeks later, polyphenols (pinosylvins) were present in considerable quantities in lesions in the dominant trees, but still not detectable in lesions in the suppressed trees. It is interesting that trees displaying the resistant response were stimulated to produce more ethylene by *Sirex* – *Amylostereum* attack than by mechanical injury. Moreover only slight amounts of polyphenols were observed in mechanical lesions after 3 weeks. The above observations raise the question whether suppressed trees produce less ethylene, and hence a less resistant response, largely due to the environment or to the genotype. There are indications from this work that genotype plays the most important role in determining the host's capacity to respond to injury.

In studies on resistance to kino vein formation in *Eucalyptus regnans* both genetic and environmental factors have significant influence (Nelson and Hillis 1978a) (Section 5.9.3). The bark of clones of *Hevea braziliensis* that are most susceptible to various forms of rubber tapping contains also higher concentrations of ethylene than do the resistant clones (Sivakumaran et al. 1983). In other situations the responses of the tree to fungal lesions may be influenced by the fungi concerned as they can have varying capacities to enhance ethylene production by the host tissues (S.F. Yang and Pratt 1978).

5.6 Type of Extractives

The term extractives covers a large number of compounds of different classes which can be extracted from wood or bark with polar and nonpolar solvents. Usually extractives are considered to be those compounds which are soluble in organic solvents. But there are also water–soluble carbohydrates and their derivatives as well as polymerized polyphenols which require a small amount of alkali to dissolve them.

The chemical compositions of the extractives of various zones in the same tree can differ from each other. Furthermore, as heartwood is a "conservative" taxonomic tissue it can provide extractives that are characteristic of the family, genus, and even the species to which the tree belongs (Swain 1963). For example, the Cupressaceae are the sole source of tropolones such as a β–thujaplicin (Gardner 1962). The extractives composition has been used for taxonomic purposes with a number of genera as with *Pinus* and other conifers (Erdtman 1963), *Prunus*

(Hasegawa 1958), *Acacia* (Clark–Lewis and Dainis 1967, Tindale and Roux 1974), *Nothofagus* (Hillis and Orman 1962), and *Eucalyptus* (Hathway 1962, Hillis 1972a). Nevertheless, varieties of species containing significantly different wood extractives from the normal have been found in genera such as *Cinnamomum*, *Pinus*, *Pterocarpus*, *Ocotea*, and *Acacia* (Hillis 1966, Clark–Lewis and Dainis 1967).

Not only may the composition of the extractives in different zones of the same tree differ, but the extractives in different tissues in those zones may also differ. The extractives in the resin canals can have markedly different compositions from those of the nearby parenchyma (Back 1960b, Mutton 1962, Kimland and Norin 1972).

Detailed summaries of the composition of extractives can be found in specialist sources such as Hillis (1962a), Sandermann (1966), Zinkel (1981), Fengel and Wegener (1984), and Rowe et al. (in preparation).

Extractives which increase in amount on heartwood formation may be divided into the following major classes.

5.6.1 Galactans and Cyclitols

These compounds are rarely found as heartwood extractives and then only in conifers.

The arabinogalactans isolated from the heartwood of *Larix* species (Côté et al. 1966) are water-soluble, heavily branched polysaccharides based on residues of arabinose and galactose. The molecular weight of the arabinogalactans isolated from various larch species varies from 29 600 to 58 500.

The cyclitol, pinitol (a monomethyl ether of d-inositol), was first discovered in the exudate of *Pinus lambertiana* by Berthelot in 1855 (Ballou and Anderson 1953). A 30-fold increase in amount of this and another cyclitol, myoinositol, occurs when sapwood is converted into heartwood (Anderson 1952). Pinitol is found also in the heartwood of other Haploxylon pines, as well as additional cyclitols in *Sequoia sempervirens*, *S. gigantea*, and *Picea rubens* (Ballou and Anderson 1953, Anderson et al. 1968). Average yields of up to 9.8 percent have been obtained from the heartwood of *P. lambertiana* stumps (Anderson 1953).

5.6.2 Terpenoids

The terpenes and their derivatives are a large group of compounds which are widespread in trees and found very largely in softwoods. They are light-colored or colorless, insoluble in aqueous solvents, and a number are fragrant. The wide range of terpenoids present in the volatile oils of the oleoresins in the woods of a wide range of species has been reviewed by Kurth (1952). The genera include *Abies*, *Pseudotsuga*, *Picea*, *Larix*, *Pinus*, *Boswellia*, and *Cedrus*, as well as various genera of the Taxodiaceae, Cupressaceae, Araucariaceae, Podocarpaceae, Santalaceae, Lauraceae, and other families.

The terpenoids are built from a number of five carbon isoprene units (Sandermann 1966, Zavarin 1970, Goodwin and Mercer 1983, Croteau and Johnson 1985). The monoterpenes contain two isoprene units linked from the head of one unit to the tail of the other. The sesquiterpenes are built up from three units similarly linked, the diterpenes of four units, the sesterterpenes of five units and triterpenes of six units. The terpenes are hydrocarbons, usually linked intramolecularly to form one or more rings. The terpenoids have also functional groups such as hydroxyl, carboxylic acid, carbonyl etc. (Fig. 5.5).

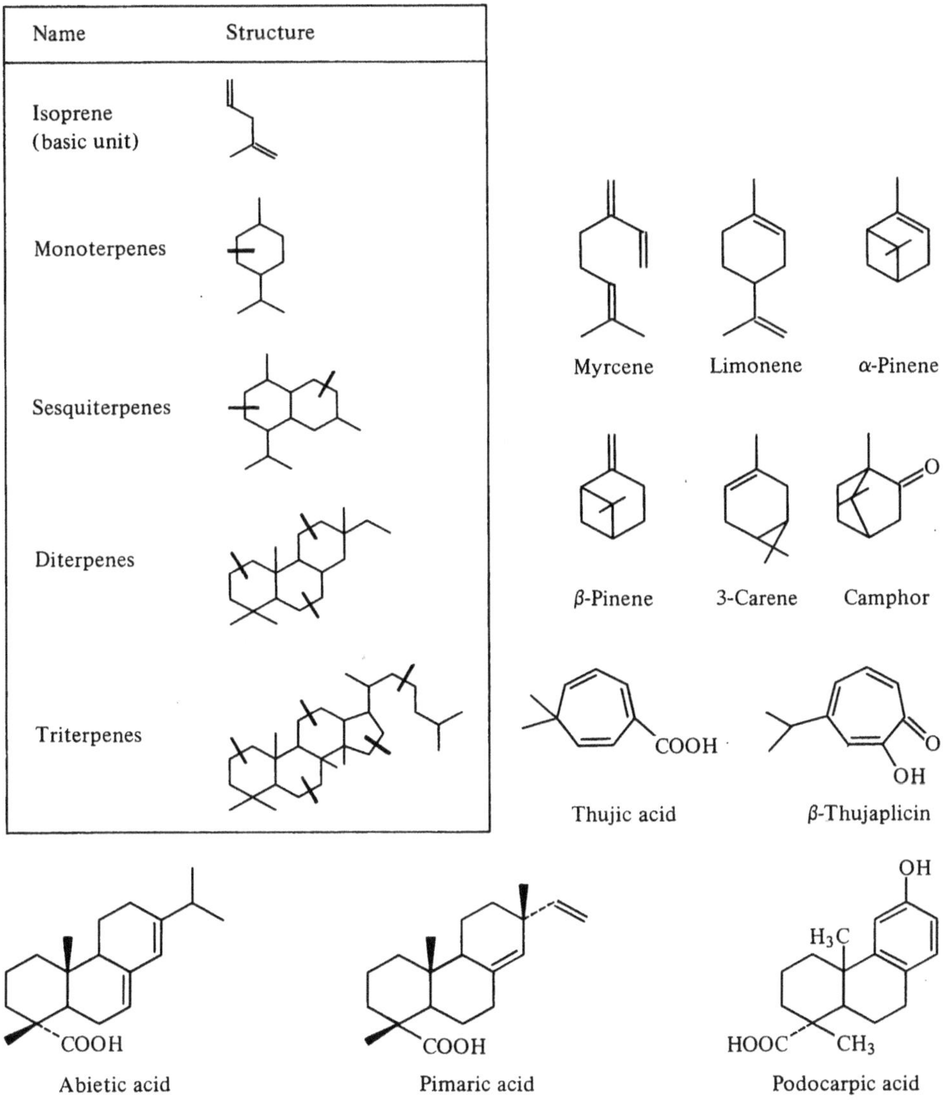

Fig. 5.5. Formulae of terpenoid compounds. The thick bars of the formulae in the box are the point of "head-to-tail" junction of isoprene units

All types of the monoterpenes can be found in the volatile fraction of many softwoods and particularly of *Pinus* species. The most common are α- and β-pinene, and often β-phellandrene, limonene, and myrcene. The hardwood *Cinnamomum camphora* contains the monoterpene camphor. Sometimes the hydrocarbon n-heptane is a major component of wood resin rather than monoterpenoids, as in the cases of *P. jeffreyi* and *P. sabiniana*. The composition of the volatile monoterpenes in different species and varieties varies much more than other terpenoids.

The tropolones are considered as derivatives of monoterpenes (Lai 1971) and consist of structures involving 7-membered rings. Tropolones such as thujic acid, β-thujaplicin and thujaplicinol (Fig. 5.5) are found only in the Cupressaceae family (Gardner 1962).

Sesquiterpenes and sesquiterpenoids of increasing complexity exist in *Shorea* (Bisset et al. 1971), *Pinus*, and *Pseudotsuga* species. They have been found rarely in hardwoods although they have been isolated from *Ulmus*, and *Celtis* species, and *Santalum album*.

The oleoresins synthesized by the epithelial cells surrounding the resin canals of softwood species often contain relatively high percentages of diterpenes and diterpenoids. The most common members of this group are the resin acids which are mostly tricyclic compounds, such as the abietic, pimaric, and palustric acids (Fig. 5.5). This ring system is apparently very stable, as diterpenoids have been found in fossils. The resin acid fractions in pine sapwood and heartwood can vary from each other (Conner et al. 1980a,b,c) and the amounts are higher in the heartwood and particularly the inner heartwood (e.g., Porter 1969, Hemingway and Hillis 1971, Lloyd 1978).

Most of the triterpenoids exist as steroids and have been isolated from both the softwoods and more particularly the hardwoods. The triterpenoid amyrin has been obtained from the latex sap of *Hevea* species. Saponins are glycosides of triterpenoids and are present in several tropical hardwoods.

Compounds containing more than 6 isoprene units are known as polyprenes of which rubber, gutta percha, and caoutchouc are examples. They differ in their chain conformation with the isoprene units in gutta percha being arranged in the trans-form and those in caoutchouc arranged in the cis-form. Only a few species contain caoutchouc in the xylem; lignum vitae (*Guaiacum officinale*) is the most notable of these (Sandermann et al. 1963).

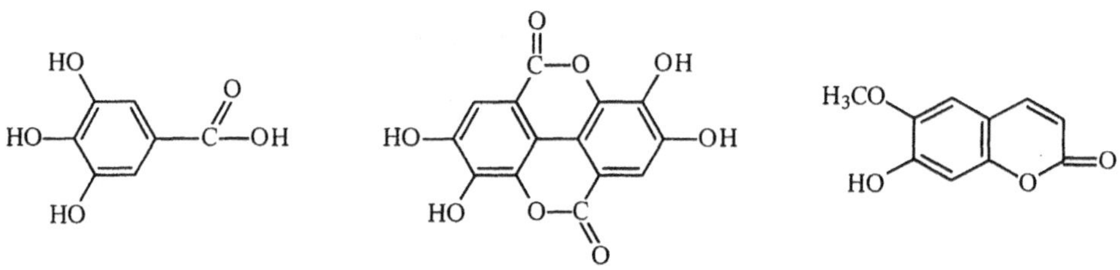

Gallic acid Ellagic acid Scopoletin

Fig. 5.6. Formulae of C_6C_1 and C_6C_3 compounds

5.6.3 Fatty Acids and Related Compounds

More than 20 long-chain fatty acids have been identified in wood. In general they are saturated (e.g., stearic), monoenic (e.g., oleic), dienic (e.g., linoleic), trienic (e.g., linolenic) acids with 16–22 carbon atoms although shorter and longer ones have been detected. Most of the total fatty acids are composed of stearic (C_{18}), oleic (C_{18}), linoleic (C_{18}), and 5,9,12-octadecatrienic acids. The percentage of free fatty acids is considerably higher in heartwood than in sapwood (Lindgren and Norin 1969, Porter 1969).

Fats are esters of fatty acids with glycerol, and triglycerides are more common than mono- and di-glycerides both in softwoods and in hardwoods. Waxes are esters of fatty acids with saturated straight-chain alcohols, with 16–28 carbon atoms such as docosanol ($C_{22}H_{45}$). Waxes are usually present in smaller amounts than the fats.

5.6.4 Phenolic Compounds

Phenolic compounds are second only to carbohydrates in abundance and are the most widespread of the secondary components. Tissues containing them in large quantities have been used for centuries to convert hides and skins into leather in the process of "tanning". Also it was found that these tissues stained blue with iron salts so that all vegetable extracts which gave this color became known as tannins and this term has been used widely, and often erroneously, even when their use is other than preparing leather. True vegetable tannins are water-soluble, phenolic compounds having molecular weights between 500 and 3000 and, besides giving the usual phenolic reactions, have special properties, such as the ability to precipitate alkaloids, gelatin, and other proteins (White 1957). Much of the botanical information concerning the occurrence of tannins is misleading as it is based on tests that are insufficiently specific.

The value of phenolic compounds in the study of the formation of heartwood and similar tissues lies in their chemical stability, ease and speed of identification, widespread distribution, and structural variability. There are many thousand polyphenols, but rarely is the whole repertoire of them in the extractives of different tissues determined. That repertoire may dramatically alter in tissues subjected to different environmental conditions. The major classes are as follows.

5.6.4.1 Simple Phenols and Phenolic Acids

Simple phenols are rare in wood and cannot be classed as heartwood extractives. Phenolic acids (C_6–C_1 compounds) can be present in large amounts in heartwood either as free acids (notably gallic acid or its dimers digallic acid and the hexahydroxydiphenic acid) or esterified with glucose, polyols, and other phenols. These are grouped as hydrolyzable tannins and further subdivided into gallotannins, which yield gallic acid on hydrolysis, and ellagitannins, which yield largely ellagic acid from the diphenic acid (Haslam 1982a, Hillis 1985b) (Fig. 5.6). Gallotannins are found in the heartwood of *Castanea* and some *Quercus* species and large amounts of ellagitannins exist in the heartwoods of some *Eucalyptus* and *Quercus* species. The C_6–C_2 compounds are rarely, if ever, found in heartwoods.

The C_6–C_3 compounds such as derivatives of hydroxycinnamic acids occur only in small quantities in woody tissues. However, bicyclic derivatives, the coumarins, can be found in considerable quantities in lesions such as scopoletin (Fig. 5.6) in the sapwood of *Prunus domestica* affected by the fungus *Stereum* (now *Chondrostereum*) *purpureum* (Hillis and Swain 1959), as well as in other tissues affected by fungi (Hughes and Swain 1960).

5.6.4.2 Lignans

These compounds consist of two phenylpropane units linked in different ways and they can be found in considerable quantities in the heartwoods of both softwoods and hardwoods. Pinoresinol, matairesinol, α-conidendrin, hinokiresinol, secoisolaricinol and thujaplicatin are examples (Fig. 5.7). Large quantities of secoisolariciresinol relatives have been isolated from knots of *Araucaria angustifolia* (Anderegg and Rowe 1974).

5.6.4.3 Stilbenoids

Both the stilbenes and flavonoids are derived from an intermediate involving the ester group of a hydroxycinnamic acid and a nucleophilic attack by an anion of acetyl-CoA resulting in a six carbon polyketide side chain. The side chain is folded topologically in different ways, and different mechanisms of cyclization are applied to result in stilbenoid (C_6–C_2–C_6) or flavonoid (C_6–C_3–C_6) compounds (Kindl 1985) (Fig. 5.7, 5.8).

Stilbenes and stilbenoids are present in the heartwoods of both softwoods and hardwoods (Erdtman 1958). They have been used as chemotaxonomic markers in the *Pinus, Eucalyptus, Nothofagus*, and other genera. The stilbenes in the Pinaceae, such as pinosylvin, are mostly less substituted than those in the hardwoods (Gorham 1980). Glycosides of stilbenes are rarely present, particularly in heartwood. However, resveratrol glucoside occurs in much greater amounts than its aglucone in the outer heartwood of *Eucalyptus wandoo* but vice versa in the inner heartwood; both are absent in the sapwood (Hathway and Seakins 1959).

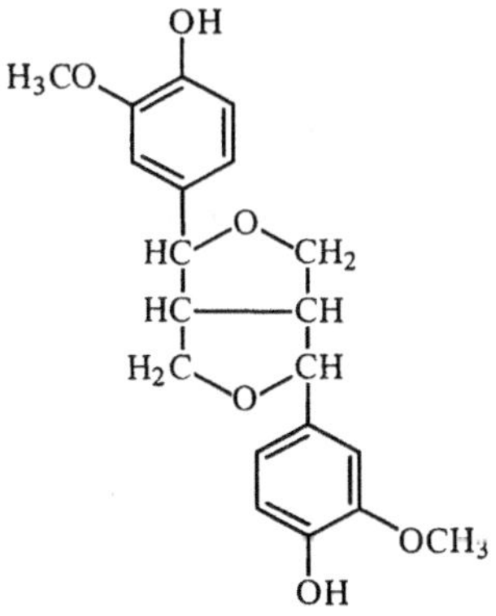

Pinoresinol

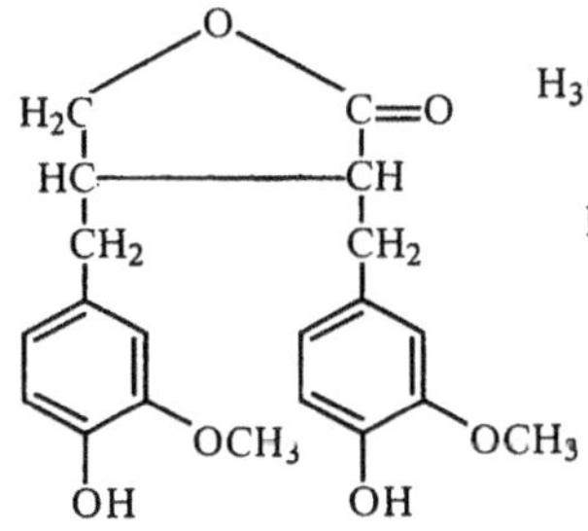

Matairesinol

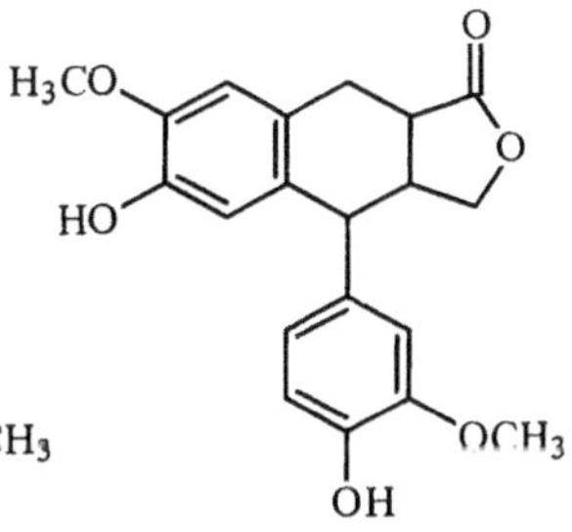

α-Conidendrin

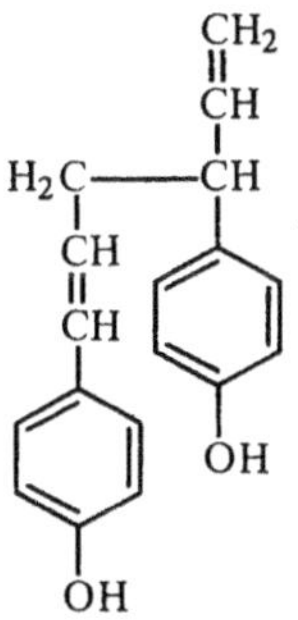

Hinokiresinol

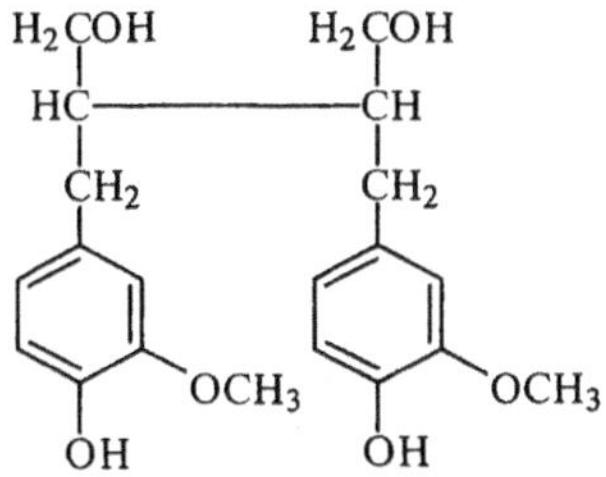

Secoisolariciresinol

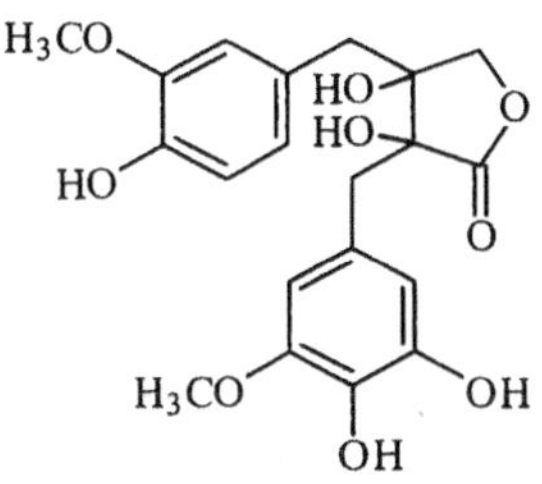

Thujaplicatin

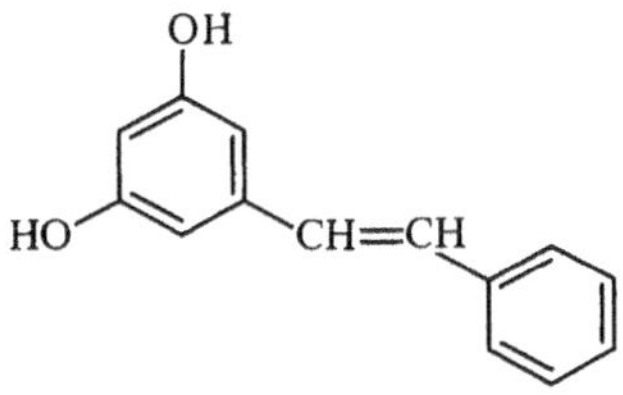

Pinosylvin

Fig. 5.7. Formulae of lignans and the stilbene pinosylvin

5.6.4.4 Flavonoids

The flavonoids comprise many thousands of C_6–C_3–C_6 compounds subdivided into flavones, flavanes, flavanones, flavanonols, and isoflavones. They occur either as glycosides or aglycones. Large amounts of flavonoids can occur in heart-

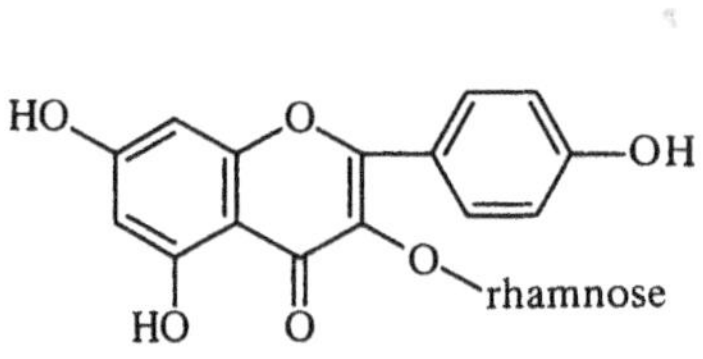

Kaempferol-3-rhamnoside

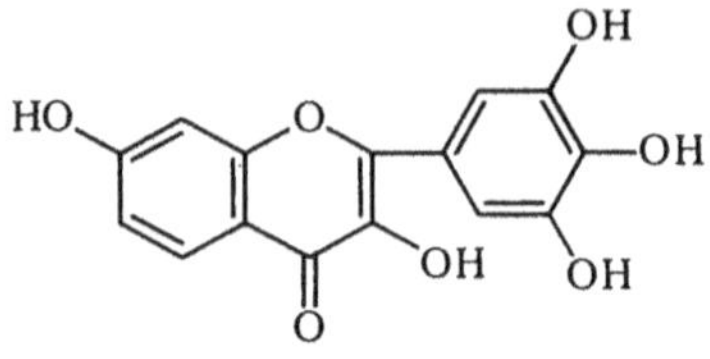

Robinetin

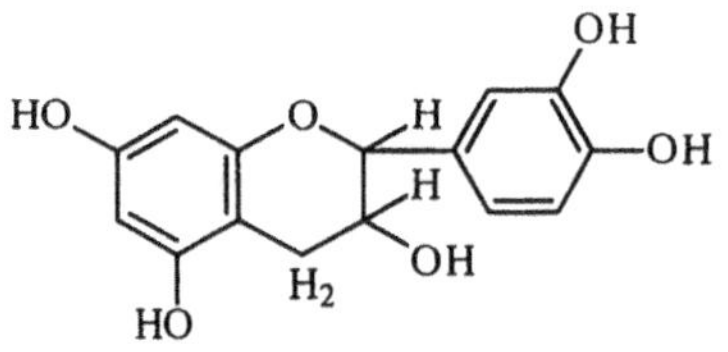

Catechin

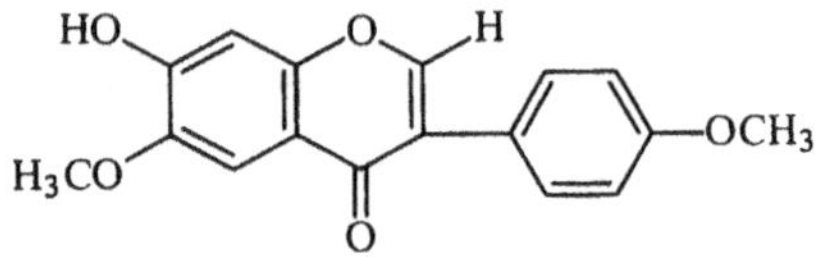

7-Hydroxy-6,4′-dimethoxyisoflavone

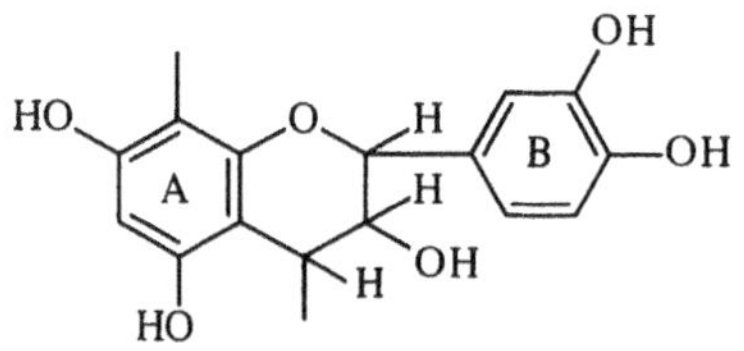

A Unit of Polymerised
Procyanidin

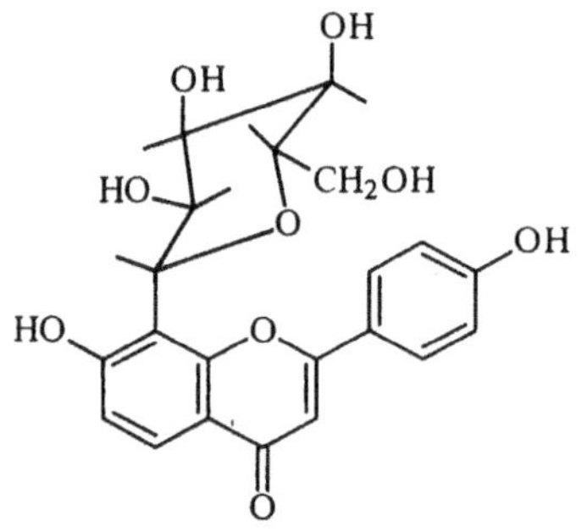

Bayin

Fig. 5.8. Formulae of flavonoids

woods but mainly as polymers with monomeric aglycones existing as a small percentage of the extracts. There are instances where monomers exist in a pure form or where a mixture or a small number occur as a major part of the extracts.

The polymers based on units of catechins (flavan-3-ols) and proanthocyanidins (flavan-3,4-diols) are known as condensed tannins (Haslam 1982b, Hillis 1985a,b) (Fig. 5.8). The linkages are through the strongly nucleophilic centers at the 6- and 8-positions in the A ring and the 4-position of the heterocyclic ring. The polymers precipitate as insoluble phlobaphenes when heated with hydrochloric acid. Condensed tannins occur more frequently in wood than the hydrolyzable tannins and can exist in large quantities in some tropical and subtropical hardwood species. The mono- and dimeric flavonoids (and other monomers) occurring in tannin extracts are termed "phenolic nontannins".

The large number of flavonoids based on the C_6–C_3–C_6 is due not only to the degree of saturation of the heterocyclic ring but also to the variation of hydroxylation of the rings. Most flavonoids are hydroxylated at the C_5 position. However, flavonoids in the heartwoods of the Anacardiaceae family and some of the Leguminoseae lack a C_5 hydroxyl. Such compounds are intensely fluorescent under ultraviolet light and provide a means to detect heartwood or to study the formation of extractives.

Although they may be present in the sapwood of some species (Hergert and Goldschmid 1958), glycosides of flavonoids are rarely found in more than trace amounts in heartwoods. However, a naringenin glycoside has been found in significant amounts in the newly formed heartwoods of *Prunus yedoensis* and *P. verecunda*, although the amount was lower than that found in the adjacent sapwood (Hasegawa 1958).

The color of the red, yellow, brown, and blue heartwoods is due to flavonoids and related components. Changes in color can be influenced by treatment with alkali metal salts, acid, and exposure to sunlight.

5.6.4.5 Quinones

A range of aromatic quinones are present as substituted benzoquinones in *Acacia melanoxylon* (Leistner 1985), as various naphthoquinones and anthraquinones and their derivatives in the heartwood of *Tectona grandis* (Sandermann and Simatupang 1966) or dalbergiones in *Dalbergia* species (Dietrichs and Hausen 1971) (Fig. 5.9). They are responsible for the strong colors, high durability, and dermatitic properties of some woods. They are largely biosynthesized from acetate units (Sandermann 1966, Leistner 1985).

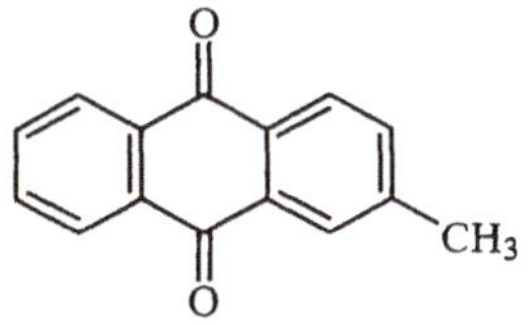

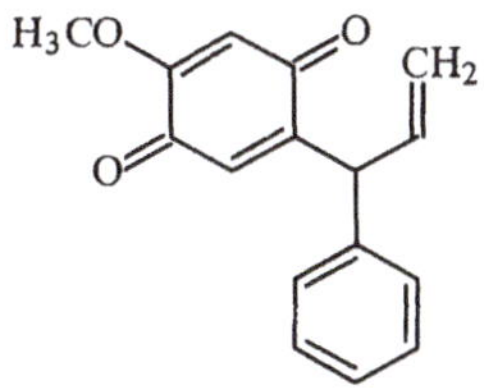

Fig. 5.9. Formulae of quinones Tectoquinone 4-Methoxydalbergione

5.6.4.6 Polymerized Polyphenols

The phenolic extractives of heartwoods are very largely in a polymerized form and with constitutions which are almost unknown. Presumably slow oxidation and polymerization occurs over many years in the heartwood, and nonenzymic oxidation has been shown to occur in several species (Goodwin and Goddard 1940). Some indication of their constitutions has been obtained from a study of the changes which can take place with monomeric compounds found in highest concentration in the outer heartwood. In recent years this knowledge has been extended by the elucidation of the constitution of bi-, tri-, and tetra-meric components. For example, in *Schinopsis quebracho-colorado*, two monomeric flavans, (+)-catechin and pro-fisetinidin (a trihydroxyflav-3,4-diol), considered to be precursors of heartwood polyphenols, are present in high concentration at the heartwood periphery and then decrease over a short distance in the heartwood (H.G.C. King and White 1957, Roux and Evelyn 1958, 1960). In *Prunus yedoensis* and *P. verecunda*, catechin present at the heartwood boundary decreased slightly in the most recently formed heartwood but much more in the inner regions (Hasegawa 1958).

Only polymerized pro-fisetinidin polymers were present in the heartwood of *Schinopsis quebracho-colorado* and the molecular weight increased as the center of the cross-section was approached. The rate of polymerization in the heartwood of this species is more rapid than in that of *Acacia* species (Roux and Evelyn 1958, 1960).

5.6.4.7 In Different Tissues

Attention has been drawn to the nature of exudates, formed by the cambial tissues, which differs markedly from that of the extractives in neighboring tissues (Chapter 2.6.2). The crystalline deposits in shakes can have different compositions from the mixture of components in adjacent tissues. Sapwood of *Prunus domestica* affected by the fungus *Stereum* (now *Chondrostereum*) *purpureum* contains considerable amounts of the coumarin scopoletin, which is present in only very small amounts in undamaged sapwood (Hillis and Swain 1959) and absent from normal heartwood of other *Prunus* species (Hasegawa 1958). When *Trametes versicolor* attacked *Prunus jamasakura*, large amounts of an abnormal constituent the lignan iso-olivil were formed, but the flavonoid components of normal heartwood were absent (Hasegawa and Shirato 1959). The fungus *Peridermium pini* induces the formation of pinosylvin dimethyl ether not found elsewhere in *Pinus sylvestris* as well as large amounts of resin (Westfelt 1966). *P. radiata* sapwood affected by *Amylostereum areolatum* lacks the flavonoids found in the heartwood but contains two fungicidal pinosylvins in a ratio different from that in the heartwood (Hillis and Inoue 1968). The sapwood of *Abies grandis* contained small amounts of flavanols (such as catechin and proanthocyanidins), whereas these were absent from the heartwood which contained the lignans matairesinol, hydromatairesinol and α-conidendrin. However, balsam woolly aphid infestation, which results in significantly greater amounts of heartwood, does not alter the composition of the heartwood extractives (Puritch 1977).

Normal metabolism can result in different types of components in adjacent tissues (Hillis 1985a). The major portion of the monomeric polyphenols in the cambium of *Eucalyptus astringens*, *E. marginata*, and *E. sieberi* was ellagitannins (polyphenolic acids esterified with glucose) although catechins and proanthocyanidins were present (Hillis and Carle 1958b, Hillis 1960b). Sapwood of *E. astringens* contains catechins and pro-anthocyanidins in a much greater proportion to the ellagitannins than in the cambium. In the heartwood, appreciable amounts of stilbenes are also present (Hillis and Carle 1962), and the bark contains appreciable amounts of pro-delphinidins (Hillis 1954). On the other hand, while *E. marginata* contains catechins and pro-anthocyanidins in the sapwood and heartwood, the bark contains ellagitannins. Gallic acid and gallotannins are the major components in the sapwood of *Schinopsis* species, whereas flavonoids of the fisetinidin type are the major components of the heartwood extractives (Roux and Evelyn 1960). In a more detailed contrast, the sapwood of *Calophospermum mopane* contains a low concentration of flavonoids of the phloroglucinol-catechol type. The heartwood also contains small amounts of these flavonoids but much large amounts of those of the resorcinol-catechol type (Drewes and Roux 1966). The sapwood of *Thuja plicata* contains small amounts of catechins and pro-anthocyanidins and trace amounts of lignans. On the other hand, the heartwood contains more than 100 times the amount of lignans (Swan et al. 1969). In *Pinus* species, small amounts of C_6–C_1 phenolic derivatives occur in the sapwood, whereas characteristic flavonoids and stilbenes are present in the heartwood in larger amounts (Hillis and Inoue 1968). In contrast, large amounts of polymerized pro-anthocyanidins are formed in the bark but not elsewhere in the tree.

5.7 Amount of Extractives

Just as the type of extractives present varies, so also does the amount. Usually there is only a few percent in the heartwood from species growing in temperate zones, but the amounts can be much larger in some species from tropical and subtropical regions. In addition, the concentration can be highest in certain parts of the tree such as the heartwood, bark or necrotic regions.

In cross-sections of a log or branch the heartwood can often be detected visually by its darker color, which is caused by the larger amount of extractives present. These observations are supported by extractions with suitable solvents as shown in Table 5.3. As expected from observations of the intensity of color, the amount of extractives is usually higher in mature hardwoods than in conifers. Also the amounts vary between the species in these two groups. Published figures such as those given in Table 5.3 are useful in indicating the amount of extractives which can be expected, but their usefulness is limited. The samples analyzed have often been taken from large trees, frequently of sizes that are now available only rarely, and in addition the location of the sample in the tree or of the tree often have not been stated. In addition some solvents used do not remove all the extractives present in certain species. Hence the value of the figures for comparison of different properties is restricted.

5.7.1 Position of Sample in the Tree

The amount of extractives can vary according to the vertical as well as radial position from which the sample was taken. This situation is illustrated by the classical figure of Sherrard and Kurth (1933) of the aqueous extractives content in

Table 5.3. Extractives content (%) of various woods

Species	Solvents and tissue (SW = Sapwood, HW = Heartwood)						
	Hot water		Ether		Ethanol		0.5% NaOH
	SW	HW	SW	HW	SW	HW	HW
a) Softwoods							
Abies balsamea			0.95	0.74–1.18			
Larix occidentalis	0.3	9.4–21.7			2.8	3.1– 7.5	
Picea rubens			0.6–1.2	0.8– 1.5			
Pinus echinata			2.6–3.8	3.9–13.3			
Pinus palustris			1.4–2.7	3.6–20.3			
Pinus palustris		7.1		6.3			22.4
Pinus ponderosa			3.2–4.8	2.7– 9.9			
Pinus radiata			0.5–1.2	2.0– 6.8			
Pseudotsuga menziesii		6.5	0.4	1.0			16.1
Sequoia sempervirens		9.9		1.1			20.0
Tsuga heterophylla			0.2–0.5	0.2– 1.0			
b) Hardwoods							
Acer saccharum			0.2–0.3	0.2– 0.3			
Acer saccharum		4.4		0.25			17.6
Betula lutea			0.3–0.9	0.3– 1.2			
Betula lutea		4.0		0.6			19.8
Eucalyptus crebra		6.8–20.2				5.0–18.6	25.4–3
Eucalyptus delegatensis		2.4– 6.5				1.6–16.5	14.8–2
Eucalyptus obliqua		7.9–26.6				10.1–29.1	20.1–4
Eucalyptus regnans		4.8–15.3				1.3–16.3	12.9–2
Eucalyptus sieberi		2.6–15.5				4.3–17.7	17.2–2
Fagus grandifolia			0.2–0.3	0.4– 0.6			
Quercus alba			0.5–0.6	0.6– 0.7			
Quercus densiflora		5.6		0.8			24.0
Rhus pentaphylla	8.5	28.6					
Schinopsis quebracho-colorado	6	28.5					
Tilia americana		4.1		2.0			23.8

cross-sections from an old, slow-grown *Sequoia sempervirens* showing an increase from the pith to the periphery of the heartwood and from the top to the butt log (Fig. 5.10). A similar pattern has been observed in other species such as *Pseudotsuga menziesii* (Kennedy and Wilson 1956, Gardner and Barton 1960, Hale and Clermont 1963), *Tectona grandis* (Da Costa et al. 1958, Rudman and Da Costa 1959) and an *Acacia* species (Hillis 1960). While the extractives in many species show this overall increase, the content can increase at different rates during various periods so that streaks or bands of differing color intensity are apparent. An examination of 480 *Pinus echinata* trees of different ages growing under a variety of conditions showed that the age of the tree influences extractives content more than any other variable, and that, while the environmental conditions contribute to the variation in content, no one environmental variable is a significant factor (Posey and Robinson 1969). There are situations also where the extractives con-

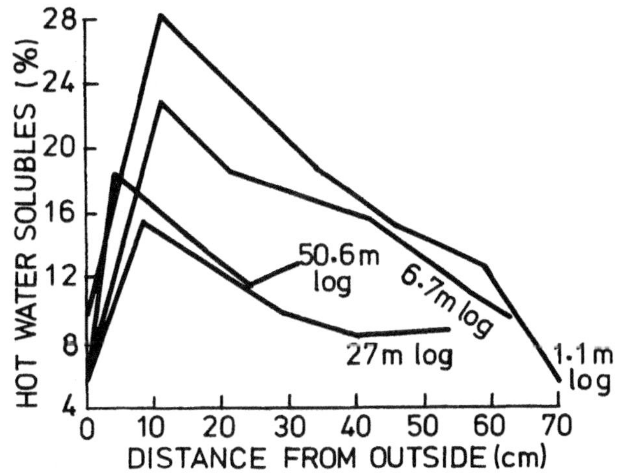

Fig. 5.10. The distribution of extractives in cross-sections of *Sequoia sempervirens*. The measurements are the height in meters at which the log samples were taken (after Sherrard and Kurth 1933)

tent increases, and even doubles, in the heartwood in the higher portions of the tree as with arabinogalactan in *Larix* species (Côté et al. 1966, 1967, Côté and Timell 1967).

The nature of the solvent in relation to the extractives should be considered with some species. Membranes lining the cell wall may hinder the extraction of ether-soluble extractives with ether (Erdtman 1943). Some eucalypt woods, and particularly those that are red-colored, require dilute alkali (one-tenth molar) to remove extractives from the cell lumens. Changes in the nature of the extractives by long exposure to mild acetic acid conditions in the heartwood (which gradually increases in acidity), or to enzymes, can affect their polymerization and extraction with neutral solvents. Consequently the lower amounts of extractives in the inner heartwood may be due not only to the smaller amounts formed but also to the degree of polymerization and insolubility. There is a progressive degradation of arabinogalactan in living trees of *Larix* species (Côté et al. 1967). With *Quercus rubra* the proportion of extractives insoluble in methanol is greater for the inner than for the outer heartwood (Nelson 1975). Interestingly, *Celtis* and *Ulmus* species had more obvious and thicker deposits on the pit membrane in the inner than in the outer heartwood (Wheeler 1983).

Together with the increase in amount of extractives as a group when heartwood is formed, there is the expected increase in new formation of individual components. For example, Hasegawa (1958) found a marked increase in the amounts of the two flavonoids, pinocembrin and isosakuranetin in *Prunus yedoensis* and *P. verecunda* on either side of the heartwood boundary. In *Thuja plicata*, a very large increase in the amount of lignans occurs at the heartwood boundary (Swan et al. 1969).

The variation of amounts of individual components in the extractives can show a similar variation within the tree. In a study of 269 *Pinus sylvestris* trees Erdtman et al. (1951) found that the average contents of the stilbenes pinosylvin and its monomethyl ether at the periphery of the butt heartwood, and in the middle and the inner portions were 0.96, 0.86, and 0.77 percent respectively. A

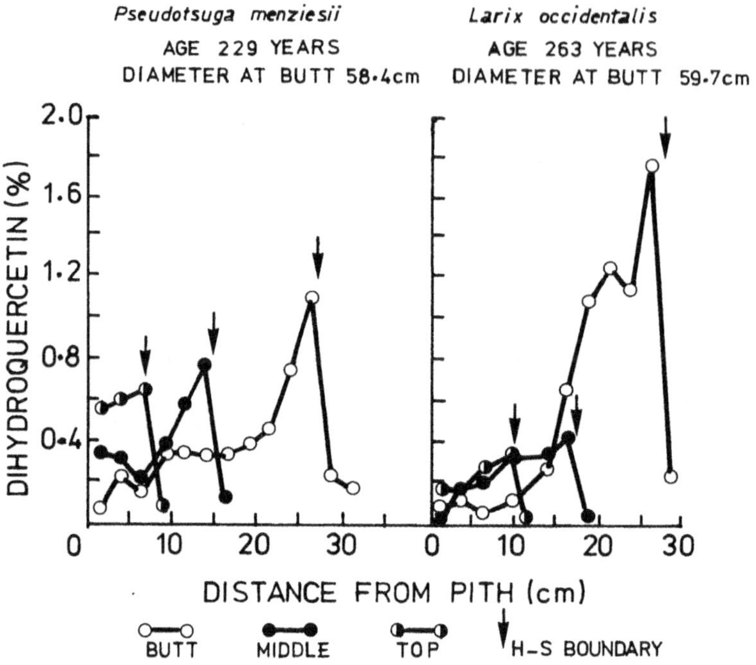

Fig. 5.11. The distribution of dihydroxyquercetin in cross-sections of *Pseudotsuga menziesii* and *Larix occidentalis*. (Gardner and Barton 1960)

small proportion of the trees had an uniform distribution of the pinosylvins across the heartwood. The content of the pinosylvins was not affected consistently by the vertical position of the sample. The distribution of the flavanoid dihydroquercetin in cross-sections taken at three different levels in *Pseudotsuga menziesii* and *Larix occidentalis* showed a pattern (Fig. 5.11, Gardner and Barton 1960, see also Hancock 1957) similar to that of water-soluble extractives in *Sequoia sempervirens* (Fig. 5.10) with the low amounts at the outer parts of the cross-section being found in the sapwood. The percentages of arabinogalactan in cross-sections of several *Larix* species show a similar, and nonuniform, increase from the center of the tree to the heartwood-sapwood boundary with higher concentrations with increasing age (Chochieva et al. 1959, Côté et al. 1966, Côté and Timell 1967). A similar distribution of proanthocyanidins has been observed in *Eucalyptus marginata* (Hillis 1956), *E. grandis* (Bamber et al. 1969) *Pseudotsuga menziesii* (Hancock 1957), an *Acacia* species (Hillis 1960), and *A. mearnsii* (Roux and Evelyn 1960), and of resin in *Abies amabilis* (Isenberg 1936) and *Picea abies* (Engström and Back 1959). However, in *Tectona grandis* the amount of tectoquinone decreased from the inner to the outer heartwood although the amount of caoutchouc increased (Sandermann and Dietrichs 1959).

The heartwood of roots and, especially, of branches of *Larix laricina* contains more arabinogalactan than does that of the stem, even when a short distance from the latter (Côté and Timell 1967). The heartwood of branches and roots of *Picea abies* contains higher concentrations of lignans (4–6 and 2–3 percent, respectively) than does the heartwood of the stem (0.1 percent) of the same tree (Ekman 1979).

In bark the amount of tannins (polyphenols) and extractives is greatest at the bottom rather than at the top of the tree, for example in *Acacia mearnsii* (Sherry 1971), *Pinus radiata* (Humphreys and Martin 1956, Sands 1975), *Quercus pedunculata* (Hathway 1959), *Q.levis* (Rogers et al. 1950), and *Castanea sativa* (Clarke et al. 1942).

There can be considerable differences in extractives content between trees (e.g., in *Pseudotsuga menziesii*, Gardner and Barton 1960); the heartwood from large, slow-grown trees is darker in color and contains more extractives than that from fast-grown trees with younger wood. Nevertheless, there are exceptions. No trends in the radial distribution of extractives was found in some cross-sections of *Pseudotsuga menziesii* (J.R. Campbell et al. 1965), *Pinus ponderosa* (Anderson 1946), and *P.radiata* (Buckland et al. 1954).

In all the *Larix occidentalis* and one-half of the *Pseudotsuga menziesii* samples examined, the largest amounts of dihydroquercetin were found at the periphery of the heartwood in the cross-section from the butt. In the other half of the *P. menziesii* samples, the maximum occurred at the heartwood periphery of the top section. In all cases, the lowest average content for the heartwood was in the middle section (Gardner and Barton 1960). The distribution found with other classes of components in other genera (Hillis 1962b, Côté et al. 1966) shows that the most common pattern is a decrease in relative amount with increasing height above the ground. The variations in the contents of arabinogalactan and the other extractives (obtained with ethanol:benzene 1:2) in a 300-year-old *L.occidentalis* follow each other quite closely with only a few exceptions (Fig. 5.12). There is an inverse relationship between the relative amounts of both these constituents and the width of the average growth ring within each group of 5 rings. (There was a poor correlation between the formation of arabinogalactan and ethanol-benzene extractives in *L. gmelinii* (Teratani et al. 1976)). The increase of 0.3 to 21.0 percent in arabinogalactan content from sapwood to heartwood was much greater than the 2.80 to 7.52 percent of other extractives. A 26-year-old *L.decidua* that had grown quickly showed an increase of only 0.60 to 7.68 percent in arabinogalactan content on heartwood formation after 8 years. Arabinogalactan was only a portion of the polysaccharides extracted with water from larch sapwood.

Most of the arabinogalactan was in the springwood or earlywood portion of the growth rings in the heartwood of various *Larix* species (Côté et al. 1966, Côté and Timell 1967, Teratani et al. 1976), and this was attributed to the larger lumen size of the earlywood tracheids. Also it has been shown that the maximum concentrations of several flavonoids occurred in the middle of the earlywood zone of a growth ring of *Pseudotsuga menziesii*, while the minimum values were at the beginning and conclusion of seasonal growth (Squire et al. 1967). These flavonoids were confined to the ray tissues. On the other hand, the amounts of resinous extractives were larger in the latewood than in the earlywood and were probably affected by the higher proportion of resin canals in the latewood (J.R. Campbell et al. 1965).

Less resin has been found in *Picea abies* heartwood than in sapwood, while in *Pinus sylvestris* the opposite trend was observed. An abrupt change in the com-

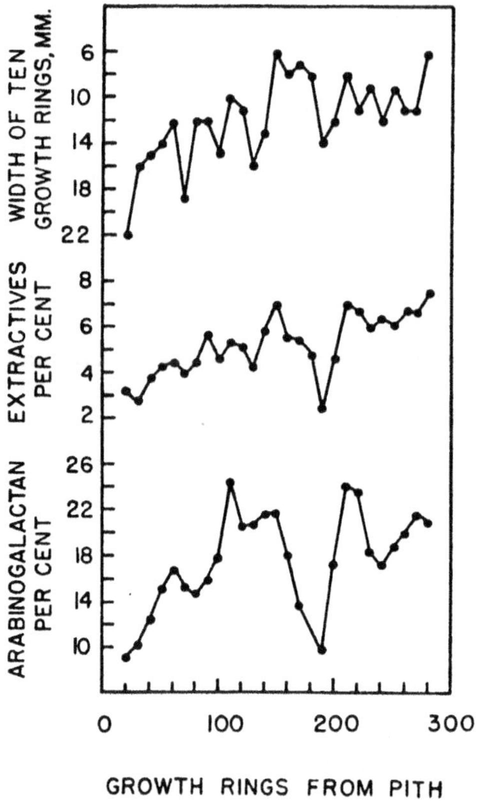

Fig. 5.12. Variation in arabinogalactan and extractives content with distance from the center of a *Larix occidentalis* and with the width of the ten growth rings present in each of the sections used. (Côté et al. 1966)

position of the extractives occurred in the narrow transition zone between sapwood and heartwood where the triglyceride content decreased rapidly and fatty acids increased (Pensar 1967, 1968, 1969). A similar situation has been found in *Pinus radiata* (MacDonald and Porter 1969, Hemingway and Hillis 1971, Lloyd 1978). There was little change in the composition of the resin acids across a cross-section, although the proportion of them increased substantially in the heartwood. The proportion of saturated fatty acids was higher in the heartwood than in the sapwood. The amount of petroleum-soluble extractives in *P. radiata* and the proportion of resin acids in them increases in the inner heartwood. The extractives in the earlywood tisues of *Picea abies* have a somewhat higher content of fatty acids than the corresponding latewood tissues (Kimland and Norin 1972), and the degree of saturation of these fatty acids appears to be higher (Pensar 1967). In addition, the amount of extractives in the earlywood of the sapwood was higher than in the latewood. On the other hand, the composition of the extractives was the same in the earlywood and latewood of *P. radiata* but the amount in the latewood of a 45-year-old tree was higher than in the earlywood (Lloyd 1978). This was especially so in the inner heartwood and particularly for resin acids, and this is attributed to an enrichment with extractives long after heartwood is formed (Harris 1965). If the epithelial cells surrounding the transverse resin canals rupture, the resin (which can be under pressures of several atmospheres in the sapwood) can flow into the latewood tracheids as their pits do not aspirate, as do those of the earlywood, on heartwood formation. This results in the resin-enriched areas in the interior of old pine trees.

In addition to the reports of increased amounts of extractives at the heartwood periphery there are other cases where the amount has increased appreciably inside the boundary. In the case of *Tectona grandis*, caoutchouc (but not other components) increased (Sandermann and Dietrichs 1959). With *Thuja plicata*, the maximum amounts of different lignans and tropolones occurred 50 to 100 annual increments inside the colored heartwood boundary (Swan et al. 1969, Swan and Jiang 1970). The lignans were found to be more highly hydroxylated in this region. Also the tropolones were not synthesized in the heartwood of very young trees. In *Cryptomeria japonica*, whereas one component reached maximum amounts in the transition zone adjacent to the heartwood, the other two major components were found in largest amounts in the third growth ring inside the heartwood periphery, as were also the total methanol extractives (Nobuchi et al. 1985). These differences cannot be explained by the variations in amount sometimes observed across various heartwoods.

5.7.2 Effects of Rate of Growth

In many cases when comparisons are made of dark-colored heartwoods, such as those of eucalypts, those from fast-grown trees can be seen to be paler than those grown more slowly. When the extractives content was related to the rate of growth of *Tectona grandis*, a marked increase in ether- and methanol-soluble extractives was apparent when the growth rate fell below a certain figure (Rudman and Da Costa 1959), although this was less evident in a later study (Da Costa et al. 1961). Similarly the narrow growth rings in the heartwoods of *Larix occidentalis*, *L.decidua*, *L. laricina*, and *L.gmelinii* contain more arabinogalactan than wide ones (Côté et al. 1966, Côté and Timell 1967) and a similar situation has been found with an *Acacia* species (Hillis 1960). Fast-grown *Picea abies* have lower extractives content than slow-grown trees (Kimland and Norin 1972). On the other hand, the between-tree variation in the amount of total extractives in *Juglans rubra* and *Quercus rubra* was not related to the rate of radial growth (Nelson 1975), and neither was there a relationship with *Eucalyptus albens*, *E. bancroftii*, *E. dealbata*, *E. goniocalyx*, *E. macrorhyncha*, and *E. sideroxylon* (J. Wilkes personal communication). The amount of extractives in these fast-grown eucalypts may have been influenced by the high proportion of juvenile wood present.

Attempts have been made to correlate the width of the growth ring and the amount of extractives present. These have not been successful due to the formation of the heartwood extractives several years after the growth ring had been formed in the cambial layer. Also some studies have been conducted on young trees before mature wood has been formed and the differences in the physiology of juvenile and mature wood formation are not yet understood.

The greater width of growth rings in tension wood, or in compression wood, than those in normal wood of the same log cross-section offers a means of comparison that can avoid between-tree variations. Smaller amounts of extractives have indeed been found in the compression woods of *Larix laricina*, *L.gmelinii*, and other *Larix* species (Chochieva et al. 1959, Tsvetaeva 1959, Côté et al. 1966, Côté and Timell 1967, Teratani et al. 1967) and of *Sequoia sempervirens* (Sherrard and Kurth 1933) and in the tension woods of *Eucalyptus goniocalyx* (Schwerin 1958), *Angophora costata*, and an *Acacia* species (Hillis 1960, Hillis et al. 1962). One of the explanations for the lower amounts of extractives in tension wood areas is the reduced availability of carbohydrates for centripetal translocation from the cambial region owing to their greater consumption in that region for the formation of thicker cell walls and the "gelatinous" layer of tension fibers. Similarly the above-normal lignin content in compression wood tracheids could result in lower extractives content in compression wood areas. This speculation is supported by the association of tension wood with lower contents of starch and sugar in the sapwood and of extractives in the heartwood of *Angophora costata* (Hillis et al. 1962). Also these differences diminish with increasing height in the tree (Hillis et al. 1962, Côté et al. 1966), which occurs concurrently with decreasing severity of reaction wood formation. The extent to which the changes in hormonal balance associated with reaction wood formation also affect extractives formation remains to be determined as well as the amount of carbohydrate produced in the crown and consumed in cell wall production. Increases in ethylene levels, which are associated with extractives formation at the heartwood periphery, have also been observed in zones of tension wood formation (Nelson and Hillis 1978c). The whole question of the control of photosynthate allocation (i.e., to the greater production of wood cells, to cells with thicker cell walls, to extractives, or to food storage) requires further study (Hillis 1968a).

5.7.3 Effect of Site

The effect of growth sites, environmental conditions, and genetic variation on the amounts of heartwood extractives cannot be clearly separated. The amount of pinosylvin and its monomethyl ether was found to be the largest in the heartwood of *Pinus sylvestris* from southern Sweden and the least in central Sweden. Trees from the northern part of that country contained the pinosylvins in amounts just below the average for the whole country (Erdtman et al. 1951). Similarly *Castanea sativa* trees grown in northern Europe contained 7–8 percent tannin in the heartwood, whereas those grown in southern Europe contained 10 percent or more. The barks of *Rhizophora mucronata, Bruguiera gymnorrhiza, Quercus sessiliflora*, and *Q.penduculata* trees grown in different localities contain significantly different amounts of tannin (Hillis 1962b). The bark of *Acacia mearnsii* trees grown in high rainfall areas contain less tannin than those from lower rainfall areas (Sherry 1971).

The significant within-species differences in the color characteristics of *Liriodendron tulipifera* and *Prunus serotina* have been related to the geographic areas in which the trees had grown (Sullivan 1967). In a quantitative study of the differences in wood color of individual trees and groups of trees of *Juglans nigra* grown at specific locations, the relationship of heartwood color to soil properties was greater than it was to either tree age or to diameter-growth rate (Nelson et al. 1969). The important nutrients in these soils were phosphorus, calcium, magnesium, and nitrogen. Forest-grown *J. nigra* had a darker-colored heartwood (and less sapwood) than open-grown trees (Phelps and McGinnes 1980a). The effects of planting site on the monoterpene composition of cortical oleoresin of *Pinus elliottii* were negligible, as was the planting site/provenance interaction (Gansel and Squillace 1976).

5.7.4 Genetic Differences

Although early studies found that the broad-sense heritabilities of extractives yields from the wood of 5-year-old grafted and other *Pinus elliottii* were low (Mergen et al. 1955, Einspahr et al. 1964), later work on clonal plantation indicated stronger genetic controls (Franklin et al. 1970). These predictions of genetic gains, based on family in addition to mass selection, were for 20 percent in oleoresin yields with a correlated genetic gain in the yield of extractives of about 12 percent. Other reports of high genetic gain in yield of oleoresin have also been made (Einspahr 1972). The heritability of oleoresin yield and the juvenile wood/mature wood correlations are so high as to make selections for yield one of the easiest traits to work with in the juvenile stages (Franklin et al. 1970, Franklin and Squillace 1973, Squillace and Gansel 1974). Variation in the response of oleoresin formation in *P.elliottii* after paraquat treatment seems to be largely due to environmental rather than to genetic effects (Squillace and Moyer 1976).

An extensive study of the cortical oleoresin from several clones and families showed the monoterpene composition to vary considerably between trees, mostly as a result of genetic factors (Mergen et al. 1955). The broad-sense heritabilities of the four major components (α-pinene, β-pinene, myrcene and β-phellandrene) were very strong and the narrow-sense heritabilities moderately strong; no effects of tree age were evident (Squillace 1971). Trends in the monoterpene composition of turpentine from *P. elliottii* have been associated with the distance from the live crown so that for genetic purposes it is desirable to sample near the base of the crown (Franklin 1976). Based on the behavior of α- and β-pinene relative to 3-carene in the xylem oleoresin of pine hybrids, β-pinene appears to be under direct genetic control, while α-pinene responds in a complementary fashion (Strauss and Critchfield 1982).

The considerable variation of extractives content in the wood of trees from various provenances of *Eucalyptus globulus* results in a related variation in yield of pulp of standard quality. The amount of sodium hydroxide consumed (as an indication of extractives content) to produce 100 parts of pulp varied between 35.6 and 74.2 parts, according to the provenances of the sample (Turner et al. 1983). The alkali requirements were related to the rainfall of the origin of the provenance but to no other environmental factor.

5.7.5 Crystals

Distinctive crystals of organic compounds, as well as those of inorganic compounds (Chattaway 1955a, 1956) such as calcium oxalate, exist in certain genera but have received little attention. Whereas the inorganic crystals are mostly formed in the cambium and outer sapwood, the crystals of organic compounds do not appear until the parenchyma cells are at the time of death, i.e., at heartwood formation. The crystals in eucalypt heartwood first appear in the ray cells, usually where they are contiguous with a vessel (Chattaway 1953). Even though the tissues may contain considerable amounts of extractives, crystals are not always present, and their occurrence is patchy and sometimes associated with fungal hyphae. The crystals are ellagic acid in *Eucalyptus* species and this compound, which is very insoluble under nonalkaline conditions, has a marked tendency to form super-saturated solutions. When these solutions are suitably seeded (such as with fungal hyphae) clumps of ellagic acid crystals will form. Frequently, crystals occur within a few millimeters of the outer limit of the heartwood and when fungal hyphae are present it appears they provide only a passive focal point for crystal formation. Crystals visually similar to ellagic acid have been found in the heartwoods of several genera in 19 families, but their chemical constitution has not been established (Chattaway 1953).

Whereas ellagic acid crystallizes from a complex mixture of extractives because of its insolubility, crystals of other compounds appear because of their selective formation and in a high degree of purity. Some of the vessels in the heartwood of *Intsia bijuga* contain characteristically yellow crystals of pure robinetin (3′,4′,5′7-tetrahydroxy-flavonol), whereas adjacent vessels may be empty or contain a mixture of flavonoids or stilbenes (Hillis 1977, W.E. Hillis, J. Ilic, Y. Yazaki 1977 unpublished data). Some vessels of *Castanospermum australe* contain pure bayin (a flavone C-glucoside). Crystals of pterostilbene have been observed in the heartwood of *Pterocarpus soyauxii* (Sandermann and Dietrichs 1957). The crystalline white deposits in the lumens of many cork cells of the bark of *Pseudotsuga menziesii* are almost pure dihydroquercetin (Krahmer and Wellons 1973).

Although not always in crystalline condition, floccosoids (Gerry 1943) contain high concentrations of lignans. Floccosoids can be frequently seen as white flecks with the naked eye in the transition zone and heartwood of *Tsuga heterophylla*, unsymmetrical in shape and randomly scattered. The lumens of the tracheids, but not the ray cells, in these regions are filled with α-conidendrin (Barton 1963). A more detailed study (Krahmer et al. 1970) showed that very small amounts of other lignans were present in the white opaque floccosoids and that the clear colorless floccosoids contained mainly hydroxymatairesinol, which is a minor component in the other type of floccosoids. Crystals in shakes and cavities are discussed in Chapter 2.6.2.9.

5.8 Reagents for Heartwood Detection

When freshly felled, the heartwood of many species can be readily delineated from sapwood by its drier (particularly with softwoods) and darker appearance. Also the cut surface of sapwood of some pines (e.g., *P. elliottii*) is soon covered by globules of resin. The delineation can be confirmed by staining the living parenchyma cells in the sapwood with the vital stain tetrazolium chloride (see, e.g., Kile and Wade 1975). There are still a number of important species containing heartwoods that are very similar in appearance to sapwood, particularly when dried. Distinction is even more difficult when there is an intermediate wood merging between the two major regions. A number of reagents can distinguish most heartwoods and be used for both freshly cut and dried specimens that have not been affected by fungi, decay, or preservatives.

Due to the paucity of extractives, aspirated pits or tyloses in sapwood, solutions of stains are more readily absorbed into sapwood cross-sections than when applied to heartwood. Tyloses are often, but not always, present in significant proportions in the vessels of the heartwood of hardwoods and can be detected on a clean-cut end grain with a hand lens.

The presence of starch indicates that the wood is sapwood but, due to seasonal and other demands, the amount of starch varies. Consequently the absence of starch does not indicate heartwood. Starch gives a deep blue color with iodine. Sugars can be present in the sapwood but not the heartwood and may be detected with limited success with Fehling's or with Benedict's solutions.

The amount of acid (mainly acetic acid) present increases significantly when heartwood is formed in species such as *Quercus robur, Ulmus glabra, Castanea sativa*, and *Pseudotsuga menziesii* (Arni et al. 1965 a,b) and also increases continuously towards the pith in *Larix occidentalis* (Côté et al. 1967). This acidity originates in the living tree from the acetyl groups of hemicelluloses, and with acetic acid being autocatalytic for the reaction (Stewart et al. 1961), the heartwood of old trees can become notably acid. In some eucalypts, such as *E. diversicolor*, pH values of 2.6 or less have been found. The differences in pH values between sapwood and heartwood are usually not large, e.g., pH 5.13 and 5.06 for *Fagus sylvatica*, 5.09 and 4.83 for *Picea abies*, 4.4 and 3.5 for a *Quercus* species (Sandermann and Rothkamm 1959), 4.75 and 3.65 for *L. occidentalis* (Côté et al. 1967) and 5.94 and 3.78 for *Ps. menziesii*. As the pH scale is logarithmic these differences represent larger differences in hydrogen ion concentrations. The pH of the heartwood of some tropical species can become less acidic than the sapwood, as in *Dalbergia melanoxylon* with pH 8.0 and 5.8 respectively (Sandermann and Rothkamm 1959). The acidity of stored undried samples can increase (Packman 1960) particularly if not sterilized, for example with propylene oxide (Arnie et al. 1965 a,b). Apart from the widespread acetic acid in heartwoods, some species contain acidic extractives, such as the strongly acidic plicatic acid in the heartwood of *Thuja plicata*, with acidities as high as pH 2.46 (Sandermann and Rothkamm 1959). The increase in acidity on heartwood formation can provide a basis for detection of the sapwood/heartwood boundary by means of pH indicators. One such stain that has been found useful for eucalypts is a 0.2 percent solution of dimethyl yellow in ethanol, which colors heartwood red and the sapwood yellow. Similar pH indicators of methyl orange, bromcresol green, bromphenol blue, or alizarine red S in water or ethanol have been used for softwoods to obtain contrasting colors.

Different reagents have been developed to accentuate the color of the extractives. Most notable among these are the diazo compounds which conjugate with polyphenols to give a range of often vivid colors. The most sensitive and versatile of these – benzidine – is no longer manufactured owing to its potent carcinogenic properties (Anonymous 1981b). Solutions (0.5 percent) of aromatic amines such as *o*-anisidine (Stalker 1971), *p*-nitroaniline, sulfanilic acid, tolidine or *p*-amino-N N-dimethyl aniline (Cummins 1972) in 24 percent hydrochloric acid are mixed immediately before use with an equal volume of 10 percent of sodium nitrite and then applied to cross-sections. A range of stabilized diazotized amines, which require only dissolving in a 20 percent solution of sodium acetate before use, are available from dye manufacturers. Sprays containing diazotized sulfanilic acid are followed with 20 percent sodium carbonate. The heartwood responds with a brighter, stronger, and usually reddish color than the sapwood with these dyes. In most cases the transition zone responds with a significantly paler color. When a 5 percent aqueous solution of sodium nitrite is painted on cross-section of *Eucalyptus obliqua* and *E. bicolor,* the initial pink shown by the heartwood changes to purple, and then deep blue in 30 minutes (A.J. Watson personal communication).

Polyphenols also react with various iron salts so that the heartwood gives a brighter blue coloration; 2–10 percent aqueous solutions of ferric chloride or nitrate, ferrous sulfate, or ferrous ammonium sulfate are used for this purpose (Kutscha and Sachs 1962). In a number of genera the amount of flavanols increases markedly when sapwood is transformed to heartwood, and flavanoids containing a phloroglucinol nucleus become brilliant red on the application of a 10 percent solution of vanillin in hydrochloric acid (Bate-Smith and Swain 1953), or dihydroflavonols in *Pseudotsuga menziesii* with zinc-hydrochloric acid (G.M. Barton 1973 personal communication). When *Tsuga heterophylla, Picea sitchensis*, and *P. abies* are dipped for 1–3 seconds in solutions of 45–50 percent perchloric acid, the heartwood becomes green in 20 seconds (Eades 1958). A brief heating at 100 °C after cross-sections have been sprayed with a 30 percent solution of phosphomolybolic acid results in darker colored heartwoods (Sandermann et al. 1967).

Some extractives fluoresce strongly in ultraviolet light and offer an additional method to confirm the heartwood boundary. Examples of such extractives are the 5-deoxyflavonoids found in heartwoods of the Anacardiaceae and of *Acacia* species which fluoresce brilliant yellow, and the stilbenes in *Pinus*, *Morus* and some *Eucalyptus* species which fluoresce a brilliant white or light blue. Some stains have been developed which not only make the heartwood more distinct but also enable the identification of species in a mixed resource of known species such as the spruce-pine-fir and other species in western Canada. Bromophenol blue (Kutscha et al. 1978) turns yellow-orange when sprayed onto freshly cut *Pseudotsuga menziesii* heartwood, Fast Bordeaux turns pine-orange with *Pinus contorta*, and the true firs (*Abies lasiocarpa* and other *Abies* species) develop red streaks when sprayed with Ehrlich's reagent (Swan 1966, 1983, Fraser and Swan 1972). The first two reagents react quickly and are used industrially to sort species.

Preservative treatment can vitiate some color tests, but one test which can still be used successfully with pines in the presence of chromium, copper, and arsenic involves staining for lignin. The characteristic lignification of resin canal tissue on heartwood formation in *Pinus* species can be clearly indicated by a double staining technique involving safranin followed by alcian blue, aniline blue, or fast green. In heartwood the resin canal tissue is stained red and in sapwood blue or green depending on the stain used (Bamber and Colley 1983).

X-ray scanning of logs of conifers and angiosperms has given a clearer picture of their heartwoods and peripheries than have sanded cross-sections (Benson-Cooper et al. 1982).

5.9 Exudates

5.9.1 Resin

The resinous exudates from conifers frequently have the same composition as the oleoresins of the same species and oleoresins contain very largely terpenoid compounds (Section 5.6.2) with very little of the fatty acids, glycerides, or waxes found in the wood resins (Mutton 1962, Kimland and Norin 1972). "Pocket resin" from *Pseudotsuga menziesii* can have a different composition from normal resin (Erdtman et al. 1968) (see also Chapter 2.6.2.1).

Resinous exudates can also contain phenolic compounds, such as p-coumaric, caffeic and ferulic acids, and vanillin from the damaged cambium of *Pinus* species (Erdtman 1958). The lignan pinoresinol has been obtained from the resinous exudate of several *Pinus* and *Picea* species (V. Meyer and Jacobsen 1924) and lariciresinol from *Larix decidua* (Haworth and Kelly 1937). Resinous exudates from hardwoods also contain lignans such as guaiaretic and the dihydroguaiaretic acids from *Guaiacum officinale* (Carnmalm 1954), olivil from *Olea europea* (Vanzetti 1936), iso-olivil from *O. cunninghamii* (Briggs and Frieburg 1937), and gmelinol from a deposit in the shakes of *Gmelina leichhardtii* (Birch et al. 1954).

5.9.2 Gum

True gums are nontoxic, odorless, colorless, and tasteless carbohydrates (Whistler 1973). They are usually water-soluble or they swell to form clear, mucilaginous solutions or dispersions but are insoluble in oils and in most organic solvents. The gels in water hinder agglomeration and the precipitation of fine particles. On heating, gums decompose without melting and usually char.

Gums are generally subdivided into (i) soluble gums such as gum arabic (from *Acacia* species and notably *A. senegal*) which dissolves in water to form viscous, sticky solutions, (ii) insoluble gums which absorb water to swell into thick jellies or jelly-like mucilages such as gum tragacanth (from *Astragalus* species), and (iii) semi-insoluble gums forming thin jellies (from *Prunus* species). They are mixtures of complex carbohydrate materials of high molecular weight involving arabinose, galactose, rhamnose, and glucuronic acid together with calcium, magnesium, and potassium cations in gum arabic (from *Acacia senegal*) and in addition other sugars in gum ghatti (from *Sterculia urens*) and gum tragacanth (from *Astragalus* species) (F. Smith and Montgomery 1959, Aspinall and McKenna 1968, A.E. Clarke et al. 1979, Davidson 1980).

5.9.3 Kino

Kinos are largely polymerized proanthocyanidins (formerly known as leucoanthocyanins) which when heated with hydrochloric acid yield one or more of the brilliant red cyanidin, delphinidin, or pelargonidin (Robinson and Robinson 1933, Robinson 1937). Even shortly after formation the proanthocyanidins are in a polymerized condition. They are light-sensitive, and further polymerization and darkening takes place so that little is known of their structure in kinos (Hillis 1962b).

Some freshly formed kinos are almost entirely polymerized proanthocyanidins, such as that from *Eucalyptus sieberi* (Hillis 1964), whereas a few contain significant amounts (less than 10 percent) of characteristic flavonoids such as dihydrokaempferol and its derivatives, flavanoid glycosides and C-glycosides, flavanones, the lignan eudesmin, and ellagic acid (Hillis 1962b, 1964, Hillis and Carle 1963, Hillis and Yazaki 1974) (see Chapter 2.6.2.3).

The proanthocyanidins can be detected specifically by their formation of a red color when heated with butanol and hydrochloric acid (in contrast to the brown or no reaction with other compounds) and this reaction has been adopted for anatomical studies (Hillis 1958). The traumatic parenchyma surrounding kino veins, which colored red with this reagent, formed a layer of only about five cells and became apparent in less than 20 days after injury (Skene 1965).

5.9.4 Latex

Pure rubber is a hydrocarbon, cis-1,4-polyisoprene, composed of 500 to 5000 units joined linearly and occurring in latex as more or less spherical globules 5 to 6 μm in diameter and stabilized by a thin absorbed film of protein and phospholipid. Rubber latex contains a higher proportion of isoprene polymers than other latexes with less than 6 percent impurities. Gutta percha from *Palaquium gutta* contains more than 85 percent of trans-1,4-polyisoprene of lower molecular weight than rubber, whereas the similar balata from *Mimusops balata* is much more resinous. Chicle, the only latex to contain both the trans- and cis-isomers (present in a ratio of 7:3), contains only about 20 percent polyisoprene. The reason for the differences in elasticity lies in the differences in the crystallinity of the polyisoprene (B.L. Archer and Audley 1973).

Rubber is biosynthesized from acetate, which in further stages develops mevalonate and then isopentenyl pyrophosphate, which forms polyisoprene on the surface of existing rubber particles. Gutta percha and chicle are probably formed by the same general mechanism (B.L. Archer and Audley 1973).

There is no evidence that any of the polyisoprenes have any biological function and the polymers are probably best regarded as inert end products of metabolism.

5.9.5 Manna

The exudate from *Fraxinus ornus, Myoporum platycarpum*, and other species (Chapter 2.6.2.5) largely contains the hexitol mannitol. Mannas from other sources can contain other carbohydrates.

Chapter 6

Biological Features

Several biological features concerned with heartwood and exudate formation exist in the sapwood, transition zone, and wound wood although their importance alters at different periods. Rather than describe the features in all regions, their description will be introduced in the early sections of this chapter and only the change in their status described when appropriate.

6.1 Sapwood

6.1.1 Wood Tissues

Fibers or tracheids, parenchyma, and vessels are the three distinct categories of cells in wood and their relative proportions vary considerably between and within genera. The major function of the several different types of fiber appears to be that of support in hardwood trees or of support and conduction in softwoods

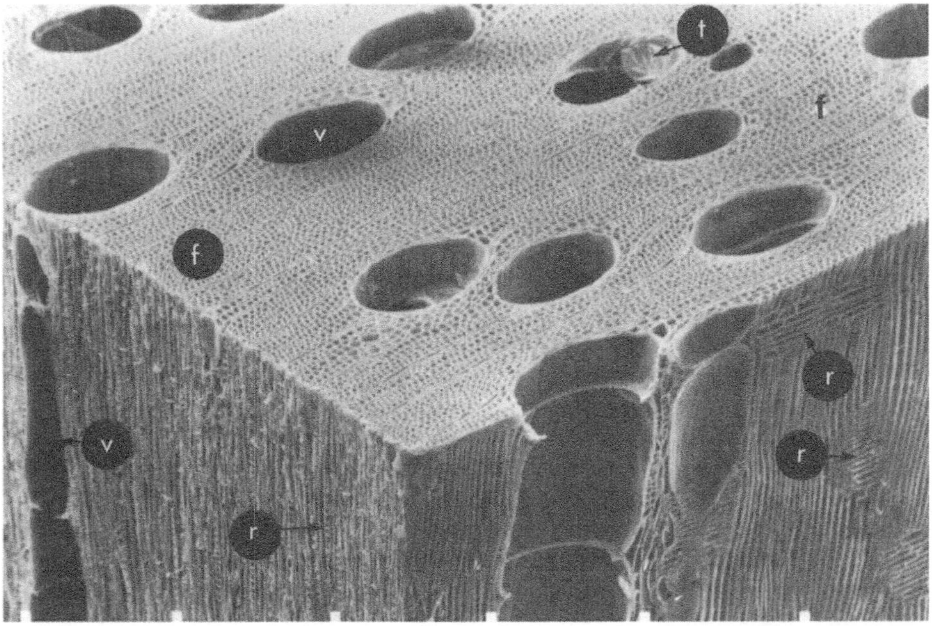

Fig. 6.1. Scanning electron micrograph of a small cube of *Eucalyptus regnans*, a diffuse porous species, showing vessels (*v*), some of which contain tyloses (*t*), ray parenchyma (*r*), and fibers (*f*) (unit distance 300 μm)

(where they are termed tracheids). Fibers lose their contents shortly after differentiation to become dead cells in the sapwood. Vessels (or pores) are unique to the hardwoods and are primarily water-conducting; the protoplasm dies shortly after differentiation. When the size of the pores do not change greatly in diameter throughout the growth ring, the wood is known as diffuse-porous (Fig. 6.1). The ring-porous hardwoods have an abrupt change in pore size between earlywood and latewood in the growth ring. Parenchyma cells usually retain their protoplasts and can remain living as long as they are in the sapwood; the proportion of them varies between 5 and 40 percent of the volume of the total tissue.

6.1.2 Lumen Volume

The extractives of some heartwoods are located in the lumens of fibers and vessels as well as those of the parenchyma (Fig. 6.2). To ascertain the influence of lumen volume on the amount of extractives present, examinations have been made of cross-sections of specimens containing reaction wood. Examination of such cross-sections with a range of anatomical features could remove the effect of variations of growth rate, site, and genetic differences present in samples from different trees. Tension wood contains thick-walled fibers some of which possess also gelatinous layers, and compression wood contains lumen volumes smaller than those found in normal earlywood. However, although the amount of extractives was lower in the tension wood of *Angophora costata* (Hillis et al. 1962) (Fig. 6.3) and of an *Acacia* species (Hillis 1960) and a *Larix* species (Côté et al. 1966) containing compression wood, no clear conclusions could be drawn on the influence of fiber lumen size. After a comprehensive survey of the literature, T.E. Timell (personal communication) found that the extractives content of compression wood in comparison with normal softwood could be more, or less, or the same, and the differences were mostly small. The amount of extractives present does not appear to be controlled by the lumen volume but rather by the amount of carbohydrate available at the time of requirement and the activity of the enzymes involved in the biosynthesis of the extractives.

6.1.3 Deposition of Extractives on Wall Surfaces

Erdtman (1943) examined the membranes lining the lumens of pine heartwoods which reputedly prevented the removal of ether-soluble extractives when the wood was extracted with ethyl ether. The membranes were soluble in ethanol and acetone and were obtained in 0.1 percent yield. Their composition closely resembled pinoresinol and similar lignans, and was consequently different from the main extractives components. Membranes exist in *Acacia catechu* (P.R. Rao and Seshadri 1948) and *Pseudotsuga menziesii* (Pew 1948). Pew (1949) isolated, in their original shape, the cell linings from the rays in the heartwoods of *Ps.menziesii, Tsuga heterophylla, Pinus taeda*, and *Quercus marilandica.*

Linings have been observed also on the lumen walls of tracheids, the pit membranes and pit chamber walls in the heartwood (but not the sapwood) of *Pinus*

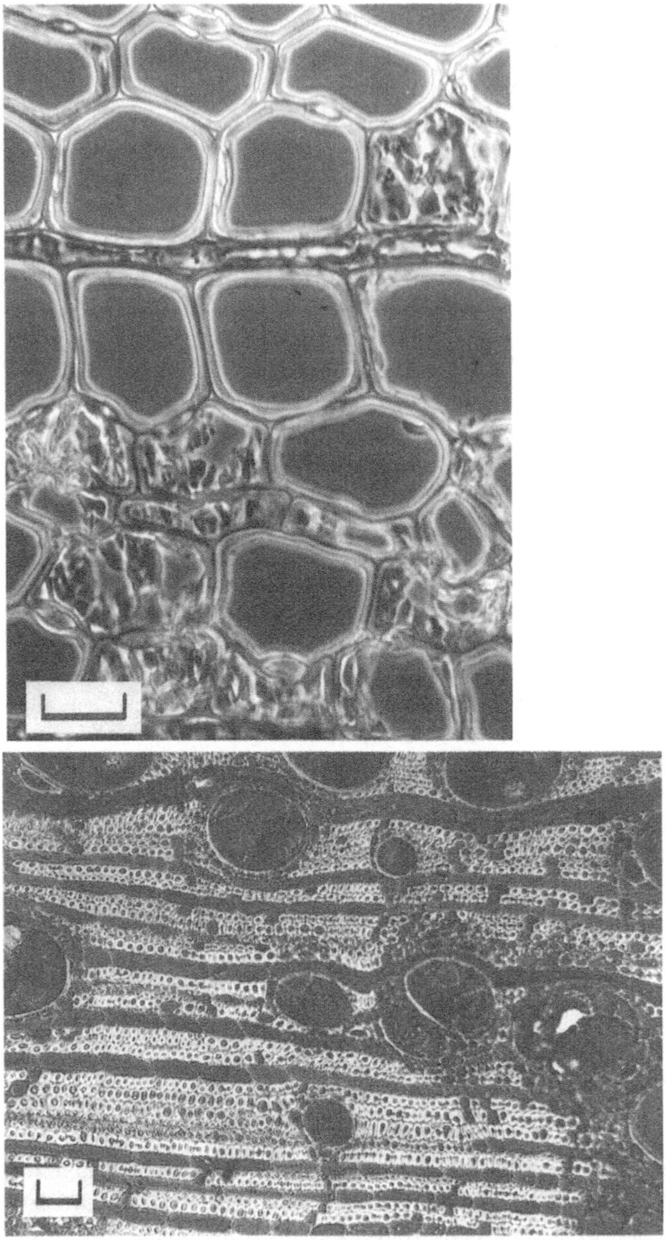

Fig. 6.2. Transverse section of heartwood of *Larix occidentalis* (*top*). Both tracheids and ray cells contain arabinogalactan (phase contrast, scale 50 μm). (Côté et al. 1966). Transverse section of heartwood of *Eucalyptus marginata* (*bottom*). Polyphenols are present in the lumens of vessels, ray and other parenchyma cells, and fibers (scale 1 mm)

Fig. 6.3. Portion of a cross-section from an eccentric stem of *Angophora costata* containing regular heartwood. The cross-section shows the pale-colored transition zone (*TZ*), a borer tunnel stopping at the heartwood periphery, kino veins and wider growth rings in the tension wood zone (*the upper portion*). There is no relationship between the heartwood periphery and the growth rings or the outer cambial region (width of specimen 22 cm)

radiata, P. muricata (Wardrop and Davies 1958), *Sequoia sempervirens,* and *Libocedrus decurrens* (Kuo and Arganbright 1980b), *Pseudotsuga menziesii, Thuja plicata, Tsuga canadensis,* and *T. heterophylla* (Krahmer and Côté 1963) (Fig. 6.4).

The pit membranes in the bordered pits of sapwood of several coniferous species have tori (Fig. 6.5) consisting of pectins, cellulose, and hemicelluloses (Bauch et al. 1968), and sometimes amorphous materials (Fujikawa and Ishida 1973, 1974) but not phenolic compounds (Bauch and Berndt 1973). The fibrils of the margo are composed of cellulose. *Pseudotsuga menziesii* (but not *Pinus sylvestris*) has sapwood tori and membranes which contain aromatic compounds. Other species with these compounds in the tori and pit membranes include *Abies* and *Tsuga* species, *Picea abies, Larix decidua, Sequoiadendron giganteum, Sequoia sempervirens, Taxodium distichum, Cryptomeria japonica, Chamaecyparis lawsoniana,* and *Juniperus virginiana.* On heartwood formation the tori were encrusted (see also Section 6.1.5.2) with additional aromatic compounds (Bauch et al. 1968, 1974, Bauch and Berndt 1973). There is evidence (Bauch et al. 1968, 1974, Bauch and Berndt 1973) to indicate that lignin is deposited on pit membranes during heartwood formation, but the similarity of many properties of lignin and polyphenols – which are all phenolic compounds – can lead to misinterpretation of results. In addition, the lack of direct contact of the pit membranes with a protoplast associated with lignification has been recognized (Bauch et al. 1974, Thomas 1975). Detailed studies of the tori of *Pinus taeda, Abies fraseri,* and *Picea rubens* (Thomas 1975) strongly support the view that during heartwood forma-

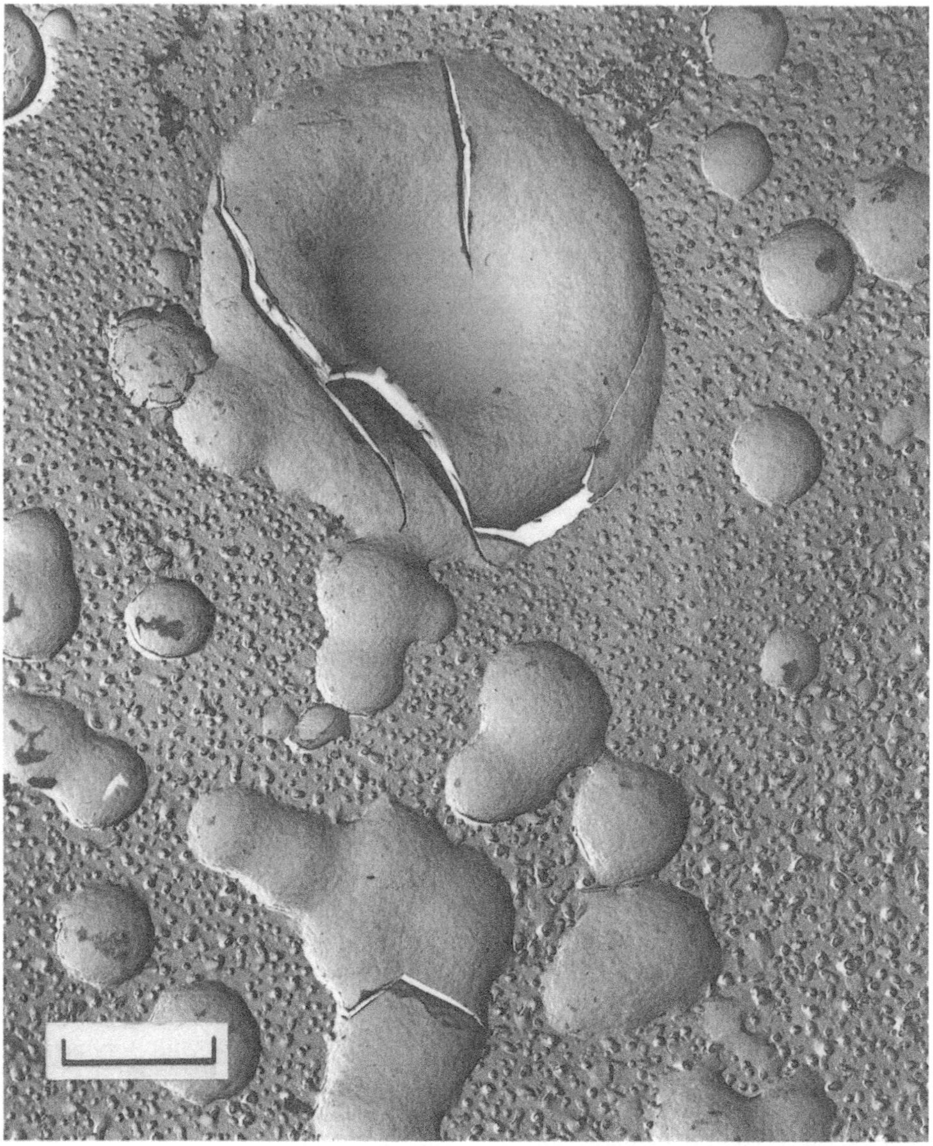

Fig. 6.4. Extractives on the tracheid walls of *Tsuga heterophylla* covering a pit aperture and the warty layer and accumulated as large deposits. (Krahmer et al. 1970, scale 5 µm. TEM by R.L. Krahmer)

tion in those species polyphenols rather than lignin are produced and deposited within the tori.

The pit membranes between vessels, rays, and tracheids have been found to be coated and infiltrated by extractives in the heartwoods of *Nothofagus fusca* (Kininmonth 1972), *Acer rubrum* (Wheeler 1982), *Thuja plicata* (R.M. Meyer and G.M. Barton 1976 personal communication), and *Quercus* species (Wheeler and

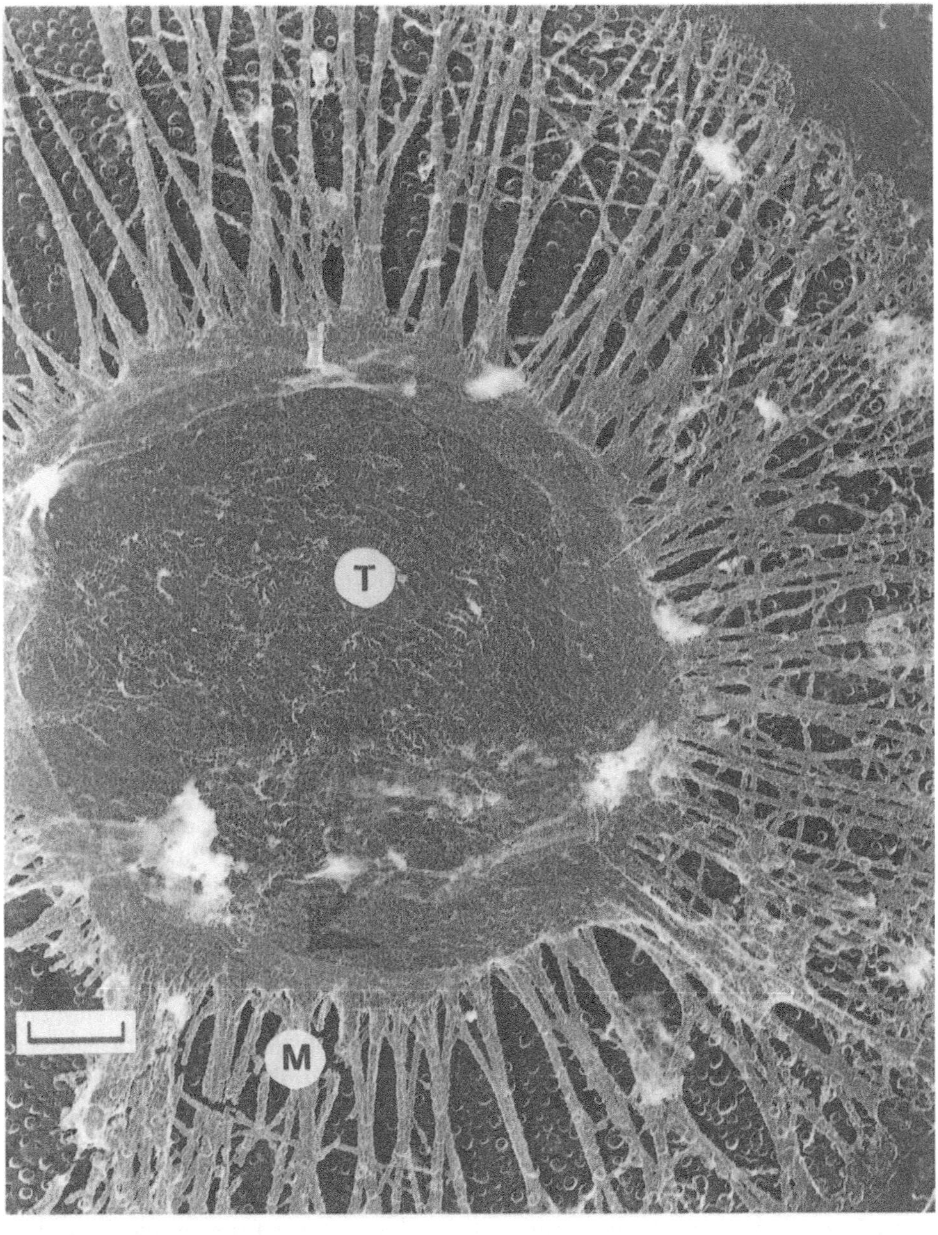

Fig. 6.5. Part of bordered pit of *Tsuga heterophylla* with tori (*T*) and margo (*M*) (scale 1 μm. TEM by R. L. Krahmer)

Thomas 1981). Moreover the lumen walls of the latewood tracheids of *Sequoia sempervirens* and *Libocedrus decurrens* were more heavily coated than those of the earlywood tracheids. The linings could be removed with organic solvents and in the case of *S.sempervirens* with water also, although in this species the extractives on the pit membranes were water-insoluble, indicating that further polymerization and oxidation of them had taken place. The presence and degree of incrustation in tracheids may be related to their distance from the ray parenchyma cells (Kuo and Arganbright 1980a).

In contrast to these findings, Morgan and Orsler (1969) presented evidence for the insolubility of some compounds in certain solvents being due to their presence in the cell wall. For example, some solvents, such as acetone, swell and penetrate the cell wall and remove extractives in a way that cannot be done by a relatively nonswelling solvent such as ether. There is little information to indicate the differences in the amount of extractives in the cell wall and in the cell lumen, the processes which control that distribution in different woods, and whether some compounds selectively remain in the cell wall rather than the lumen.

6.1.4 Impregnation of Cell Walls

The ultimate fate of the free space between and/or within the fibrillar components of the cell wall of newly formed sapwood can affect different properties of wood. These properties include uptake and transport of various ions, the movement of liquids and gases from lumen to lumen through the walls, host-parasite interactions, and various wood properties affected by extractives such as stability of dimensions, durability, pulpability, and color of pulp (e.g., Hillis 1977). In precise determinations on individual cells, Berlyn (1969) found the S2 layer (which represents about 60 percent of the tracheid cell wall) of fully hydrated latewood cell walls of the sapwood of *Pinus resinosa* contains about 25 percent free space. The water in the capillaries of the wet walls of pine and spruce woods has been estimated at more than 40 percent (on a dry wood basis) with about half the water being in the free state (J.E. Stone 1973). The inner and outer portions of the S2 layer of the cell wall are more porous than the middle portions. Furthermore because the middle lamella between the cells is less porous than the secondary wall, there would be a lower void volume (Bailey and Preston 1969, Saiki 1973). There is evidence to suggest that these pores are linked into two sets of microcapillary systems with one set running parallel to the microfibrillar orientation and the other set approximately at right angles to that orientation (Rudman 1966a).

The experimental difficulties in determining the diameter of the capillaries in cell walls have resulted in a wide range of values (Berlyn 1969, Hillis 1971). Recent determinations indicate that the cross-sections of the capillaries have diameters from 160 to 600 nm (Hillis 1971, Sawabe et al. 1973, J.E. Stone 1973). This is large enough to allow molecules of significant size to penetrate; for example the widest part of a glucose molecule is 63 nm and that of a flavanoid monomer about 150 nm. Presumably, when the membranes containing extractives rupture, those components which are small enough penetrate the cell wall through the two sets of microcapillary systems (Bailey and Preston 1969). Microspectrophotometric studies have shown the presence of polyphenolic extractives in the cell wall of different species (Bland and Hillis 1969), and improved precision in this instrument should provide detailed knowledge of their type.

Recent studies on the extractives in the cell walls of *Sequoia sempervirens* and *Libocedrus decurrens* have shown a distribution pattern similar to that of capillary volume (Kuo and Arganbright 1980b). The concentration of extractives decreases from the lumen inwards towards the S1 layer and with lower amounts in the thin S1 and S3 layers. Furthermore, the heartwood tracheid walls were more strongly impregnated with extractives than those of sapwood. Bosshard (1968) has pointed out that the regular or obligatory-colored heartwoods have lower shrinkage and swelling levels than other types because the extractives can penetrate the cell wall and cover the cellulose and hemicelluloses. The 50 percent of the components which contribute to the color of *Cryptomeria japonica* heartwood and are not extractable with methanol are apparently the oxidative products of one of the major extractives (Kai and Teratani 1977) which has coupled with the lignin. Prolonged extraction of the finely divided heartwood of *Eucalyptus delegatensis* (Hillis and Carle 1959) and several other eucalypts has failed to remove all the proanthocyanidins and ellagic acid entrained in the cell wall.

Extraction of the heartwood of *Sequoia sempervirens, Libocedrus decurrens* (Côté 1963, Resch and Ecklund 1964) and *Pseudotsuga menziesii* (Bailey and Preston 1969) greatly increases the permeability and diffusion of liquids into the cell wall. Extraction would not only remove the cell wall extractives but also cell contents and cell wall linings. The extractives (such as proanthocyanidins) of some species combine with cell wall components and prevent quantitative studies of their effect on wall properties. However, the wall extractives (between 40 and 80 percent of total extractives) of heartwoods are readily removed with water (in the case of *S.sempervirens*) and with organic solvents (*L.decurrens*) (Kuo and Arganbright 1980a). The good dimensional stability of these two species and their much lower volumetric shrinkages than other coniferous species is attributable to extractives in the cell wall. Tarkow and Krueger (1961) reported that about 74 percent of the total hot-water extractives in *S.sempervirens* were located in the cell wall. In a more detailed study of this species it was found that 50.6, 77.6 and 74.1 percent cell wall extractives occurred in the total extractives of the outer heartwood, intermediate, and inner heartwood regions respectively (Kuo and Arganbright 1980a). The corresponding values for *L.decurrens* were 39.3, 60.7, and 72.5 percent. Although considerable amounts of extractives can be present in the walls of some species, only a minor part of the arabinogalactan is in the cell wall of *Larix occidentalis* (Côté et al. 1966).

The blockage of moisture movement pathways in *Sequoia sempervirens* and *Thuja plicata* by extractives results in severe moisture gradients. The consequent checks in the woods lower their strength to a greater extent than in other coniferous woods that have lower extractives content (Petterson and Bodig 1983). The effect of extractives on the water-vapor sorption and shrinkage of a number of Mexican woods, *Pseudotsuga menziesii, Pinus* species, and *Juglans nigra*, has drawn attention to a number of the features of sorption (Wangaard and Granados 1967, Salamon and Kozak 1968, Choong 1969, G.A. Cooper 1974). However, all aspects of the process could not be explored owing to the differences in extractives composition between the species and the absence of data on the variable distribution of extractives in cell wall and cell lumen.

6.1.5 Parenchyma

Parenchyma are concerned with the storage of various materials, wound healing, and the origin of secondary substances and structures. In a few cases they assume specialized functions. Ray parenchyma serve as communication channels radially from the cambium through the sapwood. Axial parenchyma (which are largely storage tissue) can exist as the common strand parenchyma or as specialized axial fusiform parenchyma. Apotracheal parenchyma may occur singly or as bands of axial parenchyma that are not associated with vessels, whereas paratracheal parenchyma are associated with vessels in some way. Epithelial parenchyma surround resin or gum canals and the composition of the resin synthesized by these epithelial cells is markedly different from that produced by the ray parenchyma (e.g., Mutton 1962).

The rays extend inwards from the cambium towards the center of the tree and outwards through the phloem. They are made up of a number of cells of different shape, size, and physiological function largely determined by the conditions that existed at the periphery of the stem when those cells were differentiated from the cambium initials. In her classic paper, Chattaway (1951) drew attention to three phases of ray activity. First, the cell shape is determined very soon after the ray cells have been cut off from the ray initials in the cambium. Second, in the early stages of the life of a ray cell, crystals, silica, starch, and lipids may accumulate (although these substances may be found also in other parenchyma). Third, as long as the ray cells contain living nuclei, they are capable of metabolic activity, as shown by the fluctuating content of starch or by the excessive growth of tyloses into vessel cavities in response to some new and specific type of stimulation late in the life of the cell (Chattaway 1949). In addition, heartwood extractives appear to be formed in a period when the nucleus is in a disorganized condition and probably after tyloses have been formed. However, pit aperture size is a controlling factor in the appearance of tyloses (Chattaway 1949).

Even when the ray cells are of the same shape they may not all be functionally homogeneous, and some cells always take up staining reagents differently from the rest. "Furthermore, there is a functional heterogeneity not only between the different files of cells that compose a ray, but even within a single file of cells, depending upon the different vertical elements of the wood with which the individual cells of the ray are in contact" (Chattaway 1951). There are also differences within the multiseriate rays themselves and within the peripheral cells in contact with the vertical elements of the wood and the central cells which are in contact with other ray cells only. In *Acer saccharum* and some other species, specialized parenchyma cells, known as "contact cells", connect the axial and radial parenchyma systems with the vessels. In *A. saccharum* the contact cells have several large pits which face the vessels and connect them with the ray cells. These pits are the sites of strong enzyme activity (Sauter et al. 1973, Section 6.1.6).

Differences exist in the functions of ray cells in different genera. In some genera both starch and polyphenols occur together in all ray cells. In others, starch and polyphenols are segregated, with starch occurring in the vertical parenchyma and erect ray cells, while polyphenols are confined to the procumbent cells. Following the abrupt transformation of sapwood to heartwood in *Larix occidentalis*, the polyphenols and some other extractives appear mainly in the rays whereas the arabinogalactan is largely found in the tracheid lumens (Côté et al. 1966).

Some extractives are formed by a small group of cells or by specialized cells. A specialized function is shown by the epithelial cells surrounding the resin canals of conifers in the formation of extractives different from those of the parenchyma. Other specialized cells include the tanniniferous tubes in Myristicaceae, the oil cells in the Lauraceae, the vertical intercellular canals of the Dipterocarpaceae, and latex tubes in the Euphorbiaceae and in *Alstonia* species (Dadswell and Hillis 1962).

Chattaway (1951) drew her conclusions after an extensive study of woods notably including a considerable number from the southern hemisphere. That her important conclusions on the different physiology or functions of different ray cells have been little developed (or even considered) is a reflection on the experimental difficulties for their study and on the fact that research is mainly on species with relatively simple ray structures. Further studies could confirm not only the differences between different cells but the sequence in the formation of cell contents, for example the formation of crystals, formation and resorption of starch, and differences in the reaction rates of formation of secondary components. Examinations of simpler plant tissues have begun to show that biosynthesis and accumulation of a number of secondary compounds take place during definite development stages of an organism. Even closely related secondary compounds can vary greatly in their location within an organ (Wiermann 1981).

6.1.5.1 Volume

The volume of parenchyma cells varies considerably in woody tissues. In the major southern pines (Koch 1972a) the rays, including the epithelial cells surrounding the horizontal resin canals, occupy about 9 percent of the volume. The vertical interconnecting resin canals can occupy a similar volume in pines; the thin-walled epithelial cells surrounding the canals secrete oleoresin. [The oleoresin in the resin canals of *Picea abies* constitutes about 45 percent of the total nonvolatile extractives in the wood (Kimland and Norin 1972)]. Resin canals in heartwood can be obstructed by tylosoids resulting from proliferation of epithelial cells. In southern pines and some other pines, the ray parenchyma in the horizontal elements are about 1 percent or less of the volume of the wood. A detailed examination of *Picea abies* has shown that only one type of parenchymatous ray cell appears (if the epithelial cells within the horizontal resin canals are excluded). These are thick-walled, their pit area is one-tenth that of *Pinus sylvestris* (Nyrén and Back 1960); about 4 percent of the weight of the wood is made up of parenchyma cells (Back 1960a).

In a number of species, polyphenols either fill the ray cells or form a lining along the inner cell wall (e.g., Preusser et al. 1961, H.H. Dietrich 1964b, Fengel 1965). Other extractives, such as the neutral extractives of *Betula verrucosa*, occur in the same way (Perilä 1954). However, it is unlikely that the volume of the parenchyma cells influences the amount of extractives stored in the wood of pines and it definitely does not in the case of *Larix* species. Parenchyma account for about 5 percent of the volume in these species, whereas the amount of arabinogalactan can be over 25 percent of the weight of the wood with the largest portion located in the lumen of the tracheids (Côté et al. 1966).

The volume of parenchyma cells in hardwoods can be up to 40 percent or more, but the proportion of living cells is unknown. These cells offer large storage volumes, and their influence on the amount of heartwood extractives has been studied with different species.

The numbers of axial and ray parenchyma per unit weight of the heartwoods of *Quercus rubra* and *Juglans nigra* varied, as did the relative volumes of parenchyma (volume/wood volume and volume/wood weight). However, no causal relationship could be found between and within trees to the quantity of polyphenolic heartwood extractives in either species or across the heartwoods. Polyphenol biosynthesis in *J.nigra* was substantially greater on any basis than it was in *Q.rubra*. The amount of extractives formed per parenchyma cell was greater during formation of the outer heartwood than during the formation of inner heartwood in both species with *J.nigra* showing the greatest increase (Nelson 1975). Apparently the rate and duration of aromatic biosynthesis during the formation of heartwood are affected much more by physiological conditions near the sapwood-heartwood boundary than by the amounts of parenchyma participating in the biosynthesis.

6.1.5.2 Cell Wall

Parenchyma cells are usually thin-walled and commonly composed only of a primary wall. Craib (1923) found that the living ray cells in the transition zone of *Taxus baccata* did not contain polyphenols although they were detected in the walls of the tracheids. The ray parenchyma and the resin canal tissue of the Diploxylon section of pines lignify (and some cells also thicken) at the sapwood-heartwood boundary (Balatinecz and Kennedy 1967, Bamber and Davies 1969, Bamber 1972, Bauch et al. 1974, Yamamoto 1982). The Haploxylon pines and other conifers have lignified sapwood ray cells. In these studies lignin was identified by histological stains, by ultraviolet light absorption or spectra so that the proportion of polyphenols present is unknown.

During heartwood formation in *Pinus sylvestris* polyphenols and lignin are laid down within the pinoid pits as well as in the walls of the parenchyma cells. The same process apparently occurs in pit membranes of tracheids (Bauch et al. 1974). Similar changes occur with *Juniperus virginiana* but not so clearly with *Larix decidua* and *Taxodium distichum*.

Harkin and Obst (1973) found that peroxidase is exclusively responsible for the dehydrogenative polymerization of lignin. On the basis that during heartwood formation C_6–C_3 compounds are deposited in the pinoid pits in ray cells, as well as in the bordered pit membranes of Diploxylon pines, followed by polymerization by peroxidase, Bauch et al. (1974) accordingly surveyed the presence of peroxidase. In *P.sylvestris* the peroxidase in the sapwood of the nonlignified tissues disappeared at the sapwood-heartwood boundary and was apparently associated with the formation of lignin. This was further supported by the incorporation during early summer and early autumn of labeled coniferin in the walls of the nonlignified ray parenchyma cells over a width of only one or two cells at the sapwood-heartwood boundary. There was, however, no apparent incorporation in the bordered pits (Bauch et al. 1974). Incorporation of phenylalanine in the transition zone of *P.densiflora* also occurred to a much greater extent than it did in adjacent sapwood (Yamamoto 1982).

As pointed out in Section 6.1.3, polyphenols and lignin have a number of similar or identical properties which can hinder their complete identification in tissues. Furthermore although peroxidase is considered to be essential for lignification, its presence does not imply that lignification will occur. Peroxidase can also oxidize polyphenols to polymers of high molecular weight, some of which are insoluble except in alkaline solvents.

In a thorough study, Yamamoto (1982) subdivided the uniseriate ray parenchyma of *Pinus* species into three types according to the site of thickening and lignification of their walls and those of the resin canals. In type A the parenchyma are thickened and lignified in the outermost sapwood. In *P.strobus* the parenchyma cells behave as type A as do one-half the earlywood parenchyma of *P.densiflora* (Fig. 6.6). In the latter species the remaining earlywood parenchyma behave as type B and are thickened and lignified in two rings of the transition zone. The radial resin canals of these two species behave as type B and the walls are thickened and lignified in the inner growth ring of the transition zone. The latewood parenchyma of *P.densiflora* behave as type C and lignify without thickening in the transition zone. Similarly, most of the cells of the radial resin canals mature

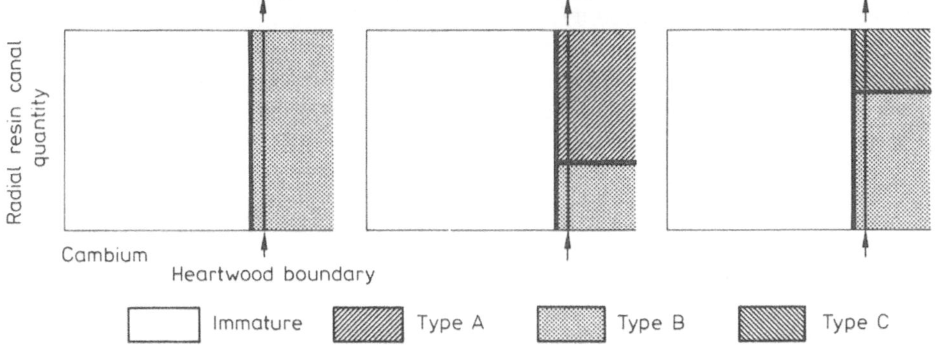

Fig. 6.6. Quantitative changes in parenchyma cells of different *Pinus* species on maturing with age. (Yamamoto 1982)

and lignify without thickening in the inner ring of the transition zone. In *P.banksiana* the situation is more complex although the cells of the radial resin canal behave as types B and C. In this species about one-half of parenchyma cells commence to thicken and lignify (type B) in the middle sapwood and the remainder of type B mature in the transition zone.

The season of maturation of the cell walls of the different types of ray parenchyma in pines varies. The thickening and lignification of the A type proceeds in the late summer to autumn period in the outermost sapwood; the cells continue to live in the sapwood and store primary metabolites. The maturation of the B type occurs in late summer to early autumn in the transition zone or in the inner sapwood. Lignification of the C type occurs during summer and autumn (Yamamoto 1982).

The nature of the ray parenchyma cells apparently differs with their position in either the outer or inner part of the transition zone. The early stage of heartwood formation occurring in the outer part of the transition zone of *Pinus densiflora* is characterized by the wall development of B type parenchyma, the commencement of lignification of C type parenchyma, the incrustation of bordered pits with polyphenols and normal cytological behavior. The last stage of heartwood formation is characterized by the completion of lignification of C type parenchyma, the final incrustation of cross-field pits, higher peroxidase activities, strong incorporation of lignin precursors, the decrease of food reserves, the lowering of moisture content, and the degradation of the nucleus (Yamamoto 1982).

Secondary wall formation in the radial resin canal parenchyma cells of Haploxylon pines occurs during heartwood formation (Bamber 1972, 1976a). The thickening and lignification of the secondary wall in these cells in *P.strobus* took place at the sapwood/heartwood boundary mainly when the cambial activity began to decline; the parenchyma then began to die until the commencement of dormancy of the cambium (Yamamoto et al. 1978). There was a short time lag between the secondary wall formation and lignification of both uniseriate rays and resin canal tissue and in addition the parenchyma cells of the radial resin canals died about 1 year earlier than those of the uniseriate rays.

In addition to the different behavior of ray parenchyma in different parts of the growth rings, there can be different cells comprising the rays of whose functions there is little knowledge. Hugentobler (1965), for example, describes in *Acer pseudoplatanus* the different lengths and extent of pitting in ray cells at different distances from the vessels, with the adjacent ray cells being devoid of starch, and with changes in the shape of the cells and of the size of nuclei across a growth ring. The shape and size of nuclei also vary with the distance from nearby vessels as well as from the cambium.

6.1.5.3 Cytology

By definition, heartwood ceases to contain living cells although under certain circumstances they have been reported present (Good and Nelson 1951). Chattaway (1952) concluded from her observations that a period of intensified metabolism occurs adjacent to the heartwood and then the protoplasmic membrane disintegrates to release the extractives in order to produce the marked differences from the sapwood. Frey-Wyssling and Bosshard (1959) considered heartwood formation to be an aging process in a living tree with the parenchyma cells undergoing irreversible changes, such as disorganization of the cells' systems, but without accompanying physiological activation.

The sequence of changes leading to the formation of heartwood can be illustrated with *Tamarix aphylla*. Starch disappears in the 7–13 growth rings from the cambium, positive tetrazolium tests (indicative of dehydrogenases) are obtained in the 14–18 growth rings only and disintegration of the nuclei finally occurs (and heartwood forms) in the 16–21 growth rings from the cambium (Fahn and Arnon 1963). This sequence indicates that an increased supply of glucose is one of the earliest steps in the demise of the parenchyma cells and possibly linked with a later changed mitochondrial activity to account for the high dehydrogenase activity.

Frey-Wyssling and Bosshard (1959) measured the morphology and dimensions of the cell nuclei to define the vitality or activity of the storage tissue cells involved in the transformation of sapwood to heartwood in *Pinus sylvestris, Larix decidua, Taxus baccata, Pseudotsuga menziesii, Sequoia sempervirens, Robinia pseudoacacia, Picea abies, Abies alba, Carpinus betulus, Tilia cordata, Fagus sylvatica*, and *Fraxinus excelsior*. These workers, and Fahn and Arnon (1963), who studied *Tamarix aphylla*, reported that the nuclei in ray parenchyma cells change in shape and disintegrate with increasing distance from the cambium. Coniferous cells close to the cambium, and in the cambium itself, possessed large and more or less round cell nuclei characteristic of very active tissues and were larger than those in angiosperms. The nuclei became radially elongated and their degrees of slenderness increased with increasing distance from the cambium. (The degree of slenderness of the cell nucleus used as a criterion of cell vitality was determined in radial sections as the quotient of the long axis and short axis of a cell nucleus). The peak in this curve was normally reached in the first 5–10 rings from the cambium. After this, the cell nuclei became shorter and the slenderness degree decreased more or less rapidly; there was also a decrease in the reductive capacity of the mitochondria. The changes did not occur at the same time in all cells in a ring, and some cells may survive for a further few years. At the heartwood boundary the cells died, the nuclei disintegrated to form debris and disappeared. Similar decreases in slenderness ratio and nuclear volume have been observed by Higuchi et al. (1964) with *Chamaecyparis obtusa, Cryptomeria japonica, Thujopsis dolabrata, Magnolia obovata, Pterocarya rhoifolia, Betula maximowicziana*, and *Illicium religiosum* sampled in middle to late summer. The nuclei of the ray cells of the angiosperms were much smaller than those of the conifers, and in the inner parts of both types of sapwood the nuclei of all parenchyma cells were or became globular. Similar observations have been made on a wide range of species sampled in northern Japan in mid-autumn (Fukazawa and Higuchi 1965) and by Nobuchi et al. (1982) with *Cryptomeria japonica*. The slenderness ratios and nuclear volumes of the earlywood of the sapwood of *Larix, Abies*, and *Picea* species were greater than those of the latewood bands (Fukazawa and Higuchi 1965). A gradual decrease in the nuclear area from the outer sapwood to the transition zone exists in *Bridelia retusa* (Nair et al. 1981). These observations do not support the proposal of Chattaway (1952), which would require the ability for strong nuclear activity.

Results similar to those of Frey-Wyssling and Bosshard (1959) were obtained by Hugentobler (1965), who examined during the four seasons the hardwoods *Quercus robur, Fraxinus excelsior, Carpinus betulus, Acer pseudoplatanus, Juglans regia, Robinia pseudoacacia*, and *Prunus avium*. The peak of the slenderness degree was reached within five growth layers with *Fraxinus excelsior, Acer pseudoplatanus*, and *Carpinus betulus*. However, the situation is made more complicated than originally observed in that the ring-porous or semi-ring-porous species, such as *Juglans regia, Robinia pseudoacacia, Quercus robur*, and *Prunus avium*, which are characterized by a much smaller sapwood zone, showed a different behavior to the other three species. After an initial rapid decrease of the slenderness degree, there occurred a marked increase in the slenderness ratio and a tendency to form bigger and longer cell nuclei and an increased ratio of nucleoli volume to nuclei volume in the transition zone between sapwood and heartwood. There was a distinct shrinkage of the cell nuclei and disappearance of the nucleoli, then the surface of the nuclei became irregular, collapsed in parts, degenerated, and lay against the walls of the parenchyma in the heartwood. Decomposition of the nucleus was invariably preceded by hydrolysis of the starch and a marked increase of colored deposits (Hugentobler 1965), and it was suggested that enhanced metabolic activity at the sapwood-heartwood boundary depended on the time of the year and possibly occurred in late summer. The DNA content of the cell nuclei also revealed characteristic changes in this zone and at the moment when the nuclei were about to disappear (Lairand 1963, Higuchi et al. 1964).

The slenderness degree of the cell nuclei of the upright ray cells across the sapwood of *Beilschmiedia tawa* was clearly different from that of the procumbent ray cells and only the latter are apparently engaged in the transformation of sapwood to heartwood (Bosshard 1968). There was an increase of the slenderness degree of the nuclei of the procumbent cells (possibly an indication of increase in vitality) but little change with the upright cells, which appear to correspond with the ray tracheids of the conifers. Lairand (1963) examined the sapwood and heartwood of *Pinus sylvestris* and *Larix europea*, the sapwood and ripewood of *Picea excelsa*, and the sapwood of *Betula verrucosa* and *Populus tremula*. Although only a portion of the quantitative data was reported, he found that elongated nuclei predominate in the middle parts of the rays and retain their shape (without becoming rounder) until they disappear. Also, rounded nuclei exist in positions where there is little reason to expect them to lose some of their vitality. However, cells in the rays that intersect a resin duct in pine possess nuclei that are much larger than normal in both length and width. Further evidence of the differences between genera has been adduced with *Acacia catechu, Bauhinia tomentosa*, and *Bridelia retusa* which have distinct heartwoods and – at an unspecified period in the year – ray parenchyma cells which have nuclei with areas and slenderness ratios showing a strong negative correlation with the distance from the cambium (Nair and Chavan 1983). *Lagerstroemia speciosa* has a distinct heartwood, but only the nuclear area shows a strong negative correlation with distance from the cambium but not the slenderness ratio; on the other hand in *L. indica*, which does not form a distinct heartwood, the area of nuclei in the ray parenchyma cells shows no relation with the distance from the cambium, whereas the slenderness ratio shows a positive relationship. *Bauhinia purpurea*, unlike *B.tomentosa*, has neither heartwood nor nuclear relationships. Increases in nuclear area at the heartwood periphery has been reported for *Albizia lebbeck*, *Prosopis juliflora*, and *Ougeinia oojeinensis* (Nair and Chavan 1983).

As a part of their extensive studies, the Kyoto school grouped 20 coniferous species according to the survival curves of the sapwood ray parenchyma cells (Nobuchi et al. 1979). All the cells of one type survive from the cambium to the transition zone between sapwood and heartwood, as with *Pinus densiflora*, *P. pentaphylla*, *Taxus cuspidata*, *Thuja orientalis*, *Thujopsis dolabrata*, and *Juniperus rigida*. With Type II a part of the ray parenchyma cells die in the inner sapwood, as with *Pseudotsuga japonica*, *Tsuga sieboldii*, *Cryptomeria japonica*, *Chamaecyparis obtusa*, and *Torreya nucifera*. With this type the necrosis of the ray cells starts generally in the upper and lower rows of the ray parenchyma cells or in the rows adjacent to the ray tracheids. In Type III, part of the ray parenchyma begins to die in the outer sapwood as with *Picea abies*, *Larix leptolepis*, *Abies firma*, and *A. sachalinensis*, and these species do not exhibit a sharp boundary between sapwood and heartwood.

Cytoplasmic organelles such as mitochondria, Golgi apparatus, endoplasmic reticulum (ER), and plastids are plentiful in the ray parenchyma of the outermost sapwood of *Cryptomeria japonica* (Nobuchi and Harada 1968, Nobuchi et al. 1976) and *Abies sachalinensis* (Hirakawa et al. 1979). Mitochondria and ribosomes were observed to be particularly abundant in the growing season when respiration and protein synthesis are expected to be high. In the inner sapwood these organelles appear to be less and the amyloplasts more abundant.

In the transition zones of *Cryptomeria japonica*, *Abies firma*, and *Pinus densiflora* the organelle content of the parenchyma cells vary. It has been concluded that these cells do not begin to age simultaneously during heartwood formation (Nobuchi et al. 1976). The boundary of the ray parenchyma cells containing nuclei coincided with the transition zone of *Robinia pseudoacacia* in spring to midsummer. After this time ray cells containing nuclei could be found in the outer heartwood (Nobuchi et al. 1984b). This co-existence of nuclei and colored substances in the ray parenchyma cells has also been found in *Cryptomeria japonica* (Nobuchi et al. 1982).

The necrosis of ray parenchyma cells varies in the sapwood of *Pinus strobus*, *P.densiflora*, and *P.banksiana* (Yamamoto 1982). With the first two species, death of all these cells in the uniseriate rays occurs suddenly in the transition zone during late summer. With *P.banksiana*, death begins gradually from the middle sapwood, increases at the heartwood periphery, and occurs mainly in autumn. The maturation and death of the radial resin canal tissue is completed during late summer. The nuclei in the epithelial cells of the vertical and horizontal resin canals of *Larix leptolepis* and *Picea jezeonsis* disappear before those of the ray parenchyma in the transition zone (Fukazawa and Higuchi 1965).

Nečesaný (1958b, 1965, 1966) examined the vitality of the sapwood ray cells from the cambium to the demarcation of the heartwood of *Fagus sylvatica*, *Populus nigra*, *Ulmus effusa*, *Acer platanoides*, *Pinus sylvestris*, *Quercus robur*, *Tilia platyphylla*, and *Robinia pseudoacacia*. He found that the electrical resistance decreased, the conductivity increased, there was a decline in the resistance of the cells to repeated plasmolysis, an increase in the rates of deplasmolysis, a decline in the osmotic value, a weakening of respiration, and increased staining with neutral red. From these observations he concluded that a gradual decline took place in the vitality of the living cells until death occurred at the heartwood periphery as a result of a natural aging process influenced by environmental conditions. He (1966) suggested that the higher vitality of cells in the border zone of false heartwood relative to that of regular heartwood caused the higher frequencies of tyloses and exceptionally high extractives content in the outer zone of the false heartwood. The rapid death of the cells through accidental causes when the latter is formed causes extractives to coagulate in the lumens and possibly results in the distinct dark border lines surrounding false heartwood. On the other hand, the slow death during formation of regular heartwood results in the extractives impregnating the cell walls. *Quercus cerris*, a species with a wide sapwood of 30 growth rings or more, can contain regular heartwood as well as false heartwood associated with frost splits. An examination of this species confirmed these views. False heartwood can form before the regular heartwood forms, or later when it envelopes the regular heartwood (Nečesaný 1968) and enables comparison of ray parenchyma cells in the one species and even in the one stem containing the two different types of heartwood. Later studies on *Eucalyptus botryoides* showed decreases in osmotic values and isoelectric points of the living ray parenchyma cells across the sapwood and a seasonal effect with higher values being obtained in the spring (Nečesaný 1973a,b). Segments of branches of this eucalypt species were allowed to die at different rates in moist air when some cellular changes followed (V. Nečesaný and W.E. Hillis 1970 unpublished data). These changes were the slenderness ratio of the nuclei in the living ray cells as well as the number and diameter of starch grains and the amounts of nucleic acids before and after extraction of ribonucleic acids. The trend in values was the same as in the large cross-sections reported previously. A clear increase in both types of nucleic acid occurred immediately before death and at the moment of formation of tyloses. The cells in the samples collected in spring aged more slowly than in those obtained in the summer and presumably because the latter had higher physiological activities.

The nuclei in the ray parenchyma can begin to disintegrate at different stages. In the inner sapwood of *Cryptomeria japonica* the upper and lower parenchyma cells in the rays show a strong tendency to lose their nuclei first (Nobuchi et al. 1982). By using Ziegler's (1968) measure of the rate of survival of wood parenchyma cells, these workers found that there were about 50 percent of living ray cells in the innermost annual rings of sapwood or the outermost layers of the transition zone of this species, with nuclei and protoplasmic organelles being absent from the remainder.

No direct relationship between the DNA (deoxyribonucleic acid) content of the nuclei of the ray cells could be found with the metabolic activity of *Abies sachalinensis*, *Picea jezeonsis*, and *Larix leptolepis* sampled in summer and autumn. Also the RNA (ribonucleic acid) in the nuclei and in the cytoplasm of these species disappeared rapidly during aging and the decrease in metabolic activity of the sapwood (Higuchi et al. 1967a, Fukazawa and Higuchi 1965).

Morphological studies have revealed differences between various ray parenchyma cells, and chemical studies further indicate differences. This aspect is important in relation to different patterns of formation of heartwood and of extractives. Yamamoto et al. (1979) have drawn attention to the possibility of two modes of heartwood formation, one of which involves the ray tissue as a whole, while the other involves the particular physiological condition of each individual cell and is not subject to seasonal control. Both modes could overlap each other. Fukazawa and Imagawa (1970) postulated two types, one of which could not form "heartwood substances".

Later, three types of ray parenchyma cells were observed in *Cryptomeria japonica*. In the transition zone, one type lacks nuclei and protoplasmic organelles, is pale in color with the colored contents spread evenly throughout the lumens. Nobuchi et al. (1982) suggested that the cell contents arise from surrounding cells during the spring to mid-summer period.

Type II cells contain nuclei, starch grains and other cell organelles with colored materials spread partially in the lumens. These cells are observed mainly in the inner sapwood (Nobuchi et al. 1982) and appear to be in the transitional stage from life to death with coloration commencing before the cells die.

Type III cells have a pattern of colored substances similar to Type II, but they are darker in color, and lack starch, nuclei, and protoplasmic organelles. The colored substances are not only darker in color but larger in amount and have been presumed to arise from the starch at the same time as the disintegration of the nuclei (Nobuchi et al. 1982). Type III cells are found at the heartwood periphery.

It is important in future work that clear differences of function or activity of individual parenchyma cells in the intermediate wood or transition zone be defined.

6.1.6 Respiration and Enzyme Activity

Senescence in many herbaceous plants is accompanied by increased respiratory activity (Varner 1961), but there are contradictions concerning this activity with aging sapwood cells and particularly intensification in the zone adjacent to the heartwood periphery. These discrepancies are probably due to the season in which samples were examined. For example, Higuchi et al. (1967a) examined the respiration patterns and cytological behavior of the conifers *Cryptomeria japonica, Chamaecyparis obtusa, Abies sachalinensis, Picea jezoensis, Thujopsis dolabrata*, and the angiosperms *Magnolia obovata, Pterocarya rhoifolia, Betula maximowicziana, Fraxinus mandshurica*, and *Quercus crispula*. The trees were examined during summer and autumn and probably before dormancy began. Respiration decreased rapidly from the cambium towards the inner sapwood and transition zone. [The same pattern had been found with *Fraxinus nigra, Quercus robur, Acer rubrum* (Goodwin and Goddard 1940), and *Picea abies* (Möller and Müller 1938)]. Respiration was much lower in the transition zone than in the cambium, and no evidence of intensified respiration was obtained. On the other hand Zelawski (1960a) confirmed the findings of Djaparidze by showing that there is intensified physiological activity in the inner sapwood rings of various *Quercus pedun-*

culata examined in late winter to early autumn. Also during storage after felling, respiration of the living sapwood cells increased considerably, particularly in the zone adjacent to the colorless transition zone. Oxygen absorption of the transition zone did not increase on storage. The respiration rate decreased with age in the sapwood of *Populus tremula* but increased somewhat at the sapwood/heartwood boundary and then decreased in the heartwood (Nekrasova 1973). Evidence has been obtained for the increased respiratory and other metabolic activity in the transition zones surrounding necrotic regions in *Pinus radiata* and *Picea abies* compared with surrounding sapwood (Shain 1971, Shain and Hillis 1972). The oxygen uptake and amount of carbon dioxide in the inner sapwood growth ring of *Robinia pseudoacacia* felled in early winter was considerably greater than in other parts of the cross-section (Höll and Lendzian 1973). Higuchi et al. (1969) found evidence to support the view that fatty acids, frequently occurring in the ray parenchyma of sapwood as glycerides in some conifers and hardwoods, are consumed as substrates for respiration and as a carbon source in the biosynthesis of secondary products.

The oxygen uptake by the transition zone of *Pinus radiata* was significantly higher than that of middle and inner sapwood in the dormant season (late autumn to early spring) and much higher in the transition zone in the dormant season than in the growing season. On the other hand, oxygen uptake by the sapwood was not appreciably different throughout the year. Even in the growing season the respiration of the transition zone tended to be higher than the middle and inner sapwood (Shain and Mackay 1973a). The presence of a dry transition zone throughout the year and the greater respiratory activity in this zone (compared with the inner sapwood zones) during most of the year suggest that heartwood formation proceeds throughout the year but mainly during the dormant season.

Tissues involved with the storage and mobilization of carbohydrates and of their transport have high activities of acid phosphatase. Sauter (1966b) investigated the activity of this enzyme in the ray parenchyma of a *Populus* species with 3 growth rings of sapwood over the different seasons, using histochemical techniques. A high activity was found during dormancy, a strong increase in early spring and the starch mobilization period, a decrease during starch deposition in summer, and a second increase during starch dissolution in late autumn and winter. The contact cells connecting the rays and the vessels, particularly on the tangential walls, showed a higher acid phosphatase activity than other cells at the times of starch dissolution and intense radial transport (i.e., in spring and in summer). Similar contact cells in *Acer saccharum* also differed from all other parenchyma cells and showed a greatly increased activity of respiratory enzymes during times of sugar secretion. Both aerobic and anaerobic respiration were much higher in the contact cells at this time (Sauter et al. 1973). The activity of acid phosphatases was much higher in these cells, but only during the time when sucrose appeared in the vessels during autumn and winter when starch was mobilized; in late spring their activity disappeared.

The concentration of acid phosphatase activity and nucleoside triphosphatase on the large pits between the contact cells and the vessels of *Populus*, *Betula*, and *Acer* species is detectable only when sucrose has to pass these specific sites (Sauter 1966b, 1972, 1983, Sauter et al. 1973). In relation to this concentration of enzyme activity it is noteworthy that high concentrations of peroxidase have been found on the pit membranes of *Pinus sylvestris* (Bauch et al. 1968) and that aromatic compounds have been observed on the tracheid-side (but not the ray-side) of pit membranes of *Tsuga heterophylla* (Krahmer et al. 1970).

During storage of *Cryptomeria japonica* sections, or during the dormant season, the respiratory breakdown of glucose through the glycolysis-TCA cycle system gradually changed to that through the pentose phosphate pathway (Higuchi et al. 1967b). In this connection it is interesting to note the effect of enzyme inhibition. Although the addition of dinitrophenol had no obvious effect on extractives formation in *Rhus succedanea* (Hillis and Inoue 1966), it increased the rate of glycolysis in the tissues of *C.japonica* and particularly in the transition zone (Higuchi et al. 1967b). The addition of arsenite and iodoacetic acid (which block the utilization of acetate in the TCA cycle, J.L. Webb et al. 1949, Beevers 1960) to sapwood tissues of *R.succedanea* resulted in a marked increase in biosynthesis of flavonoids, a group of compounds which require acetate units for the synthesis of the A-ring of their molecules (Hillis and Inoue 1966). However, the effect of arsenite on the respiration of *C.japonica* (Higuchi et al. 1967b) increased the rate of glycolysis towards the pentose phosphate pathway and hence the formation of shikimic acid and subsequently the B-ring of the flavonoids.

6.2 Transition Zone

6.2.1 Seasonal Variations

Not all heartwoods are surrounded by a transition zone (Chapter 2.4), but the variable appearance of it within and between species could be due to its seasonal formation or its seasonal transformation to heartwood. A detailed study of the transition ("white") zone in *Cryptomeria japonica* reports its variation in width throughout the year (Nobuchi and Harada 1983). One interpretation of data concerning coloration of ray parenchyma (Nobuchi et al. 1982) is that the transition zone, but not the heartwood, enlarges in autumn to winter, rather than the heartwood itself. Slight variations were found in its moisture content (see Section 6.2.3).

The transition zone of *C. japonica* was present in all seasons but there were seasonal variations in the distinctiveness of its boundary with the sapwood and in its whiteness. The variations in the latter were attributed to differences in the moisture content between the transition zone and the adjacent sapwood, with the differences being greatest in the winter months (Nobuchi and Harada 1983). On the other hand, a low moisture content in transition zones has been reported in *Pinus densiflora* and *P. banksiana* during the summer season (Yamamoto 1982).

6.2.2 Appearance

The variation in distinctiveness of the transition ("white") zone in *Cryptomeria japonica* is considered to be due to differences in moisture content (see Section

6.2.3). On microscopic examination, the region having living ray parenchyma cells can be most clearly distinguished from that having dead cells during the spring and summer season. After that period the living and dead cells overlap each other (Nobuchi et al. 1982).

When cross-sections are treated with various staining reagents (Chapter 5.8), the transition zone often responds less than either the sapwood or the heartwood and is relatively colorless. This is clearly shown in the colored illustrations of *Pinus sylvestris* and *Picea abies* treated with phosphomolybdic acid, and with bromcresol green; a 2,3,5-triphenyl tetrazolium chloride stain was also effective. A concentration of amino acids was also revealed by a stain for amino acids (Sandermann et al. 1967).

6.2.3 Water Content

The transition zone has a lower water content than the sapwood and in some instances it has a lower water content than the heartwood, as in *Taxus baccata* (Craib 1923). Yazawa and Ishida (1965b) classified species into three types according to the average moisture content of the transition zone. The first type represented by *Larix leptolepis* has a moisture content between sapwood and heartwood. The moisture content of the transition zone of the second type is similar to the low value of either the sapwood or the heartwood [e.g., *Chamaecyparis obtusa, Cercidiphyllum japonicum, Prunus sargentii, Ulmus davidiana, Abies* and *Ulmus* species, and *Dacrydium cupressinum* (see also Dadswell and Hillis 1962)]. In the third type it is lower than in both sapwood and heartwood [*Cryptomeria japonica* (see also Nobuchi et al. 1985), *Quercus mongolica, Ostrya japonica, Acer mono*, and in *Quercus pedunculata* in different seasons of the year (Zelawski 1960a,b)].

Only slight seasonal changes of moisture content occurred in the radial direction in the transition ("white") zone of three trees of *Cryptomeria japonica* sampled throughout the year (Nobuchi and Harada 1983). The changes were less than those occurring in the heartwood and particularly the sapwood. There were no changes with height in the stem in contrast to those in the heartwood. The transition zone is more readily detected in the winter because of its paleness, which is related to differences in moisture content.

The moisture content of the transition zone of *Pinus strobus, P.densiflora*, and *P.banksiana*, which is less than that of the sapwood, but a little more than that of the heartwood, remains about the same throughout the year with the first two species, but shows a significant decrease during summer with *P.banksiana* (Yamamoto 1982). This may be associated with heartwood formation in the latter species.

The moisture content of different parts of the transition zone in *C.japonica* varied between 130–80 percent in comparision with 150 percent or more in the adjacent sapwood and 200 percent or more in the adjacent heartwood; the moisture content decreased to 100 percent in the inner heartwood (Nobuchi and Harada 1983, Nobuchi et al. 1984b). Earlier work on trees of the same species felled in late autumn indicated that the sapwoods had average moisture contents of 198 percent, the transition zones 60 percent, and the red heartwoods 69 percent (Yazawa and Fukazawa 1956). The transition ("dry-wood") zones of *Pinus radiata* trees had a moisture content of 40 percent in comparision with the adjacent sapwood of 170 percent of the weight of the oven-dry wood (Harris 1954a). The moisture contents of the transition zones of *Quercus pedunculata* and *Q. robur* are less than those of both sapwood and heartwood (Zelawski 1960a, Ebermann and Stich 1985).

A marked decrease in moisture content in the transition zone adjacent to sapwood occurred together with the beginning of the death of the ray parenchyma cells in *Cryptomeria japonica* (Nobuchi et al. 1985). The transition zones surrounding some wound woods (Section 6.4.1) and paraquat- treated pine wood (Section 6.4.2) also have moisture contents below that of surrounding tissues and even of *P. radiata* heartwood (Sioumis and Lau 1976).

Most of the discussions concerning the relative amounts of water, carbohydrate, and salts in the various zones pay attention to their radial movement. Future consideration must also be given to axial movement. There is evidence from various injuries and injections of various chemicals that axial movement upwards can be substantial, particularly in the inner sapwood of some species.

6.2.4 Pit Aspiration and Tylosis Formation

Because of the strong association between heartwood formation and the occurrence of physiological drought (Harris 1954b), Harris (1954a) investigated some 200 *Pinus radiata* trees with an age of 27 years and grown in New Zealand. The mean percentages of aspirated pits in the earlywood bands across the sapwood, the middle of the transition zone (which Harris described as "drywood"), and the heartwood were 30, over 90 and 96 respectively. Dominant, co-dominant, and suppressed trees behaved similarly. The percentage of pits aspirated in the latewood seldom exceeded 20. The percentage of pits aspirated was not directly related to the age of the growth ring or its distance from the cambium, but increased layer by layer to that point at which the transition zone is first formed when about 50 percent of the pits were aspirated. Under this condition, passage of water through the system is almost impossible, so that there is a barrier which prevents water in the outer part of the stem from diffusing inwards (except through the living cells of the rays) (Harris 1954a). In other samples of the same species grown in Australia, a gradual, consistent increase in aspiration did not occur across the sapwood and instead, within one growth ring of the sapwood adjacent to the transition zone, a marked increase to about 90 percent aspiration of the pits occurred (J.A. Kininmonth 1976 personal communication). A similar situation has been reported in *Cryptomeria japonica*, where the percentage of aspirated pits rose sharply at the border of sapwood and transition zone from 10 to 60 and with a further slight increase in the heartwood (Nobuchi and Harada 1983). Although most *P. radiata* heartwoods have a high percentage of aspirated pits and are impermeable to neutral aqueous solutions of chemicals, occasionally specimens may show patches of heartwood into which solutions have penetrated. Also some specimens of *Pseudotsuga menziesii* contained heartwoods that were more permeable than sapwood (Kozlik 1982).

In addition to pit aspiration, the incrustation of the pits with extractable materials and particularly at the sapwood-heartwood boundary can further decrease the permeability of the wood to fluids (Krahmer and Côté 1963, Liese and Bauch 1967). Of those species studied, incrustation was greater with *Thuja plicata* and *Tsuga canadensis* than with *T. heterophylla* and *Pseudotsuga menziesii*. Incrustation of bordered pit membranes increased abruptly in the transition zone or before heartwood formation in *Pinus strobus, P.densiflora*, and *P.banksiana* (Yamamoto 1982) (Fig. 6.7).

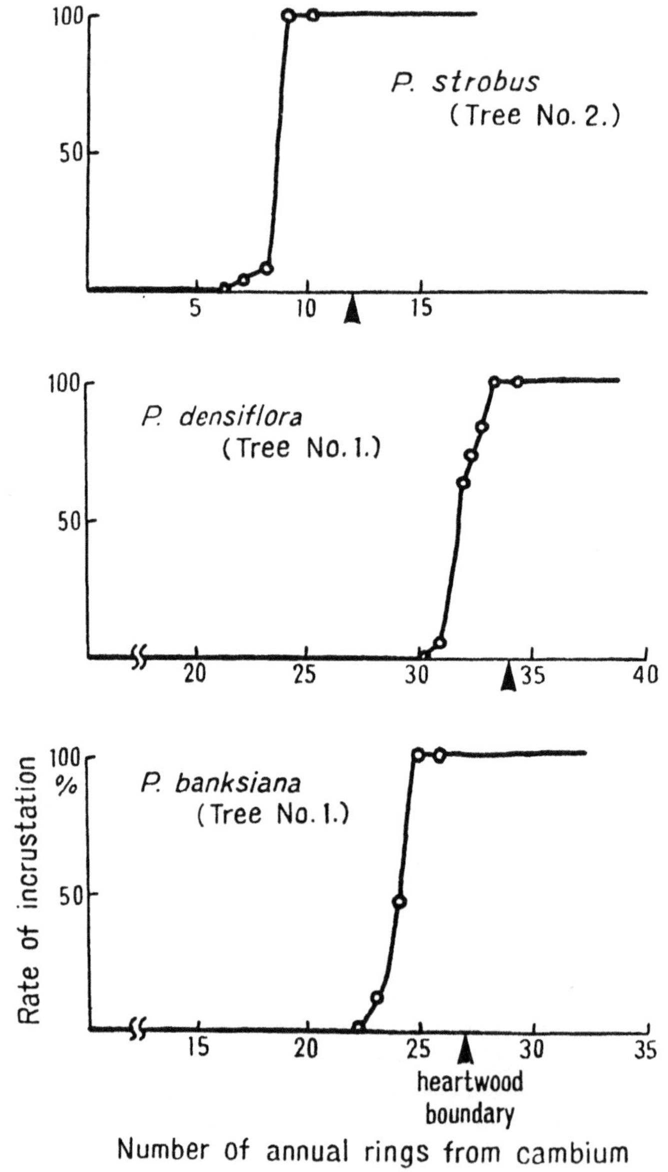

Fig. 6.7. The rate of incrustation of the bordered pit membranes in tracheids from sapwood to heartwood. (Yamamoto 1982)

Permeability dropped sharply at or close to the heartwood periphery and some improvement in permeability could be achieved on extraction. In addition to the pit aspiration at the periphery of the heartwood of *Pinus sylvestris*, there is also a much greater impregnation of the torus by osmiophilic substances (probably polyphenols) than that in sapwood (Fengel 1970). A similar situation has been found with the incrustation by polyphenols of the pit membranes of *Abies alba* (Parameswaran and Bauch 1975). Furthermore, these substances completely incrust and seal the membranes of the bordered pits between ray tracheids and between ray and longitudinal tracheids in *P.sylvestris*.

Aspirated pits generally arise in softwoods when a tracheid wall lies between one tracheid containing gas and one containing water. The hydrostatic tension in the inner sapwood possibly results in the drier transition zone when the permeability of its parenchyma increase (Shain 1979). It is unlikely that the metabolic processes in the transition zone would consume sufficient water to reduce the water levels to the amounts present.

The balloon-like structures of tyloses develop when parts of ray parenchyma cells project through pit pairs into adjoining cell lumens (Koran and Côté 1965). Their formation in the vessels of hardwoods (Foster 1964) commences when the water content decreases, and if the pit opening is larger than about 10 µm (Chattaway 1949, Jurášek 1956), so that the protoplast can be forced through the unthickened and extensible intercellular membrane into the vessel lumen. Apparently tyloses grow by synthesis of wall substances (Shibata et al. 1980). Extractives only form when the pit opening is smaller (Chattaway 1949). Although tyloses are commonly found in the heartwood of a wide range of angiosperms, they can occur also in some sapwoods in small regions (e.g., *Quercus phellos*, *Robinia pseudoacacia* and *Fraxinus excelsior*) or in injured zones. Zycha (1948) found that oxygen (but not fungi) caused tyloses to form, and in *Fagus sylvatica* the moisture content must be higher than 50 percent (Jurášek 1958). Tyloses appear abruptly in the transition zones of those eucalypts which contain them also in the heartwood (Bamber 1976b), and extractivities or colored materials develop later (Nečesaný 1973a). On the other hand, it is reported (Nair et al. 1981) that the degeneration of the nucleus and appearance of tyloses and polyphenols occur simultaneously in the transition zone of *Bridelia retusa*.

In many angiosperm genera, the formation of tyloses is an important feature of the changeover of sapwood to heartwood and they are abundant in *Castanea*, *Quercus* (but not *Q. rubra*), *Robinia*, and *Juglans* species. They are present in moderate numbers in *Fraxinus*, *Populus*, *Salix*, and *Plantanus* species and much less frequently in *Acer, Prunus*, and *Betula* species. Tyloses form in the outer sapwood of *Robinia pseudoacacia* and the conditions controlling their formation may differ from those found in transition zones. The most recent work shows that their development commences at almost the same time in these two regions, but the development is delayed during the period of conspicuous heartwood formation (Nobuchi et al. 1984b).

The seasonal development of tyloses can differ in different climatic regions. Tyloses in the outer sapwood of *Robinia pseudoacacia* growing in a cold temperate zone began to form in late summer, but their growth was suspended until after the next growing season (Ishida et al. 1976). In a warm temperate zone tyloses in the same species began their growth in early autumn, attained their maturity in early winter, and originated not from ray parenchyma cells, but from axial parenchyma cells (Fujita et al. 1978). Consequently tylosis formation, which is a characteristic feature in heartwood formation in many angiosperms, is a normal metabolic change due to seasonal growth and affected by climatic conditions.

The presence of aspirated and occluded pits and/or tyloses in dried wood can affect the penetration of neutral solutions. The uptake of polyethylene glycol by the pale-colored transition zone of *Eucalyptus camaldulensis* was appreciably lower than that of the tissues on either side and the volumetric shrinkage of this zone, and its content of water-soluble materials was also appreciably higher (Tischler 1976).

6.2.5 Cytology of Parenchyma Cells

Chattaway (1952) concluded from anatomical findings that the living cells of sapwood go through a phase of increased activity before they die. Frey-Wyssling and Bosshard (1959) proposed that an altered, rather than intensified, metabolism produces colorless polyphenols in the inner sapwood and in the outer transition zone. However, the disintegration of the nuclei in the ray parenchyma cells of *Cryptomeria japonica* begin in the transition zone after the decrease in moisture content found in sapwood. The percentage of living cells decreases gradually across the transition zone to zero at the outer heartwood boundary (Nobuchi and Harada 1983, Nobuchi et al. 1984a, 1985).

An increase in protein content (see Section 5.4) and frequently an increase in water-soluble vitamins have been found in the sapwood adjacent to the heartwood in *Pinus strobus, Larix decidua, Fagus sylvatica*, and *Robinia pseudoacacia* (Ziegler 1964, 1968). As the vitamins tested are co-enzymes, the findings indicate an increased physiological activity in the inner sapwood.

The enzymes of the tricarboxylic acid cycle and of oxidative phosphorylation are located in the mitochondria, which appear to be involved in the early stages of heartwood formation (Frey-Wyssling and Bosshard 1959). Wardrop and Cronshaw (1962) found in sections from trees of *Eucalyptus elaeophora* that those cells rich in phenolic substances contained few or no starch grains, and that polyphenols may also develop in vesicles within amyloplasts in ray parenchyma as starch disappears. [Whenever starch was mobilized in the ray cells of a *Populus* species, acid phosphatase activity was frequently observed to be localized on the starch grains (Sauter 1966b)]. There is also an accumulation of both starch and polyphenols in the ray cells of mature sapwood of *E.melliodora* and *Acacia mearnsii*, nuclei are present, and the polyphenols occur in small vacuoles which later fuse together (R.C. Foster 1970 personal communication, Fig. 6.8). In the transition zone of *Quercus iberica* where the concentration of shikimic acid was high and where intensive formation of ellagitannins took place, mitochondrial activity could not be found (Nutsubidze and Abashidze 1980). The polyphenols of *Fagus sylvatica* are localized in vacuoles of parenchyma cells, but when they are present, nuclei are absent (Dietrichs 1964b). During senescence and death of the parenchyma cells in *Melia azedarach*, a more or less simultaneous disintegration

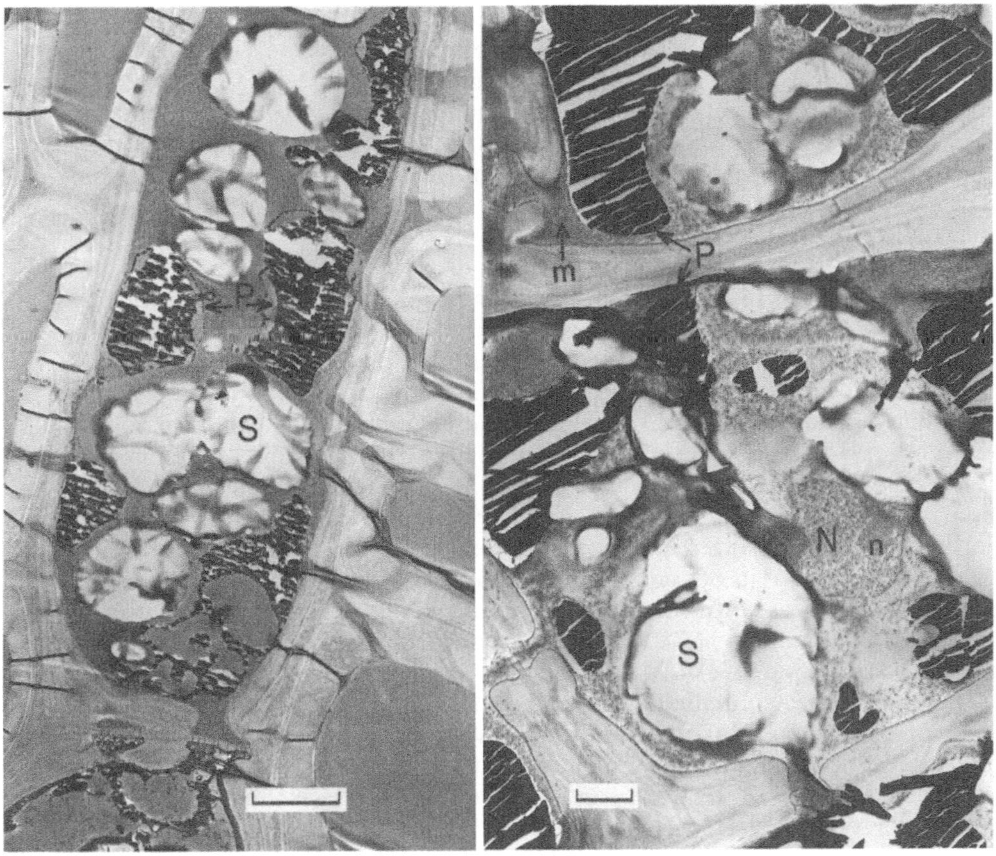

Fig. 6.8. A sapwood ray cell enclosed by latewood fibers and adjacent to the heartwood of *Eucalyptus melliodora* containing both polyphenols (*P*) and starch (*S*) (*left*, scale 4 μm), and details of the inclusions as well as the cytoplasmic structure with nucleus (*N*), nucleolus (*n*) and mitochondria (*m*) (*right*, scale 2 μm) (TEM by courtesy of R.C. Foster)

of the nucleus, mitochondria, and starch occur. The cracks appearing in the starch grains to allow the entry of amylolytic enzymes result in degradation of starch and its conversion mostly into lipids which later appear to be converted to polyphenols (Pandalai et al. 1965). The polyphenols first appear around the periphery of the lipid droplets and vacuoles, but their origin is obscure. The nuclei of the ray parenchyma cells of *M.azedarach* are absent at the sapwood-heartwood boundary at the time of phenolic accumulation. Although there is increasing hydrolysis of starch in the sapwood of *Abies alba* with increasing distance from the cambium, together with an intensive production of phenolic vacuoles and their fusion with one another, no relationship could be established between the two phenomena (Parameswaran and Bauch 1975). In *Abies alba*, granular osmiophilic particles appear to be synthesized on the inner side of the unit membrane surrounding the vacuole. The ribosomes lining the outer surface of this membrane could also participate in the production of these polyphenols because they decrease in number with the enhanced accumulation of polyphenols in the vacuoles and they disappear in the heartwood (Parameswaren and Bauch 1975). The nuclei in the ray parenchyma cells in the transition zone of *Cryptomeria japonica* ceased to have the spherical form seen in the sapwood, vacuolation was apparent, and some organelles were pushed to the periphery of the cytoplasm (Nobuchi and Harada 1968). Nevertheless some cells were relatively rich in organelles and metabolically active (Nobuchi et al. 1976).

Nobuchi and Harada (1985) have shown distinct cytological differences between the axial and ray parenchyma cells, as postulated many years previously by Chattaway (1951 personal communication). In the outer sapwood, the ray parenchyma cells of *Cryptomeria japonica* were found to be rich in protoplasm and to contain little metaplasm. On the other hand, the axial cells contained few protoplasmic organelles but large vacuoles containing material that was osmiophilic and presumed to be polyphenolic. Apparently the cells have different functions. The ray parenchyma are actively concerned with storage and mobilization of materials, whereas the axial cells are passively concerned with storing waste materials. Conspicuous changes in the parenchyma cells took place in the transition zone, commencing at its boundary with the sapwood. The percentage of living ray parenchyma decreased with the increase in the quantity of extractives (Nobuchi et al. 1985), and although the volume of vacuoles in the ray cells increased considerably, the amount of osmiophilic substances did not. On the other hand, more of these substances accumulated in the vacuoles of the axial parenchyma cells.

The experimental difficulties in obtaining samples at different times from transition zones hinder the determination of the sequence of the cytological formation of polyphenols. However, an indication of their development can be obtained from other tissues such as the mycorrhizal association of the Hartig net of *Pinus radiata* roots and from tissue cultures. About one-half of the acetone-soluble extractives of the mycorrhizae of pine consist of a mixture of five polyphenols (Hillis et al. 1968). Foster (Hillis 1968b) found that the formation of polyphenols takes place in specialized areas of the cytoplasm which become dilated and form tanniniferous vesicles. Polyphenols were at first deposited as granular material at the periphery of the small vesicles but later became aggregated into discrete, densely staining bodies. Two different types of polyphenol vesicle could be distinguished at an early stage, but while both types of vesicle may occur within the same tissue they did not usually form in the same cell. In some cases all the vacuoles in the cell were tanniniferous, in others tanniniferous and nontanniniferous vesicles occurred in the same cell. The vesicles extended and slowly coalesced until the major portion of the cell volume contained tanniniferous vacuole material. During the development of the tannin vesicles in the root cells, the cytoplasm contained abundant organelles. Proplastids became filled with starch so that each amyloplast could contain three or more starch granules. Where extensive polyphenol deposition took place there was some reduction in the starch content of the amyloplasts. In no case in this species did tannin form within the amyloplast but always in the tanniniferous vesicles. There was some indication that polyphenol materials might be incorporated into the cell wall during development.

Tissue cultures have been studied to clarify the mode of formation of "tannins" (polyphenols) within the cell and whether it is mutually exclusive of that of starch. Heartwood polyphenols were formed in tissue cultures of *Pinus resinosa* (Jorgensen and Balsillie 1969) and accumulated within actively metabolizing cells of *P.elliottii* tissue cultures (Baur and Walkinshaw 1974). In the latter species large accumulations were found in vesicles within cell vacuoles which differentiated from smooth endoplasmic reticulum. This took place without affecting the function of the cell or causing its degradation. One type of osmiophilic compound appeared to lead to cell death and degradation, when these occurred, more readily than another type. The polyphenol particles in the tissue cultures of *Picea glauca* developed initially within small vacuoles, possibly derived from endoplasmic reticulum, which increased in size through enlargement and coalescence of the vacuoles (Chafe and Durzan 1973). Although the sites of development differ from those in *Eucalyptus elaeophora* (Wardrop and Cronshaw 1962), there was an inter-relationship between polyphenols and starch within the one cell. Cells containing massive starch accumulation revealed little polyphenol formation, and conversely cells with large quantities of polyphenols contained only a few starch grains (Chafe and Durzan 1973). Apparently all stages in between could exist.

6.2.6 Respiration

Respiratory activity of the dry transition zone of *Pinus radiata* was greater than the inner sapwood zones during most of the year but particularly during the dormant season (Shain and Mackay 1973a). In contrast, the oxygen uptake by the middle and inner sapwood was not significantly different in either the growing or dormant seasons. There have been similar increases in respiration rates in *Robinia pseudoacacia* (Höll and Lendzian 1973) and a *Populus* species (Nekrasova 1973). (The lower respiration rates found in the transition zones of other species (Section 6.1.6) could be due to the time of the year during which samples were collected, but it is unlikely that the higher rates of respiration would have been influenced by higher temperatures (Zelawski 1960b). These facts, together with a similar pattern of ethylene production (see also Chapter 5.5 and Section 6.4.1), point to a connection between the formation of heartwood and the localized formation of extractives.

6.2.7 Enzyme Activity

Even if the enzyme activity in the transition zone, or adjacent to the heartwood boundary, is low there is direct evidence it is adequate to convert labeled precursors to extractives. Labeled glucose administered to the cambial region of *Eucalyptus sieberi* has been converted to labeled extractives at the heartwood periphery 19 days later (Hillis and Hasegawa 1963). Similarly labeled extractives were formed by the transition zone of *Cryptomeria japonica* (Higuchi et al. 1969).

The physiological situation in the transition zone (or at the heartwood periphery) is complex. It requires extensive studies on the enzymes of the pentose phosphate pathway and those related to carbohydrate, fat, and nucleic acid metabolism and their relation to the reserve materials in the sapwood. This information is gradually being accumulated from studies on various species.

There are several indications of increased activity of some enzymes in the zone between sapwood and heartwood. Concentrations of phenol oxidizing enzymes have been observed at both the outer and inner peripheries of the transition zone of *Eucalyptus polyanthemos* (Hillis and Yazaki 1973b) (Fig. 6.9). Activation of peroxidase and polyphenol oxidase and new synthesis of peroxidase have been noticed in several diseased tissues (Farkas and Kiraly 1962), and polyphenol oxidase activity in sweet potato is increased by cutting (Uritani and Muramatsu 1952, Shannon et al. 1971). Accordingly, the increase in enzyme activity at the heartwood periphery could be due to de novo synthesis and/or introduction of an activator. In this context it is noteworthy that a series of water-soluble vita-

Fig. 6.9. End of log of *Eucalyptus polyanthemos* 1 week after cutting. The white transition zone (*TZ*) surrounding a red heartwood is enclosed by thin, brown-black bands which appeared on exposure to air (scale 5 cm)

mins (thiamine, riboflavin, nicotinic acid amide, pyroxidine, and biotin) increased in the sapwood near the heartwood of *Pinus strobus* (Ziegler 1964, 1968). All the vitamins are co-enzymes of a number of key enzymes, suggesting that physiological activity is increased in the inner layers of the sapwood.

Lairand (1963) found that although the nucleus had disappeared in earlier growth rings, there was a sharply increased peroxidase activity in the layers adjacent to the heartwood of *Pinus sylvestris* and *Larix decidua* where it was concentrated in the torus (see also Bauch et al. 1968). (It is interesting to note that in aspen, which forms only ripewood and not heartwood, there is a steady decrease in activity of peroxidase from cambium to near the center of the tree, where it disappears). Wardrop and Cronshaw (1962) also found a marked peroxidase activity in sapwood adjacent to the heartwood of *Eucalyptus elaeophora*, as did Dietrichs (1964a,b) with *Fagus sylvatica*. Following a decrease in peroxidase activity from the cambial region to the middle sapwood of *Quercus robur*, there was an increase of enzymatic activity in the inner part of the sapwood without significant change in nitrogen content (Ebermann and Stich 1985). A slight increase in peroxidase activity at the heartwood periphery has been shown by staining various wood discs (Harkin and Obst 1973). In *Acacia nilotica* some peroxidase activity was found in the ray and axial parenchyma cells at the transition zone, whereas the activity of succinate dehydrogenase was intense, indicating active mitochondrial involvement and actively respiring cells. The latter activity was higher in the axial than in the ray parenchyma cells and sometimes even in the vessels (Shah et al. 1981a). Succinate dehyrogenase was significantly active only in the transition zone of *Melia azedarach* (Baqui et al. 1979). Adenosine triphosphatase (ATPase, which is implicated in cellular transport, membrane permeability, and high-energy consuming processes in cellular metabolism) and lipase were also active in the transition zone. The absence of increased succinate dehydrogenase activity in the transition zone of *Bridelia retusa* was attributed to the sampling periods (mid-summer and mid-winter) (Nair et al. 1981). Kondo (1964), Matsukuma et al. (1965), and Sameshima et al. (1967) have reported increased activity of a catechol oxidase, maltase, lipase, and amylase in the transition zone of *Cryptomeria japonica*; there was also an increase in starch content in this zone (Sameshima et al. 1967). The marked acid phosphatase activity in the middle sapwood of *Bridelia retusa* was associated with the hydrolysis of the starch present and the increased amount of lipids in the transition zone (Nair et al. 1981). In tropical trees of Venezuela (Fink 1982) the acid phosphatases were highly active in the paratracheal parenchyma and particularly in the pits close to the vessels. This activity continued over a major portion of the year, as would be expected with the continuous growth of tropical trees.

Peroxidase activity in the inner sapwood of *Juglans nigra* peaked during the late growing season to early dormant season, and in a period before the maximum emanation of ethylene (see Chapter 5.5). Peroxidase activity in the outer and middle sapwood samples peaked at higher levels in the dormant season (Nelson 1978). The middle and inner sapwood samples and transition zone of *Prunus serotina* showed very low peroxidase activity throughout the year (Nelson 1978). The peroxidase activity in the transition zone of *Pinus densiflora* and *P.banksiana* was higher than in the sapwoods throughout the year (Yamamoto 1982).

Kondo (1964) has pointed out in his review of the topic that all enzyme systems, and particularly those in the cell wall, are not necessarily deactivated when the nuclei disintegrate and the normal physiological functions stop. He found the activities of an enzyme hydrolyzing sucrose to glucose and fructose in different tissues to be in the following order:
intermediate zone $\gg$ sapwood > heartwood > control (sterilized wood). Höll (1972,1973) showed in *Robinia pseudoacacia* that aldolase activity was highest at the heartwood periphery and amylase activity higher in the youngest sapwood and at the periphery of heartwood (Höll and Lendzian 1973). Malic dehydrogenase (MD) and glucose-6-phosphate dehydrogenase (G6PD) also have enhanced activity near the sapwood/heartwood boundary of *Pinus radiata* (Shain and Mackay 1973a). The histological tests for these enzymes showed the G6PD activity to be considerably greater in the transition zone than in the adjacent inner sapwood, whereas less difference was noticed with MD activity. These findings suggest that the pentose shunt activity increased more dramatically in the transition zone than did that of the citric acid cycle (Shain and Mackay 1973a). A similar situation has been observed in the transition zone of *Cryptomeria japonica* (Higuchi et al. 1967). These activities would be expected in tissues synthesizing polyphenols containing flavonoids with A and B aromatic rings. Phenylalanine ammonia lyase was not detected in the inner sapwood or transition zone of *Chamaecyparis obtusa* and *Cryptomeria japonica* during summer (Higuchi et al. 1967a) and was weakly present in late autumn (Higuchi and Fukazawa 1966). Low levels, or absence, of phenylalanine ammonia lyase activity was found in some tissues of *Eucalyptus* species which were able to biosynthesize polyphenols (Hillis and Ishikura 1970).

The high respiratory activity in the transition zone of *Robinia pseudoacacia* tested in early winter (Höll and Lendzian 1973) was not matched by a high carbon dioxide uptake in a sample tested in the same period (Höll 1974). The high concentration of carbon dioxide in this region could be due to either its production or a diminution of its refixation. A study of *Tilia cordata* and *Fraxinus excelsior* (examined in early summer) showed that the phosphoenolpyruvate carboxylase activity, and replacement of compounds for the tricarboxylic acid pathway, are limited by the concentrations of other enzymes required for the reactions (Höll and Meyer 1977). During the dormant season, activity of malic dehydrogenase was significantly higher in the transition zone adjacent to the heartwood of *Pinus radiata* than in the middle and inner sapwood; there was also a substantial increase in the activity of glucose-6-phosphate dehydrogenase in this zone, particularly during the dormant season (Shain and Mackay 1973a). Similar increases were found with phenol-oxidizing enzymes in the transition zones of *Juglans nigra* (Nelson 1978). Again the activity of peroxidase was highest in the sapwood 1 cm from the heartwood of *Quercus robur* trees during the winter and lowest during the spring (Ebermann and Stich 1985).

In comprehensive histoenzymatic studies on cross-sections of *Acacia auriculiformis* undertaken at bi-monthly intervals, Baqui and Shah (1985) found an involvement in heartwood formation of not only the inner sapwood enzymes but those of other regions also. The enzymes studied were acid phosphatase, adenosine triphosphatase, glucose-6-phosphatase, lipase, glucose-6-phosphate dehydrogenase, succinate dehydrogenase, and peroxidase as well as the distribution of starch. The latter was very largely present in the axial rather than in the ray parenchyma cells and particularly during the wettest and hottest period. In addition to its association with lignification of the cell wall, peroxidase was also strongly present in the axial and ray parenchyma cells in the inner sapwood and at the heartwood periphery. The activities of acid phosphatase and glucose-6-phosphatase were localized in the ray and axial parenchyma cells and associated with degradation of the starch. Glucose-6-phosphate dehydrogenase, which can be related to an increased activity of the pentose phosphate pathway, was most intense during the hot and wet period of the year in the inner sapwood and at the heartwood periphery and more so in the axial than the ray parenchyma. Succinate dehydrogenase was most active in the same period although early in this period it was mostly in the ray parenchyma.

The osmiophilic compounds observed on the margo of ray tracheid bordered pits in the inner sapwood of *Pinus sylvestris* were considered to be enzymes by Fengel (1970).

6.2.8 Formation of Extractives

The levels of C^{14} across cross-sections of *Quercus* species have shown that, following the commencement of atom bomb testing, very pronounced incorporation of labeled carbon into the wood of growing trees occurs at the place where the sapwood was changed into heartwood. Some of this radioactivity was incorporated into the cell wall (Cain and Suess 1976, Cain 1980). The radioactivity in the heartwood of *Thuja plicata* was removed with acetone (Fairhall and Young 1970).

Labeled glucose or sucrose administered to the cambial region of *Eucalyptus sieberi* in early summer (Hillis and Hasegawa 1963) and of *Prunus yedoensis* in late autumn (Hasegawa and Shiroya 1965) was translocated through the rays to the heartwood periphery within 19 and 12 days respectively, where labeled polyphenols were also formed. Metabolism was different in the outer and inner sapwoods and the transition zones of *P. yedoensis*. Whereas shikimic acid was the major derived acid in the outer sapwood, quinic acid was the major one detected in the transition zone. The ability to form polyphenol glycosides was greater in the outer sapwood than it was in the inner sapwood and the transition zone, but the degree of flavonoid formation was much larger in the transition zone than in the inner sapwood (Hasegawa and Shiroya 1965, 1966, 1968). A number of amino acids had higher radioactivities in the transition zone than in the inner sapwood. It is evident that the biosynthesis of polyphenols, which involve both the shikimic acid and tricarboxylic acid pathways, takes place in the transition zone.

Putative precursors that had been radioactively labeled were fed to cross-sections of *Cryptomeria japonica, Chamaecyparis obtusa, Fagus crenata, Quercus mongolica*, and *Magnolia obovata* felled in late autumn and early spring. All tissues except heartwood had sufficient biosynthetic ability to convert the precursors into extractives in situ. Whereas acetate and glucose were incorporated into all fractions, mevalonic acid and phenylalanine were converted mainly into terpenoids and phenolic compounds, respectively (Higuchi et al. 1969).

Harris (1954b) first drew attention to the formation of polyphenols in the inner layers of the transition zone (called by him the "drywood zone") in *Pinus radiata*. The polyphenols first appeared during the flush of growth and had become incorporated in heartwood by early summer. The outer layers of the transition zone of *Eucalyptus polyanthemos* formed a methylellagic glucoside, whereas the inner layers formed in addition proanthocyanidin polymers (Hillis and Yazaki 1973b). In a study of the cytology and extractives formation in *Cryptomeria japonica*, Nobuchi et al. (1984a, 1985) found that one polyphenol was present in greatest amount in the transition zone, while the other two major polyphenols were more abundant in the heartwood. The amount of extractives in the transition zone increased at the same time as the decrease in moisture content and the percentage of living cells. The largest amount of extractives was found in the outer heartwood after all cells had died. In *Acacia nilotica* the polyphenols are localized only in the cell lumens and were concentrated in the ray and axial parenchyma cells of the inner transition zone, whereas they were mainly in the vessels of the heartwood (Shah et al. 1981a).

Oil droplets and osmiophilic materials (probably polyphenols) were observed in the tracheids at the half-bordered pit-pairs between the ray parenchyma cells and the tracheids of *Cryptomeria japonica* but not at the pit-pairs between axial parenchyma cells and tracheids (Nobuchi et al. 1984a, Nobuchi and Harada 1985). An earlier study (Krahmer et al. 1970) of the contents of rays and tracheids of *Tsuga heterophylla* showed that whereas lignans were present in the occluded tracheids, these polyphenols were absent from the adjacent ray cells. In one section, deposits were found on both sides of a half-bordered pit, but aromatic compounds were absent in the ray deposits, whereas the tracheid contents possessed a UV absorption curve typical of lignans. Conversion to these polyphenols must take place close to the pit membranes separating these cells. Yamamoto (1982) has illustrated a pit chamber in *Pinus densiflora* adjacent to a tracheid, both sides of which were full of extraneous material. Parameswaran and Bauch (1975) were unable to decide whether or not the polyphenols in the pit membranes of *Abies alba* were derived from those in the sapwood ray cells or whether there was a relationship with the heartwood polyphenols. However, it appeared certain that the polyphenols in the lumens of the heartwood ray cells were different in nature from those in the pit membranes. There are indications that the polyphenols in the cell walls of *Acacia melanoxylon* can differ from those in the lumen (Bland and Hillis 1969).

In *Pinus sylvestris*, where incrustration of the aspirated pits occurs, the membranes of the bordered pits between ray tracheids and between ray and longitudinal tracheids are thickened by extraneous materials. These formations indicate that not only are reactions taking place there but that other substances are being transported there for these reactions. Fengel (1970) has illustrated the osmiophilic materials that can be found in the lumen of longitudinal tracheids and how they could originate from the bordered pit chamber between that tracheid and a ray tracheid. Similar findings were reported earlier by Dietrichs (1964) with *Fagus sylvatica*.

With the transition of sapwood to heartwood, osmiophilic substances impregnate pit membranes, as has been mentioned, and are also deposited on parenchyma cell wall. In addition, they appear to migrate between adjacent ray parenchyma cells in *Pinus sylvestris* via plasmadesmata in the tangential walls. Fengel (1970) proposed that in the transition zone close to the sapwood of this species, vacuoles of the living ray cells enter the pit chambers and migrate or diffuse through the compound middle lamella and from there also move into the tracheid lumens. Subsequently on the heartwood side of the transition zone, the diffused substances react or condense to form the coloring material in the cell wall. Consequently heartwood formation takes place inside as well as outside the cell wall. A similar situation appears to occur in *Abies alba* (Parameswaran and Bauch 1975).

6.3 Heartwood

6.3.1 Seasonal Formation

To enable comparisons between the northern and southern hemispheres, the discussions in this chapter have related different seasons, rather than monthly behaviour, to the various stages of heartwood formation. However, few records of the mean temperatures of the various seasons are given in publications and even fewer of the major rainfall periods in relation to heartwood formed.

Yamamoto (1982) points out that the thickening and lignification of ray parenchyma and their necrosis in the transition zone and ultimate heartwood formation in *Pinus banksiana, P. densiflora, P. strobus* (Fukazawa et al. 1980), and *Larix leptolepis* occur in northern Japan from mid-summer to autumn (Yamamoto et al. 1978) and without any development in the intervening period (Hirai 1951). On the other hand, in the warm temperate zones, heartwood forms in trees mainly during the dormant periods of growth or in winter, when presumably the temperatures are sufficiently high (above 5 °C) to enable the required cellular reactions to occur.

Heartwood formation in larch (*Larix* species) commences in late summer to late autumn and at the beginning of dormancy (Hirai 1951). Studies on the production of ethylene, respiration of xylem tissues, and enzyme activity in *Pinus radiata*, *Juglans nigra*, and *Quercus rubra* indicate that heartwood formation occurs at the time of cambial dormancy (Harris 1953, 1954a, Shain and Hillis 1973, Shain and Mackay 1973a, Nelson 1978).

Lignification of the resin canal tissue of pine also occurs at the heartwood boundary (Bamber 1976a), and this process has been found to take place from mid-summer to early spring (Yamamoto et al. 1978, 1979). During late summer and early autumn, the surface area and volume of the cell nuclei and nucleoli are largest at the heartwood periphery of *Quercus robur*, tyloses form, starch is hydrolyzed simultaneously and immediately following tylosis formation, the parenchyma die, and the polyphenols diffuse into cell walls (Hugentobler 1965). Tyloses commence to form in the earlywood of *Robinia pseudoacacia* in late summer

(Ishida et al. 1976). Nobuchi et al. (1982, 1984a,b) concluded from a study of the cytology and coloration of the extractives that heartwood formation in *Cryptomeria japonica* and *Robinia pseudoacacia* grown in a warm temperate region of Japan commences in mid-summer and continues in the autumn and winter seasons. With the latter species the heartwood region had expanded considerably to about 70 percent of the annual ring by early autumn, and continued to late winter long after growth ring formation had ceased.

Heartwood formation can involve a number of features such as the formation of ethylene, activation of certain enzymes, and respiration. Climatic differences may affect the initiation and conclusion of heartwood formation, and one or more of these different aspects under different conditions (Fukazawa et al. 1980). Tylosis formation, an integral part of heartwood formation in some hardwoods, is also affected by climate (Nobuchi et al. 1982).

6.3.2 Appearance

Generally the color of heartwood increases from the pith to its periphery, and this corresponds to the amount of extractives present (Hillis 1962b). There are often concentric bands darker in color than surrounding bands [e.g., *Quercus bicolor* (Wardell and Hart 1970)] even when no association with discoloration and decay can be found. Analysis can show these bands to contain more extractives than the lighter- colored ones.

Dark-colored heartwoods sometimes show a brighter, outer heartwood band or zone such as has been reported in *Eucalyptus marginata* (Hillis 1956) (Fig. 2.11). Normally the heartwood extractives of this species have a low solubility in water, but in the outer zone with its distinctly pinker color the extractives are soluble in water and they can migrate or diffuse in variable amounts into the outer transition zone. In those cases the heartwood periphery becomes diffuse and indistinct. A similar behavior has been recorded with *Cryptomeria japonica* (Nobuchi et al. 1982).

6.3.3 Respiration and Enzyme Activity

The lower decay resistance of inner heartwood of many trees could be due to the small amounts of extractives formed on sapwood-heartwood transformation and/or the degradation of polyphenols by oxidation with enzymic or other involvement. Usually, fresh heartwoods darken on exposure to air or when injured in the living tree, and the changes are often considered to be nonenzymic in nature, as dried heartwood also darkens in use. The slight oxygen uptake which occurs in the fresh heartwoods of a number of species is not affected by boiling (e.g., Goodwin and Goddard 1940, Higuchi et al. 1967a). Nevertheless phenol-oxidizing enzymes have been implicated in the formation of brown stain in the heartwood of *Pinus lambertiana* (Stutz 1959). When aspects of the presence of enzymes in heartwood are considered, questions are raised as to whether (a) the enzyme(s) is produced by microorganisms inhabiting heartwood (b) the enzyme(s) is durable, of host origin, and has remained active after necrosis of the host parenchyma or (c) the enzyme(s) is produced by host parenchyma which remained alive after heartwood formation. Shain and Mackay (1973b) addressed these questions in an examination of fresh heartwood of *P.radiata*. They found two phenol-oxidizing enzymes (the *o*- and the *p*-diphenol: O_2 oxidoreductases) with strong evidence that they were of host origin remaining active after necrosis of host parenchyma. Various tests showed that fungi and bacteria were absent in fresh heartwood and that, as dehydrogenase was absent, the parenchyma were dead, and that only certain enzymes can survive after their necrosis on heartwood formation. A similar situation was reported in *Picea abies* where necrotic sapwood, preceding infection by *Fomes annosus*, contained a phenol oxidase probably produced by the host (Shain 1971).

In the discussion of enzyme activity in the transition zone, it was evident that peroxidase activity increased in a number of species. Peroxidase (and amylase) activity has been reported (without evidence) in the heartwoods of *Acer campestris, A.platanoides, Pinus sylvestris, P.nigra austrica, Picea excelsa, Taxus baccata*, and *Quercus robur* (Ebermann and Stich 1982). In the last species peroxidase activity decreases more rapidly in the heartwood than does the nitrogen content (Stich and Ebermann 1984, Ebermann and Stich 1985). Peroxidase was detected in heartwood of different species stored for 1 year or more (Ebermann and Stich 1982).

A possibly more direct measure of the assessment of the metabolic intensities is the concentration of adenosine triphosphate (ATP), the most important primary energy carrier of the cell. Höll (1973) found that the ATP content decreased across the sapwood of *Robinia pseudoacacia* (felled in the spring) and disappeared in the heartwood. On the other hand it remained at a high level in *Tilia cordata* and *Betula verrucosa* (felled in winter) which did not form heartwood.

6.3.4 Location of Extractives

Some dried wood specimens have shown marked and abrupt increases in extractives content in all cells at the heartwood periphery (Fig. 6.10). The dry conditions of the samples have prevented detection of any transition zones that may have been present or any notable differences in visual appearance which could indicate different composition. Knowledge of the precise location of different classes of extractives in a tissue would improve the understanding of the modes of biosynthesis. The red coloration of *Cryptomeria japonica* on heartwood formation is due to the extractives in the axial parenchyma, which are formed before the yellow and orange components in the ray parenchyma (Nobuchi et al. 1976, Kai and Teratani 1977, Kuroda and Shimaji 1983). Deposits are also formed around the cross-field pittings on the lumen side of latewood tracheids. The red substances appear to be distributed throughout the tracheid walls, whereas the black substances of discolored heartwood may be distributed only in the secondary walls of this species.

The osmiophilic substances, presumed to be polyphenolic in nature, in *Pinus sylvestris* are considered to migrate from the parenchyma pit membranes through the compound middle lamella to the bordered pits and from there into the cell lumina. Subsequent reactions result in the formation of typical substances in the cell walls of heartwood (Fengel 1970).

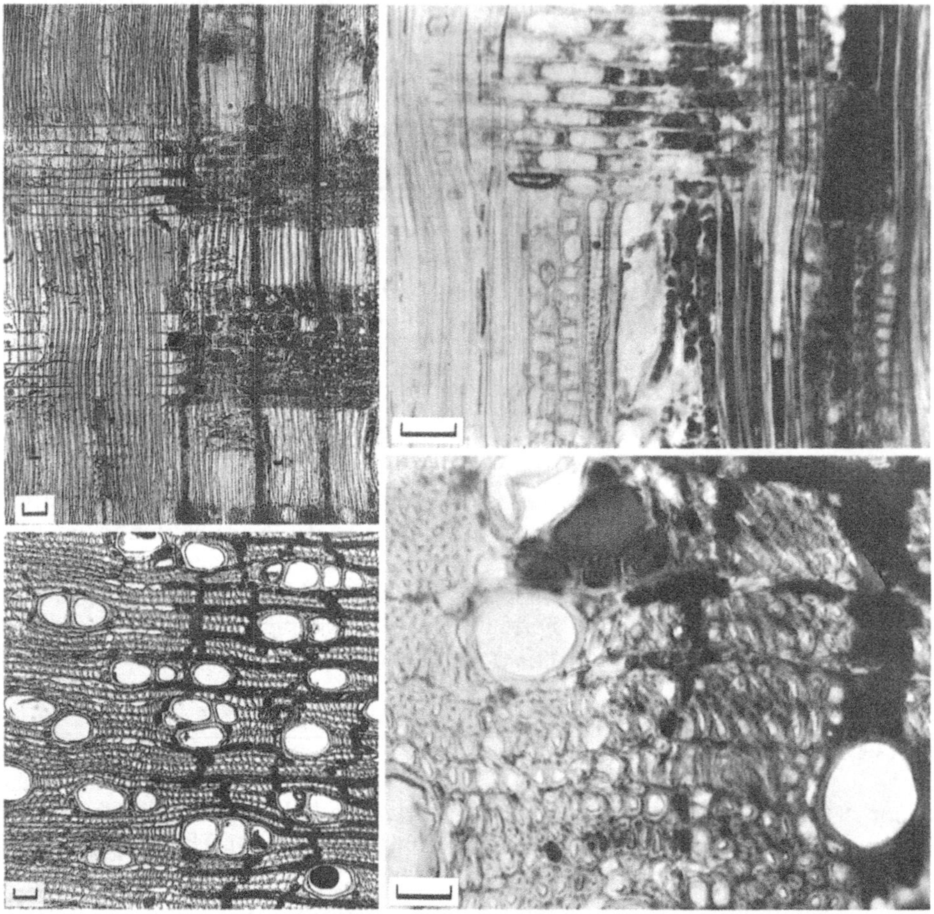

Fig. 6.10. Radial (*top*) and cross (*bottom*) sections of *Excoecaria parvifolia* (*left*, scale 100 μm) and an *Acacia* species (right, scale 100 μm) showing abrupt appearance of polyphenols at the heartwood periphery. The polyphenols can be seen in the rays, vertical parenchyma and vessels (Courtesy of R.K. Bamber, and J. Ilic)

6.4 Wound Wood and Chemically Affected Wood

6.4.1 Wound Wood

Wound wood is formed in response to mechanical or to microbiological damage. Its formation is related to a dynamic mechanism in the host to minimize further damage. Several terms have been used to describe the same response by the tree, such as pathological heartwood (Büsgen and Münch 1929, Chattaway 1952), protection wood (Jorgensen 1962) discolored wood (Shigo and Hillis 1973), and reaction zone (Shain 1967, 1979). The extent of discoloration in injured wood is greatest in the axial direction compared with radial and tangential directions (e.g., Shain 1967, Wardell and Hart 1970, Nobuchi et al. 1984b). Discoloration spreads further in the summer than in the autumn season (Lyr 1967) and in some cases to a greater extent in the inner than in the outer sapwood.

In his last investigation Yazawa (Yazawa et al. 1967) induced "artificial heartwood" in 41 species in northern Japan. This was done by boring radial holes into the trees and forcing in oxygen. These wound woods formed extensions in both the lateral, tangential, and particularly the axial directions. In a number of species the discolorations were much greater in the inner than in the outer part of the sapwood and particularly with *Fagus crenata, Betula maximowicziana, Cercidiphyllum japonicum, Tilia japonica, Ulmus davidiana, Fraxinus mandshurica, Cryptomeria japonica*, and *Picea abies*. This behavior differed from that of *Fagus sylvatica* in which the discoloration decreased in the innermost part of the sapwood (Zycha 1948). It was concluded (Yazawa et al. 1967) that discoloration of the tissues surrounding bore-holes was not due to oxygen alone.

Wound and discolored wood can involve a wide range of compounds including alkaloids. Injured wood of *Liriodendron tulipifera* can become highly discolored from violet, blue, green, and brown to black. The discoloration is most likely due to the metabolism of a number of aporphine alkaloids by invading microorganisms as the lignans syringaresinal and its dimethyl ether would not be amendable to change. The compounds present in discolored wood and heartwood can differ markedly both quantitatively and qualitatively. With one exception, nonphenolic aporphine alkaloids predominated in heartwood whereas phenolic aporphines predominated in discolored sapwood. Significant amounts of pigments were found in the latter tissues but not elsewhere (Chen et al. 1976).

The wound wood of *Pinus densiflora* contained the characteristic pinosylvins of heartwood and that of *Cryptomeria japonica* similarly contained sugiresinol and hydroxysugiresinol (Higuchi et al. 1967c). Mechanical wounds made into the cambial region of *P. resinosa* formed characteristic pinosylvins after 4 weeks in late autumn, after 9 weeks in mid-winter but not in summer (Jorgensen 1961). On the other hand, the formation of pinosylvin in woundwood of *P.sylvestris* begins in the spring (when it is clearly visible 3 weeks later) and ends at the beginning of winter (Lyr 1967). Similarly, the formation of kino veins in the cambial regions of eucalypts develops to a much greater extent in early than in late spring (Hillis 1964). Lyr (1967) found the absence of rapid desiccation important for the biosynthesis of pinosylvins, and this is achieved in the tree by exudation of resin on wounding. Resinification of wounds in pines is less in summer than at other times so that wounds are inadequately covered, and this is apparently the reason why the mechanical wounds made in the tree by Jorgensen (1961) failed to produce pinosylvins in summer. When the wounds made in *P.resinosa* in late summer were filled with solid carbon dioxide, the characteristic pinosylvins were formed. Under laboratory conditions these pinosylvins formed in partly desiccated sections most rapidly at 25 °C and over a period of 7 days, by which time the cells were dead. When solid carbon dioxide was applied to fresh sections at 25 °C, the pinosylvins appeared after 3 days (Jorgensen 1961). As discussed elsewhere (Section 6.4.3), these effects were probably due to traumatic ethylene. (It is interesting that the polyphenol phaseollin was formed in french bean hypocotyls when they were touched with solid carbon dioxide at separate sites along their length, but not when they were completely frozen (Rahe and Arnold 1975)). Artificial injuries into the sapwood of *Picea abies* resulted in an increase of α-pinene and decrease of β-pinene (Schuck 1982), and infested bark of *Abies grandis* had higher myrcene and 3-carene contents (Russell and Berryman 1976). Only the extent of heartwood formation was increased when *A. grandis* was infested with the balsam woolly aphid (*Adelges piceae*) and no change occurred in the amount or nature of its extractives (Puritch 1977). When the stems of young *Pinus taeda* were infected by *Cronartium fusiforme*, the resultant gall wood contained greater amounts of different extractives than the noninfected wood (Rowan 1970).

In a critical examination of the dynamic responses of sapwood to injury and infection, Shain (1979) considered separately the pale-colored transition zone contiguous with functional sapwood and the polyphenol-enriched reaction zone adjacent to infected wood. Both these zones are formed in sapwood in advance of infection by decay fungi. Polyphenols (and potasium, calcium, and magnesium salts) in the stem and roots of *Picea abies* and *Pinus* species are usually concentrated along the edges of the lesion forming the wound wood (reaction zone) (Coutts and Dolezal 1966, Shain 1967, 1971, Johansson and Theander 1974). The periphery of the wound wood in *Pinus taeda* is always ahead of the advance of the hyphae of *Fomes annosus* and the amount of different extractives is appreciably higher than that in the heartwood (Shain 1967) as was found also by Prior (1976) with *P. nigra*. When *P. radiata* is heavily oviposited by *Sirex noctilio*, the moisture content of the zone affected drops from 200 to 60 percent of oven-dry weight within 14 days. This occurs in advance of the extending hyphae of the symbiotic fungus *Amylostereum areolatum* (P.H.B. Talbot 1977). Both the fungus and the mucus injected during oviposition by *Sirex* contribute to the drying (Kile and Turnbull 1974b). When *P. radiata* was inoculated with the decay fungus *A. areolatum*, the two fungitoxic pinosylvins were detected within 2 days (Shain 1979) or with *Diplodia pinea* within 6 months (Coutts 1970). The ratio of pinosylvins to each other differed considerably in the affected sapwoods and in the heartwood (Hillis and Inoue 1968, Coutts 1970, Shain and Hillis 1972) and in favor of the more toxic pinosylvin. The same stilbenes were formed when the sapwood or the seedlings of *P. contorta* were attacked by *Dendroctonus ponderosae* and its associated fungus (Shrimpton 1973) or inoculated with *Europhium clarigerum* (Shrimpton and Watson 1971). A toxic sesquiterpene isolated from culture filtrates of *Fomes annosus* results in the formation of these stilbenes at some distance from its point of injection into the sapwood of *P. taeda* (Bassett et al. 1967). It is noteworthy that when sapwood blocks were removed from a *P. radiata* tree and inoculated with *A. areolatum* alone, stilbene synthesis was not stimulated (Coutts 1969). Death of radiata pine generally is ascribed to the combined attack of the *Sirex* wasp and its symbiotic fungus (P.H.B. Talbot 1977). Sapwood on the pith side of cambial zones of 2 *P. radiata* trees that had been damaged 4 years previously contained the less toxic pinosylvin monomethyl ether and one of the heartwood flavones (Hillis and Inoue 1968). It is evident that the conditions of injury affect the composition of the extractives, as indicated also by the accumulation of pinosylvin dimethyl ether and other extractives in the sapwood of *P. sylvestris* infected with *Peridermium pini* (Westfelt 1966).

When *Prunus domestica* (var. Victoria) is attacked by the fungus *Stereum* (now *Chondrostereum*) *purpureum*, the resultant wound wood contains a considerable amount of the coumarin scopoletin, a compound present in the sapwood in only very small amounts (Hillis and Swain 1959). Large amounts of the lignan isoolivil have been isolated from the wood of *Prunus jamasakura* infected by the fungus *Coriolus versicolor*. This abnormal constituent was accompanied by some of the flavonoids found in uninjured sapwood, but the major flavonoids were absent (Hasegawa and Shirato 1959). In the reaction zone surrounding *Fomes annosus* infestation of the sapwood of *Picea abies* considerable amounts of hydroxymatairesinol were isolated as well as other lignans and 4-methylcatechol (Shain and Hillis 1971, R.A. Anderson et al. 1975). Isocoumarins have been found in the sapwood of *Pinus taeda* infected with *Ceratocyctis minor* (Hemingway et al. 1977). The biaryl compound aucuparin as well as a pentacyclic triterpene has been isolated from the dark-colored interface between healthy *Malus pumila* sapwood and the stained wood colonized by *Chondrostereum purpureum* (Kemp et al. 1985). Pyrufurans were formed in *Pyrus communis* by the same pathogen and by injury (Kemp et al. 1983, 1984). Flavonoids have accumulated in the sapwoods of *Pseudotsuga menziesii* attacked by *Poria weirii* (Barton 1967), and of *Rhus succedanea* injured by certain chemicals (Hillis and Inoue 1966).

The discolored sapwood of *Cryptomeria japonica* that had been attacked by the insects *Resselilla odia* and *Epinotia granitalis* contained much larger amounts of some of the extractives than did the heartwood or wood adjacent to the heartwood that had been affected by microorganisms. One of the nonlignan extractives was present in fourfold amounts in the outer sapwood injured by the insects than in the heartwood (Takahashi and Ogiyama 1985). The most striking change in the sapwood of *Pinus contorta* attacked by *Dendroctonus ponderosae* and its associated microorganisms was the amounts of total terpenes and particularly β-phellandrene which were much higher than those found in the heartwood. The phenolic, neutral, and free acid components increased more slowly to a final concentration which was about the same as that in normal heartwood (Shrimpton 1973). The qualitative composition of monoterpenes in *Picea abies* altered considerably after injury (Schuck 1982). The ratio of the components in the group of sesquiterpenes resulting from infection of *Ulmus glabra* with *Ceratocystis ulmi* was different from that when infected with *Coriolus versicolor* and also from that infected with *Chondrostereum purpureum* (Burden and Kemp 1984). A different sesquiterpene has been found in the narrow pigmented zone between the healthy sapwood of *Tilia europea* and the inner core decayed with *Ganoderma applanatum* (Burden and Kemp 1983).

Transition zones have not been observed around all regions of wound wood or reaction zones. However, when they do occur it appears that the host parenchyma die and their physiological and metabolic processes terminate in a "programmed" fashion to form wound wood. The transition zones surrounding the sapwood of *Pinus* and *Picea* species infected with different organisms are drier than the sapwood which in turn surrounds them (Shain 1967, 1971, Shain and Hillis 1972). In addition to the greatly enhanced production of ethylene by the transition zone surrounding infections with *Amylostereum areolatum* (Chapter 5.5), there was a significantly greater respiratory activity by the lesions in pines than by surrounding tissues (Shain 1979). There was also increased activity of both malic and glucose-6-phosphate dehydrogenases in the parenchyma of this zone.

Transition zones surrounding wound woods caused by metallic borers driven into the sapwood of various species are apparently the same as those around normal heartwood. They have been observed in *Betula maximowicziana*, *Acer mono* var. *glabrum*, and particularly in *Sorbus alnifolia* but not in other species. The formation of the transition zone was apparent within 40 days in the former species when injured in mid-summer, and the zone always preceded the colored wound wood (Yazawa et al. 1967). The latter appeared larger in the interior sapwood adjacent to the transition zone surrounding normal heartwood.

During the stages leading to the death of the parenchyma, fungitoxic extractives are formed and then, with the death of the tissues in pines, the thin-walled parenchyma release oleoresin into the wound wood (Coutts and Dolezal 1966, Shain 1967). When *Pinus radiata* is attacked by *Sirex noctilio*, resin flow is one of the first observable host responses, and it precedes the synthesis of polyphenols in areas of affected sapwood. The accumulated resin limits the spread of the *Sirex* symbiont *Amylostereum areolatum* in the sapwood. Although the physical properties of the resin may have some inhibitory effects (see also J.N. Gibbs 1968, Shrimpton and Whitney 1968), there is evidence that both the volatile and nonvolatile components, of which α- and β-pinene are the most important, have chemically inhibitory effects (Kile and Turnbull 1974a, see also R.H. Smith 1961, Rudinsky 1966, van Buijtenen and Santamour 1972). The amount of fungitoxic components formed over a period of time determines the resistance of the tree to fungal decay. In pines, the pinosylvins are found in necrotic regions (Hillis and Inoue 1968), and a significant correlation has been found between the extent of infection of *Pinus nigra* inoculated with *Fomes annosus* and its pinosylvin content (Prior 1976). The components in the resin of healthy *Chamaecyparis obtusa* and of exuded resin of diseased trees are the same, but the proportions differ (Ohashi and Imamura 1973). In *Picea abies*, where the wound wood does not become soaked with oleoresin when infected with *Fomes annosus*, a greatly enhanced amount of the fungistatic lignan hydroxymatairesinol is formed (Shain 1971, Shain and Hillis 1971). On the other hand, after the sapwood of *Picea glauca* was mechanically wounded, no change in the amount or type of lignans was found in injured sapwood, but there was a considerable increase in the amount of resin acids present (Hart et al. 1975). There are further contradictions regarding the amount of the lignans, α-conidendrin and others, found in the wetwood of *Tsuga heterophylla*. One group of workers (Krahmer et al. 1970) found no difference from the adjacent normal heartwood, whereas others, particularly Schroeder and Kozlik (1972) noted a fivefold increase. The latter samples had pits occluded with extractives which presumably rendered the wetwood more impermeable than normal heartwood. The differences in composition of the wetwoods may be due the site of initiation; if this occurred in the transition zone a higher content of extractives could be expected than if it occurred in the heartwood.

Although major differences may exist during the formation of wound-initiated discoloration and heartwood, there are also a number of cytological similarities (Sucoff et al. 1967, Wardell and Hart 1970). In the formation of discoloration initiated by mechanical wounds, the response to injury is immediate, and most changes take place within a few weeks. On the other hand, the conversion of sapwood to heartwood is slow. Once discolored wood is formed it does not enlarge, unless microbiological decay sets in, whereas heartwood progressively enlarges with the passage of time. The environment in which discolored wood is formed is one of higher moisture content and concentration of oxygen. On the other hand, heartwood, in softwoods at least, develops at lower levels of moisture and oxygen. The chemical composition of the extractives formed in injured tissue can differ greatly from that deposited in normal heartwood (Shigo and Hillis 1973). The vitality of the cells in discolored wood of *Quercus bicolor* is high at their time of conversion (Wardell and Hart 1970) and, although there is debate concerning the vitality of the cells at the periphery of heartwood, there is evidence to indicate it is high at certain times.

The first detectable cytological response of the injured parenchyma of *Quercus bicolor* was the loss of starch. When trees were wounded mechanically in the summer, the starch grains in the injured zone disappeared in 8 days and in the autumn in 12 to 16 days. At this time starch began to disappear from parenchyma adjacent to the discolored wood, indicating that starch, as a raw material, was hydrolyzed at a distance from the wounding stimulus, to be translocated and synthesized into extractives at the appropriate location (Wardell and Hart 1970). The decline in starch content in the inner growth rings of sapwood adjacent to heart-

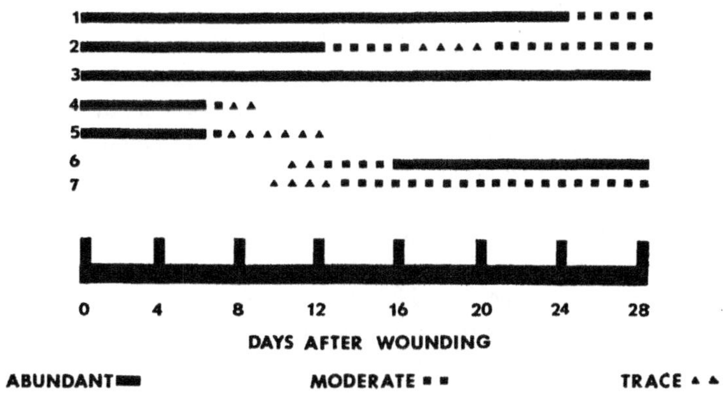

Fig. 6.11. The cytological and histochemical changes in sapwood of *Quercus bicolor* following mechanical injury : 1 nuclei; 2 lipids; 3 tyloses; 4 starch grains; 5 parenchymatous tissue able to reduce triphenyl tetrazolium chloride; 6 amorphous deposits; 7 tannins. (Wardell and Hart 1970)

wood can also been seen in other species (e.g., Hillis et al. 1962). Within 1 week after wounding, the parenchyma cells of *Q.bicolor* lost their dehydrogenase activity (as judged by their inability to reduce triphenyl tetrazolium chloride), and these cells also lacked starch. Adjacent uninjured sapwood cells lost starch as they became older and before they lost their dehydrogenase activity. This took place adjacent to the heartwood periphery in samples which were collected only in the summer and autumn seasons. The condition of the nuclei within individual cells of discolored sapwood of *Q.bicolor* varied considerably. Cells containing nuclei but lacking extraneous material in the lumen were adjacent to those in which the contrary situation existed. Generally, however, extraneous materials appeared 10–12 days after wounding and increased up to 16 days, particularly at the interface with normal sapwood. Disintegration of the nucleus appeared to take place before any deposition of extractives could occur. Wardell and Hart (1970) summarized the cytological and histological changes in the sapwood following mechanical injury in Fig. 6.11.

6.4.2 Paraquat-Treated Wood

Roberts (1973, Roberts et al. 1973) reported on the induction of "lightwood" in the sapwood of pine trees after treating them with dipyridyl herbicides, and particularly paraquat (a 1,1′-dimethyl-4,4′ bipyridylium salt) as a means to increase the availability of low cost naval stores – rosin, turpentine, and related chemicals. Near the wound site, the oleoresin content of affected wood may be greater than 40 percent of the wood (dry weight basis) compared with the usual 2–3 percent. On a total wood basis the oleoresin content can be enhanced from 100 to 150 percent (Stubbs et al. 1984) and injection of paraquat, particularly when used in conjunction with ethrel, is an economically attractive operation (Drew and Roberts 1978). Extensive coordinated work undertaken in the USA to improve various aspects of the operation has been reported in various issues of the Proceedings of the Lightwood Research Coordinating Council (R.N. Stone 1975, Esser 1976, 1977, 1978, 1979). The work has been summarized and extended by Stubbs et al. (1984) (see also Zinkel 1981), and the work on *P. radiata* has been reported by Sioumis et al. (1977, 1979).

Solutions of paraquat (or the related diquat) are applied to the cambial tissues (after cutting the bark) or injected into holes drilled into the sapwood of living trees of various pines to produce resin-soaked pinewood commonly known as "lightwood" because it is traditionally used as kindling. The highly polar paraquat cation is strongly adsorbed on the cellulose cell walls and with the continued flow of water in the sapstream the cations slowly move upwards by repeated desorption and adsorption. They are translocated at a rate that eventually becomes negligible. The paraquat ion concentrations decrease geometrically above the application site and when applied in appropriate concentrations do not reach the crown. In mature *P.elliottii*, paraquat moved 30 cm a day vertically and 0.5 cm radially (C.L. Brown and Nix 1975), but when applied to the xylem through the bark of *P.taeda*, paraquat moved only 3 cm about the area of application (Schwarz et al. 1977). It is the height of the trees, rather than age, which controls the mortality after paraquat injection, and height should be more than 10 m (Stubbs et al. 1984). Resinosis is most intense at the level of application and usually extends upwards 10 to 15 times further than it extends downwards from the point of injection. Oleoresin exudes within 2 months from the bark some distance above the treated area (Sioumis and Lau 1976). Resinosis decreases from the surface inwards and does not occur in the heartwood or the transition zone surrounding it.

It is believed that all *Pinus* species produce resinosis after paraquat injection, although the southern pines with their higher natural oleoresin content produce most resin. The response is typically greater in *P.elliottii* than *P.echinata,* and *P.taeda,* which in turn respond more than *P.virginiana* (Holton and Winston 1976, Nix 1976, Stubbs et al. 1984). The response to paraquat varies within *P. elliottii* for example (Barker and Schmid 1976, Miniutti 1976) and even in genetically identical material (Wolter and Zinkel 1976). A very high tree-to-tree variation in oleoresin enhancement has been found with increasing sample height. Increased oleoresin accumulation extends only 1.2 to 1.8 m above the wound site in some trees, whereas in others, that are comparable and growing close by, resin exudes from the bark 9 m above (Stubbs et al. 1984). On the average, paraquat treatment will increase the oleoresin content of *P.elliottii* trees from 14 to 30 kg m^{-3} or more and of *P.taeda* trees from 10 to 20 kg m^{-3}. When applied at the recommended concentrations, the growth loss of these two pines does not exceed 10 to 15 percent (Stubbs et al. 1984). Paraquat treatment of *Picea mariana* and *P. glauca* induced up to sevenfold increases in oleoresin content (Conner et al. 1980c).

When high (about 2.5 percent) concentrations of paraquat are injected into sapwood, the compound enters the axial and ray parenchyma cells and has an almost immediate effect of drying, whereby the moisture content is reduced to less than one-fifth of the surrounding tissues. The metabolism of these cells is then increased, resulting in enzyme conversion of starch and synthesis of oleoresin which is primarily resin acids dissolved in monoterpenes (C.L. Brown et al. 1976, R.N. Stone 1975, Esser 1976, 1977, 1978, 1979). The dry wood zone surrounding the wounds is mainly above the point of injection, and the paraquat moves upwards in these zones or in the trailing edge of water in the transpiration stream (Davies and Carrodus 1978). The normal pattern of oleoresin formation and secretion is significantly changed following paraquat treatment. All affected parenchyma cells, including those not directly associated with either vertical or horizontal resin ducts, begin to synthesize copious amounts of oleoresin. In *P.elliottii* the ray parenchyma produces more extractives than the epithelial cells (Miniutti 1976), although it has been noticed (Davies and Carrodus 1978) that increased numbers of resin canals were formed in the latewood in the cambial regions of *P.radiata* treated with paraquat. The affected resin ducts, in distinct contrast to normal ducts, begin to secrete or lose oleoresin centrifugally via the pits into the lumens of adjacent tracheids rather than into the duct lumen itself. Likewise, ray parenchyma cells in uniseriate rays produce excessive oleoresin and secrete or lose it initially through half-bordered pits into the lumens of contiguous tracheids (Birchem and Brown 1979). Each band of ray parenchyma is stimulated in the upward progression of paraquat to produce resin. Between the ray bands the resin flows through the normal channels (Sachs 1978). The latewood is soaked first with resin and in preference to the earlywood, possibly because the latewood pits remain open longer. When the tracheids are filled and as they lose water to the transpiration stream, the secondary cell walls are impregnated with oleoresin.

With *Azadirachta indica* lipids and polyphenols were formed in increased amounts when treated with paraquat or ethrel (Nair and Shah 1983, Shah et al. 1981b). In some cases mainly traumatic tissues were formed as in *Tsuga canadensis, Abies balsamea, Larix laricina, Pseudotsuga menziesii, Thuja plicata,* and *Picea* species, and usually with small increases of atypical extractives (Kiatgrajai et al. 1976a,b, Rowe et al. 1976, Conner et al. 1977, Sandberg et al. 1977). The formation of traumatic resin canals in the phloem of *Chamaecyparis obtusa* was accelerated by paraquat (Yamanaka 1985). Administration of paraquat to *Callitris* and *Eucalyptus* species did not increase the extractives content, and *Picea abies, P.mariana,* and *P.glauca* gave a limited response. The traumatic tissues in the above species and the wounds of *Pinus taeda, P.echinata,* and *P.elliottii* were surrounded by dry (desiccated) wood (Nix 1976, Conner et al. 1977), which was found also in *Azadirachta indica* (Nair and Shah 1983). Resin formation in response to paraquat in the latter three pine species occurred mainly in the period of lowered physiological activities (autumn to spring) (Nix 1976). However seasonal differences were noticed with *P.radiata* (Davies and Carrodus 1978) with spring often being the best period of injection, and location of the trees also appeared to play a role (Stubbs et al. 1984).

The increase in concentration of oleoresin after paraquat injection does not result from translocation of previously formed oleoresin via the resin ducts. All evidence indicates that it is synthesized de novo by cells undergoing varying degrees of senescence induced by low concentrations of the paraquat ion supplied via the xylem transpiration stream (Birchem and Brown 1979, Ryan and Schwarz 1979, Wolter and Zinkel 1984). However, as defoliation reduces oleoresin production, only some, but not all, of the oleoresin in paraquat-treated trees may be produced by current photosynthate (C.L. Brown et al. 1979).

Paraquat inducement results in a greater stimulation of turpentine formation (essentially α- and β-pinenes) than resin acid production in *Pinus serotina, P.glabra, P.echinata, P.rigida,* and *P.virginiana* (Peters and Roberts 1976, Roberts 1976). Increases of up to eight times the nonvolatile extractives (mainly free resin acids) and 20 times the turpentine content were obtained from the affected regions of *P.resinosa* compared with the control areas. There were no significant changes in the composition of the extractives of *P.resinosa* found by Wroblewska et al. (1977) or of total fatty acids (Wolter and Zinkel 1984). On the other hand, the composition of the turpentine from paraquat-treated trees shows a higher proportion of β-pinene to α-pinene than the turpentine in untreated *P.elliottii* and *P.resinosa* (Joyce et al. 1977, Zinkel and McKibben 1978, Stubbs et al. 1984), *P.contorta* has a higher or lower proportion of β-phellandrene and *Picea abies* greater amounts of 3-carene (Conner et al. 1977, Zinkel and McKibben 1978). The level of total fatty acids remains essentially the same in treated and control trees, but hydrolysis of fatty acid esters occurs. The proportion of nonsaponifiable components does not increase in paraquat-treated trees (Zinkel and McKibben 1978).

With the optimal concentrations of paraquat administered to pines the stored food reserves (starch) appear to be mobilized into soluble forms which are rapidly metabolized, followed by destruction of the integrity of the outer cell membrane and a gradual lysis of the cytoplasm with its enclosed cellular organelles, including the nucleus and other residual components of the protoplast (C.L. Brown et al. 1976, Davies and Carrodus 1978). The increased membrane permeability liberates previously compartmentalized cellular enzymes and substrates to create a large metabolic sink, resulting in enhanced oleoresin biogenesis. The biosynthetic pathway is apparently shunted to the formation of terpenoids before the death of the cell. During the period of cellular lysis, the increases in oleoresin accumulation in individual ray cells is followed by movement through the pits into adjacent tracheids until ultimately all elements of adjacent xylem become resin-soaked (C.L. Brown et al. 1976).

Bark beetles are much more strongly attracted to paraquat-treated trees than to adjacent pines identically wounded but not given paraquat. However, there is no instance where heavily attacked paraquat-treated stands caused further beetle infestation. Beetle attack is positively correlated with severity of wounding and especially the paraquat cation concentration used. It has been recommended that moderate paraquat treatments be used, as the yields of oleoresin are only 10–15 percent less than those of severe treatments and such treatments incur acceptable and often negligible losses from beetle attack, although the use of insecticides is advised (Stubbs et al. 1984).

6.4.2.1 Biochemical Changes Due to Paraquat

Administration of paraquat to *Azadirachta indica* in mid-summer resulted in enhanced activity of a number of enzymes in the axial and ray parenchyma cells at the periphery of the induced heartwood (Nair and Shah 1983). The enzymes included acid phosphatase [indicative of hydrolysis of starch and its transport (Hillis 1971) and lysis of the cytoplasm (Gahan 1981)] and succinic dehydrogenase located in the mitochondria and indicative of high respiration and metabolism (Varner 1961). At the cellular level paraquat acts essentially at a catalyst for the formation of hydrogen peroxide or hydroxy radicals by repeated recycling of the process. The products can peroxidize lipids in the membranes of the cell and cellular organelles to cause changes in membrane permeability and integrity and finally the complete disruption of the cells. In the process, starch, sugars, fatty acids, and amino acids are released and apparently also the enzymes leading to modification or stimulation of the normal biochemical activities (Birchem and Brown 1979, Stubbs et al. 1984) and the synthesis of oleoresin.

6.4.3 Ethylene-Treated Wood

There is increasing evidence that ethylene production is a common, if not universal, early response of trees to wounding and infection (Shain 1979) and to other forms of injury.

After injection of paraquat into 5-year-old *Pinus resinosa* and during midsummer a considerable increase in ethylene production occurs; it reaches a peak in 7 days and then declines. Respiration follows a similar pattern (Brown et al. 1976, Wolter and Zinkel 1976). Maximum ethylene production and respiration occurs just before a dramatic increase in oleoresin production. Even with this strong association, Wolter and Zinkel (1976) were unable to decide whether ethylene or cell necrosis is the direct cause of extractives formation, although later work (Wolter 1977, Wolter and Zinkel 1984) strengthened the importance of ethylene. Injury of living plant tissue can result in ethylene formation, and it is interesting that the pronounced resin streak around an embedded bullet in *P.ponderosa*, the study of which led to the stimulation of oleoresin production by sulfuric acid (Anderson 1955), may have been due to ethylene. The application of paraquat to the phloem of *Chamaecyparis obtusa* caused a higher production of ethylene than physical wounding alone (Yamanaka 1985).

The ethylene-releasing compound 2-chloroethylphosphonic acid ("ethrel", "ethephon") increases the oleoresin content of *Pinus elliottii, P.radiata*, and *P.resinosa* (Nelson and Hillis 1976, Peters and Roberts 1977, Wolter 1977). The complex commercial product Ripenthol (manufactured by Pennwalt Corp), which stimulates endogenous production of ethylene, induces the formation of resin-impregnated wood in the juvenile wood of *P.resinosa* but not the sapwood of mature trees (Wolter 1977). Evidence supports the view that resinosis induced by ethylene is a localized phenomenon (Wolter 1977). Oleoresin precursors do not move via the phloem, nor is there a reduction of oleoresin content in the adjoining regions. The increased extractives content caused by ethylene is dependent on in situ synthesis in the ray cells from carbohydrate sources (Wolter and Zinkel 1984).

The response of *P.resinosa* to ethylene resulted, in comparison with the controls, in a 25-fold increase in turpentine with a significant enhancement of β-pinene content, and a 14-fold increase in rosin content with an increased abietic acid but a decreased levopimaric acid content (Wolter and Zinkel 1984).

Work by Peters and Roberts (1977, Peters et al. 1978) indicated that ethrel acts synergistically with the bipyridylium herbicides paraquat and diquat, expanding their effect into more of the tree bole. However, it was concluded in later work (Stubbs et al. 1984) that ethrel-paraquat synergism with *P.elliottii* is not proven, although the addition of ethrel will usually increase oleoresin production. It is interesting that the addition of triethylamine to paraquat enhanced the yields of turpentine and resin acids (Joyce and Drew 1979).

6.4.4 Knots

A knot is a branch base that is included in the wood of a tree trunk or of a large limb. The branch bases are gradually embedded in the wood of the trunk through the annual addition of successive increments of woody tissue. As long as the branch is living, the cambium of the stem and branch are continuous and the knot is tightly enclosed in the wood. Once the branch has died, or is broken or cut off, the continuity of the cambium is interrupted and a loose knot results. Eventually the branch stub is overgrown and the embedded branch base has a generally conical shape with the small end connected to the pith. Much attention has been given to the number and size of knots found in the wood of various species grown under different conditions and their effect on the strength of structural members and on the appearance and processing of boards. Less attention has been given to the extractives in the knots and the manner in which they are formed, or their influence on the health of the tree, or their affect on the utilization of the wood containing them (Shigo 1983, Shigo and Larson 1969).

Usually, branch stubs or knots in angiosperms contain only moderate amounts of extractives. As branches die, microorganisms infect the dead tissues and begin to spread downwards to the branch base. At the position where the last area of branch tissue is enveloped by trunk tissue, a boundary of polyphenols begins to form and these compounds may be sufficient to prevent further entry of the microorganisms which continue to decay the branch so that it eventually falls off and the stub is overgrown (Shigo 1983). Sometimes the extractives convey inadequate resistance to the entry of microorganisms, particularly with fast-grown trees in humid, tropical climates, and these penetrate into the interior of the tree, resulting in discoloration and eventually decay. In other cases, the discoloration of the branch stub does not extend into the surrounding tissues and decay does not occur.

The knots in conifers, on the other hand, are highly resinified and can resist decay. The nodes of knots of *Araucaria angustifolia* collected from the forest floor contain about 30 percent "resin" which has been extracted in amounts up to 10 000 tons annually for industrial use (Anderegg and Rowe 1974). However, instead of a concentration of diterpenoids similar to those found in the wood of some *Araucaria* species, the extractives of the knots are largely a complex mixture of phenolic polar lignans and some monomolecular lignans.

The resin content of knots can be much higher than that of the surrounding tissues of *Pinus* species. Knots associated with branches which die before heartwood formation begins in the the stem of *P. radiata* may not show any immediate increase in resin content over the surrounding sapwood. Knots associated with dead branches after heartwood has started to form are much more resinous than the heartwood. In either case resinification of a knot continues after the death of the branch until a final resin content of 20–30 percent is reached in this species. The resin content of the knot is independent of the resin of the heartwood or the sapwood surrounding it (Harris 1961). The resinification of knots and stubs in pines is nearly complete 3 years after the death of the branch, although pruning accelerates the process whereby the knots will become very resinous within 18 months (Harris 1963).

Harris (1961) describes how, in normal sapwood, the resin originates in the interior of the epithelial cells bounding the resin canals and then moves into these anastomizing canals. This takes place under considerable pressure, which keeps the epithelial cells as a thin layer lining the walls of the resin canals. The resin canals of the stem continue into the growth layers of the branch. When all the living cells in the branch die, the branch itself, and the associated knot, dry out so that the moisture content may evenually fall to below the normal fiber saturation point of the wood when there is no water in the cell lumens. Resin held under pressure in the canals of the sapwood can now flow into the knot, where, in the absence of live epithelial cells to contain it, the resin will flow out and permeate the wood. Harris (1961) points out that "because the grain of the wood changes direction from the stem into a branch, resin flow into the branch is along the grain at all times. In this respect it differs from the movement of resin into the heartwood. When the cells of the heartwood die and become drier, the epithelial cells of the longitudinal resin canals within the heartwood also die along their entire length. Only the horizontal resin canals of the heartwood are still connected to living elements" Obstacles to radial movement within the stem hinder extreme resinification of heartwood. However, later Dietrichs (1973) noted that the resin is formed in the ray parenchyma cells of the dying branch of *Abies alba* and consists only of resin acids but not volatile terpenes.

6.5 Exudates

6.5.1 General

Exudates can arise when a previously uninjured tree is cut and the content of ducts of various types exude onto the cut surface of the wood (such as resins from *Pinus*, *Dipterocarpus*, and *Vatica* species) or of the bark (such as rubber from *Hevea* species, and chicle from *Palaquium* and *Payena* species). Exudates also arise from damaged cambial regions, such as kino from *Eucalyptus* species, or gum from *Acacia* or *Prunus* species. The same chemical components are also found in veins from which they may not exude unless they are of sufficiently fluid consistency.

Different exudates are formed by different procedures. In *Abies* and *Tsuga* species the epithelial cells synthesizing the resin usually die in the year of their origin, but these cells in *Pinus* species, which form much more copious amounts of resin, remain alive for several seasons (Bannan 1936). Skene (1965) points to the contrast of resin formation in the Pinaceae and kino formation in eucalypts. In the former the resin diffuses out continuously from the epithelial cells lining the vertical resin ducts and the cells function for a considerable time (Bannan 1936). On the other hand, the formation of kino involves destruction of the secretory cells so that the secretory activity of the veins and pockets is dependent on the continued activity of the peripheral cambium.

6.5.2 Resin Formation

The term resin, particularly when used in the pulp and paper industry, includes terpenoids, fatty acids and their esters, as well as other components in smaller amounts. The parenchyma of conifers contains mainly fatty acids in addition to esters, resin acids, and smaller amounts of unsaponifiable materials (e.g., Mutton 1962). Canal resin or oleoresin contain terpenes and other terpenoids (derived from condensation of C5 isoprenoid units) and smaller amounts of unsaponifiable material. Canal resin provides the exuded resin.

Many resinous exudates provide a major passive defense mechanism when many coniferous and some angiosperm trees are wounded or are exposed to fungal invasion. Some terpenoids are fungitoxic and restrain the entry or spread of fungi. Some oleoresins contain up to 50 percent of volatile terpenes which act as a solvent for the nonvolatile terpenes, so that they can flow to the wound site. The volatile terpenes evaporate at the wound to leave a resinous plug, or in some cases to allow oxidation of unsaturated terpenoids, or polymerization, so that the plug hardens to provide an effective barrier to entry of water or other agencies.

In the Coniferae, vertical resin ducts are often widely scattered in the *Larix, Pseudotsuga, Picea*, and *Pinus* species to an extent that they are considered to be a more or less normal feature of the xylem. These ducts unite with horizontal ducts in the rays at a few points so that they become canals. *Juniperus* and *Cupressus* species never form resin ducts in the wood. The diameter of the resin canals is between 30 and 100 µm in *Picea, Larix*, and *Abies* species and 50–200 µm in *Pinus* species, where the higher value refers to the vertical canal. In conifers such as *Abies, Tsuga, Cedrus*, and *Pseudolarix*, the vertical resin ducts are absent from the secondary xylem except at injuries, when they fuse with others to form a network of cavities (Bannan 1936, Fahn and Zamski 1970). Wounding stimulates the cambium to form vertical resin ducts in previously uninjured trees, as do also internal stresses. After wounding, traumatic resin ducts begin to form rapidly, for example in a *Larix* species in 7 days and in a *Picea* species in 13 days. In these and other conifers the ducts arise by schizogenous separation of the cells and are lined by a ring of epithelial cells. Resin arising from these cells appears in the ducts at a very early stage and begins to force apart the cells, even before the walls of the tracheids on either side are completely thickened and lignified (Bannan 1936, Fahn 1979). In *Abies* and *Tsuga* species the epithelial cells become thick-walled and lignified soon after they are formed and discharge their protoplasm; rarely do they function beyond the first season. The life of the epithelial cells, and the period in which their walls remain thin and unlignified, increases sometimes for a number of years with *Pseudolarix, Cedrus, Larix, Picea*, and *Pseudotsuga* species. In *Larix* species some of the cells die during the first year, while others remain alive for several years. More of the epithelial cells remain alive in *Picea* species so that the resin in the canals is under pressure in the youngest part of the sapwood. In *Pinus* species the epithelial cells are thin-walled, unlignified, and active for several years.

The differences in the character of the tissue associated with the resin ducts are paralleled by those of function. In *Abies* and *Tsuga* species a comparatively small amount of resin is produced, probably owing to the early death of the cells. The supply of resin is much more copious in *Pinus* species where the cells remain alive for several seasons. In these species the resin originates in the epithelial cells from which it passes into the lumen of the duct where it collects. When the volume of resin increases it compresses the epithelial cells and owing to the reduction of their size, their osmotic potential increases. A tendency exists for water to be drawn into the epithelial cells, and when this occurs their size increases and pressure is exerted on the resin in the duct. If the latter is ruptured, the resin is forced out, when the epithelial cells may then produce more resin to refill the duct. Obviously this procedure of emptying and refilling can be best performed in ducts lined by thin-walled cells, as found in pines (Bannan 1936). Furthermore, the amount of resin exuded when a duct or canal is ruptured can be influenced by the length of the canal. In pines the length is much larger than in the other conifers and can average from about 10 cm in the center of a cross-section to about 50 cm in the outer wood with lengths of nearly 1 m (Bannan 1936).

The traumatic tissues in pines differ from those in other conifers in that they develop both vertically and tangentially at considerable distances from the point of injury. In *P. sylvestris* wound effects extend as far as 12 m above a wound, but less in *P. halepensis* (Fahn and Zamski 1970). The ducts are of great length and their meandering course results in a more or less even distribution throughout the wood. Importantly, the pines appear to be more highly sensitive to injury than do other conifers and moreover this sensitiveness increases with the age of the tree and the cambium (Bannan 1936). An increase in vertical resin ducts all around the stem of young *P. halepensis* occurs when they are exposed to wind (Fahn and Zamski 1970). Many of the injuries made by insects result in a greater development of ducts than do purely mechanical wounds of the same size. Wounding, pressure, and the auxins, indoleacetic acid, naphthalene acetic acid, and 2,6- dichlorophenoxyacetic acid, induce the formation of vertical ducts, but radial resin ducts form after a layer of ductless wood has first been laid down (Fahn and Zamski 1970).

The genetic trait and age of the tree seem to be the most important factors which influence the rate of resin yield although these in turn can be influenced by microclimatic conditions (e.g., Posey and Robinson 1969). The contents of the resin ducts are under pressure (about 3–10 Pa but sometimes up to 30 Pa) and are very sensitive to changes in water stress. In a review and extension of earlier work (e.g., Bourdeau and Schopmeyer 1958), Vité (1961) showed in an examination of *Pinus ponderosa* that this oleoresin exudation pressure undergoes considerable diurnal fluctuations with maximum values occurring between midnight and dawn. The pressures are affected not by the age or tree class but by the physiological condition of the tree and the factors of temperature, light intensity, and humidity. The latter in turn are affected by seasonal changes (highest pressures in mid-winter and lowest in summer) and availability of soil moisture, as shown also by Zamski (1970) with *P. halepensis*. The diurnal fluctuations are more extreme on dry than on moist sites. It may also vary around and along the trunk, being higher at the lower parts of the tree. Major differences occur radially; they are usually less in the inner sapwood which is less responsive to changes in rainfall. Differences in exudation pressures in *P. ponderosa* have been related to differences in infestation by the bark beetle *Dendroctonus breviconis*. With decreasing turgidity in the epithelial cells lining the resin ducts, the rate of exudation at which the oleoresin exudes from the anastomosing capillary system of resin ducts diminishes and enables the initial bark beetle attack to succeed.

The composition of the resin acid mixture exuded from wounds after the cessation of the primary resin acid flow appears to differ in composition from the resin acids in unwounded tissues (Shrimpton 1973, Hart et al. 1975). These studies indicate that tissues surrounding wounds undergo de novo synthesis of resin acids. In support of this view it was found that a highly significant increase in the concentration of dehydroabietic acid occurred in the resin from wounded *P. syl-*

vestris seedlings and to a lesser extent in other resin acids (Gref and Ericsson 1985). Following a bark beetle attack, α-pinene was the first monoterpene to increase in concentration in the wound resin of *Abies grandis* (Bordasch and Berryman 1977), *Pinus contorta* (Raffa and Berryman 1982), *P. strobus* (Wilkinson 1980), and *Picea abies* (Schuck 1982). Higher levels of 3-carene and myrcene in wound resin have also been reported (Russell and Berryman 1976, Gollob 1980).

Resin components in the epithelial cells surrounding the resin canals probably form in the same manner as that described in paraquat and ethylene treated wood (Sections 6.4.2 and 6.4.3). There is biochemical evidence consistent with the view that terpene synthesis in general occurs within the plastids (Dell and McComb 1978). Joel and Fahn (1980) have reported that the epithelial cells surrounding the resin ducts of *Mangifera indica* become secretory before they enter into the phase of senescence and disintegration. The endoplasmic reticulum (ER) appears to be involved in the resin secretion of this species and of *Pinus halepensis*. Other cytological studies indicate that the spherosomes synthesize resin acids in both the resin duct cells and other parenchymatous cells of *P. halepensis*, and that the proportion of resin acids is much higher in the very young tissues than in older ones (Werker and Fahn 1968). The small oleoresin droplets percolate through the lipophilic plasmalemma and then through the walls of *Pinus* ducts as minute droplets (Wooding and Northcote 1965).

6.5.2.1 Anatomy of Pockets

A number of exotic *Pinus* species planted in New Zealand contain resin pockets to some extent, although often the numbers and size are small. The pockets were found to be more common in areas of low rainfall, low water retention in the soil, and low humidity in spring and summer at which time the trees can be subjected to high winds. That these conditions of water stress as well as stress due to bending are important in pocket formation was shown by their association with false growth rings. Their initiation takes place during or immediately following production of false latewood with its thick-walled cells (Cown 1973). The formation of rings during periods of drought has been recorded in other studies. A mechanical weakness exists between the false latewood/earlywood boundary, i.e., the growth horizons in which resin pockets are initiated. Different types of cell are involved in the structure of pockets with a transition from normal latewood tracheids to septate tracheids to rectangular parenchyma to callus cells. The latter cells protrude from the rays and continue to develop into the free space. The anatomy of the pockets shows that the initial split takes place in the cambial zone. Cown (1973) has proposed that the initiation of the pockets results from the rupture of cells along the false latewood/earlywood boundary immediately after the relief of water stress. Accordingly, the resultant increase in oleoresin pressure in the anastomizing resin canals at a time when the production of large-diameter, thin-walled cells is renewed can result in rupture of these earlywood cells as they are formed in the cambium. Cell rupture would be further caused by the mechanical stress of wind sway. Once initiated, resin would enter the resulting cavity and enlarge it to an extent depending on the oleoresin pressure and the amount of resin available.

6.5.3 Gum Formation

The gum from *Acacia senegal* usually exudes in the form of tears, and sometimes lumps, mainly from young branches. It also exudes from large branches and particularly from those parts that have been damaged (Ghosh and Purkayastha 1962). During the gum-forming season, gum does not appear on *Acacia* trees until about 3 weeks after injury, and the amount exuded rarely exceeds 25 g. Exudation is stimulated by stripping the outer bark or by making shallow cuts in the bark, but these treatments are not always successful. A continuous flow of 'gum-resin' from the bark of *Commiphora wightii* takes place for a long time, and intensified morphological changes in the epithelial cells commence 45 min after injury (Shah et al. 1982).

The gum of *Acacia senegal* comes from cysts which develop in the inner bark and not from the wood. The cysts form in rows surrounding damaged bark when this exudes gum. Large gum cysts form adjacent to the cambial zone due to dissolution of some of the cells in the inner phloem, and the cambium is not affected. Sometimes the wood also forms an abnormal number of small vessels. Starch is either totally absent from all the tissues of the bark and wood or is present in very small amounts in the wood parenchyma cells some distance from the cambium. The starch disappears first from the wood fibers, then from the ray parenchyma, and last from the axial parenchyma strands (Ghosh and Purkayastha 1962). Gum cysts are initiated schizogenously in the phloem parenchyma, and the very small cavity is immediately filled with a gum-like substance. The cavity gradually enlarges and the surrounding parenchyma cells, the contiguous sieve tubes, and the companion cells disintegrate and discharge gum into the cyst. In this process considerable enlargement of groups of the parenchyma cells occurs around the cyst so that conspicuous intercellular spaces are formed which are wider on the tangential than the radial face and have limited extension vertically.

The exact cause of gummosis in *Acacia* species is not known. The proposals that bacteria are responsible have received limited support, and evidence indicates that gummosis is a response to physiological disturbances caused by drought or injury (Ghosh and Purkayastha 1962). The application of formalin and hydrochloric acid or of sodium metabisulfite increases the flow of gum.

Gum duct formation occurs naturally in healthy fruit trees of *Prunus* species, and in the fruit, but gum production is intensified after mechanical or insect damage, pathogen attack, or exposure to ethylene-producing chemicals (Fahn 1979). Injection of the ethylene-releasing compound ethrel into a *Prunus* species resulted in a copious flow of gum (Hillis 1975). In a study of gum formation in the fruit of *Prunus dulcis*, it was concluded that gum is primarily the product of specialized secretory cells completely lining the gum ducts. The formation of the ducts is initiated by the schizogeneous separation of parallel rows of secretory cells. Gum is synthesized within the cells from starch and other carbohydrates and accumulates between the cell wall and the plasmalemma. It is unlikely that cell wall degradation is the primary source of the gum, both because of the dissimilarities in chemical composition between gum and cell wall polymers and because gum is present in the ducts several days before extensive cell wall degradation begins. The secretory cells are characterized by their abundance of dictyosomes and dictyosome vesicles typical of polysaccharide-secretory cells, mitochondria, and rough endoplasmic reticulum (Fahn 1979, Morrison and Polito 1985).

Injury induces the formation of gum ducts in the wood of *Azadirachta indica* and the production of a proteinaceous gum polysaccharide. The immediate derivatives of the cambium differentiate as duct initials with intense succinate dehydrogenase and acid phosphatase activities. Following an active phase of secretion of substances between the plasmalemma and the cell wall, the lysis of the epithelial cells results in widening of the duct lumen and liberation of the substances (Nair et al. 1983). Both endoplasmic reticulum and Golgi bodies are apparently involved in the secretion.

6.5.4 Kino Formation

The polyphenolic exudate kino (see Chapter 2.6.2.3) originates from pockets, and occasionally veins, in the wood of numerous *Eucalyptus* species. Recent work by Tippett (1986) has shown that species of three sections of the sub-genus *Symphyomyrtus* form veins in the phloem (e.g., *E.sideroxylon*) and not the xylem. Some of the species in a fourth section form veins in the phloem whereas in other species the veins are found in the wood. [Other species are mentioned in Chattaway (1955b) and Brockway and Hillis (1955)]. Kino results from injury to the cambial region and the connection was first recorded by Kessell (1921). Susceptibility to pocket and vein formation depends partly on the section of the *Eucalyptus* genus to which the species belong (e.g., the "bloodwoods" exude kino freely) but also on environmental conditions, branch shed, and injury (due to fire, particularly in hot, dry seasons, insects damaging the cambium, and mechanical damage). Intermittent but persistent activity of fungi can induce a series of veins in the xylem of *E. marginata* (Tippett et al. 1983). Kino is found in pockets a few cm to 2 or 3 m length, and of a lenticular or blister shape (Fig. 2.13). Veins are 2–5 mm thick (Fig. 2.14), of varying width extending for greater distances upwards from a wound than downwards; they may form up to 1 m in length in a period of 6 weeks in warm weather. The veins form a tangential layer of anastomosing passages embedded in a sheet of traumatic parenchyma cells (Irvine 1936, Jacobs 1937).

The presence of kino pockets weakens timber used for structural purposes and, although kino veins do not have the same effect, they lower the quality of timber intended for interior joinery and cabinet work (Jacobs 1955). Kino lowers the quality of pulpwood by increasing the chemical consumption to produce high quality pulp. The presence of "gum" (kino) veins is mentioned more frequently in Australian timber standards than any other aspect as a cause of degrade.

Veins form more rapidly in the warmer months of spring than the cooler. When the mean maximum daily temperature falls to about 10 °C with mountain-growing eucalypts and 16 °C with coastal species, kino vein formation is considerably retarded. However, the extent of vein formation is determined more by vigor of growth at the time of injury than by temperature, and injuries inflicted during the cold months ultimately form less extensive veins than similar injuries in the hot months (Jacobs 1937, 1955). It was considered that cracking of the bark, caused by shrinkage during fires or exposure of the cambium by injury, allows the entry of air and desiccation of the wound, which cause kino formation. When petroleum grease was applied over these injuries in order to prevent this, and hence kino formation, from taking place it was found that instead very large veins were formed, often with a profuse exudation of kino (Jacobs 1937, 1955). It is apparent from recent knowledge that the grease prevented loss of the volatile initiators of kino formation, such as ethylene, from the wound.

In a study of two provenance trials with 26 families of *Eucalyptus regnans*, no significant difference in natural kino vein formation was found (Doran 1975). However, when kino veins were induced with ethylene and related chemicals in six half-sib families of the same species grown in two plantations at different sites, family differences were found to exist (Nelson and Hillis 1978a,b). As there was also a trend for the kino veins to be larger at one plantation than another in both of these studies, it appears that environmental factors also exist. Highly significant differences in the presence of kino veins were found between different sources of *E. regnans* (the worst affected) and *E. obliqua* (Nicholls and Griffin 1978), indicating genetic differences. A later study of *E. obliqua* showed highly significant differences between provenances (Nicholls and Matheson 1980).

6.5.4.1 Anatomy of Veins and Pockets

Kino veins have been frequently observed in a number of *Eucalyptus* and *Angophora* species but their anatomy has received little intensive study although casual observation indicates that it is generally the same. During the period of active growth, the first response of the cambium of *E.maculata, E.camaldulensis*, and *E.obliqua* to injury is the induction of multiple cambial divisions to form tangential layers of traumatic or anomalous parenchyma cells which have a regular arrangement. Subsequently, there is a localized proliferation of nodules or strands of cells within the layer of traumatic parenchyma and the growth of these strands of cells pushes back the surrounding traumatic parenchyma, deforming the radial and tangential alignment of the cells in this tissue. Concentric layers of cells develop around the edges of the strands of cells, resulting from the activity of the cambial layer, and a cavity forms by the breakdown of the central cells in the strands. This results in rows of parenchyma bridging interconnecting cavities, which Foster (1973, personal communication) prefers to call "lacunae" to avoid confusion with other terminology. Some rays occur on both the bark and wood side of a vein, showing that the ray initials do not lose their identity during the formation of a vein. The parenchyma bridges may be closely spaced, producing a tangential row of lacunae as in a vein (Fig. 2.14) or they may be absent as in a pocket (Fig. 2.13). Kino begins to accumulate in these pockets or lacunae. The suberization of some of the layers of cells brings secretion to an end, the traumatic parenchyma cells become indistinct, and the cambium resumes normal xylem formation (Jacquiot and Hervet 1954, Day 1959, Bakshi and Singh 1964, Skene 1965).

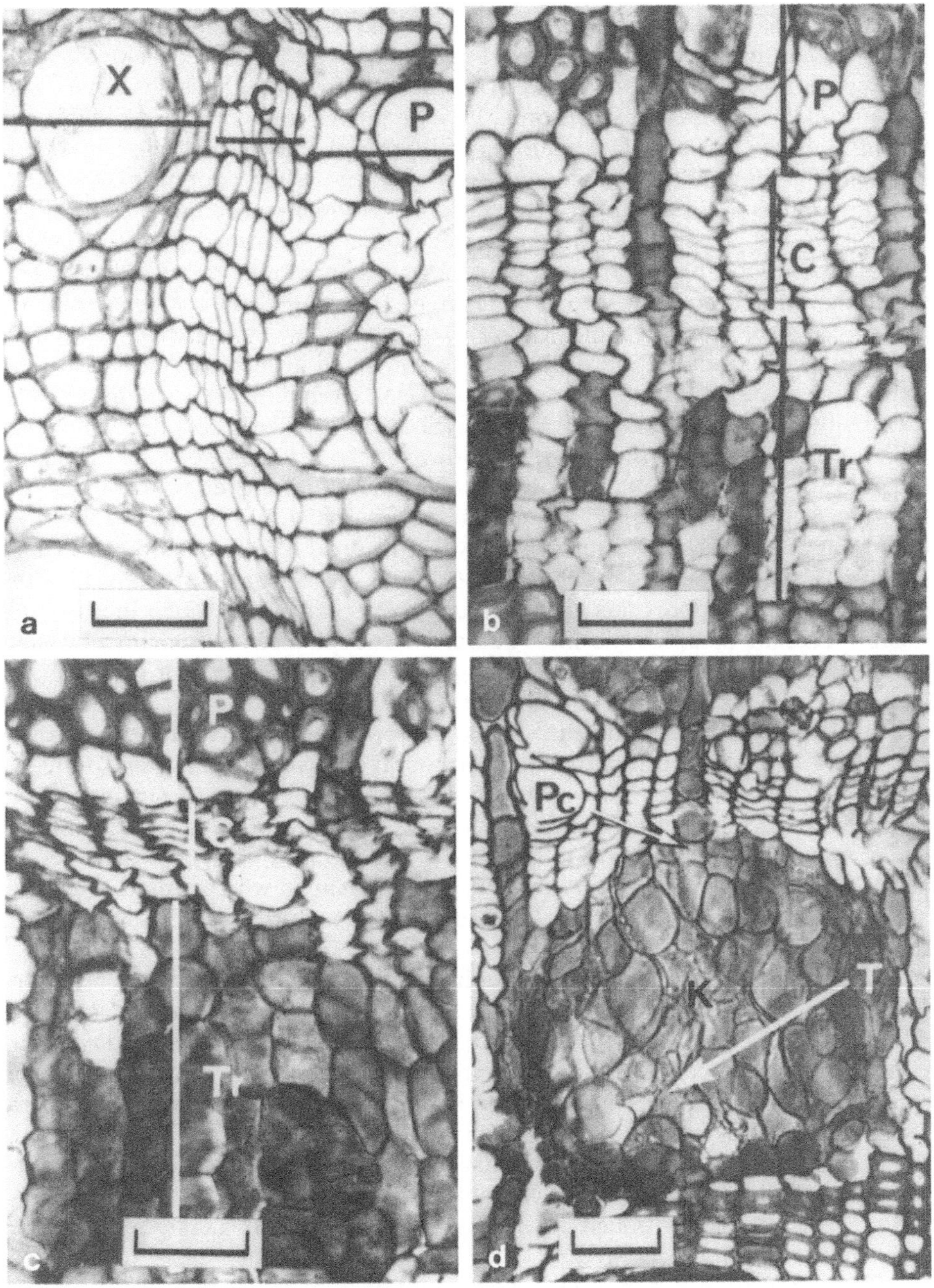

Fig. 6.12. Kino veins in *Eucalyptus obliqua*, (**a**), transverse section through the cambial region, 10 days after wounding; (**b** and **c**) transverse section through the cambial region and traumatic parenchyma, 20 days after wounding; (**d**) transverse section through a kino vein at an early stage of development, 35 days after wounding. C cambium; K kino; P phloem; Pc peripheral cambium; T torn cells; Tr traumatic parenchyma; X xylem (scale 50 μm). (Skene 1965)

Skene (1965) discerned a number of stages in kino vein development although there was considerable variation in both rate and extent of vein development in different samples. These samples were taken from *E.obliqua* trees wounded with saw cuts in late spring to mid-autumn. Development could be more rapid during the flush of growth during late winter to early spring. The first stage was observed in tangential longitudinal sections as early as 5 days after wounding, when the fusiform initials responded by transverse divisions which were completed in a further 5 days and before any traumatic parenchyma were formed (Fig. 6.12). The first anatomical changes in transverse sections were apparent about 15 days after wounding with the occurrence of a layer of thin-walled parenchyma cells of greater cross-sectional area and shorter length than the cambial cells. There was also a marked accumulation of polyphenols in small groups of these brick-like cells which 20 days after wounding were still isolated but in large groups, so that eventually only isolated cells lacked polyphenols. The cells retained the same radial seriation as when they were laid down by the cambium, despite the accumulation of large amounts of polyphenols. Later, the central cells collapse to form the lacunae of the vein (Figs. 6.13 and 6.14). Day (1959) suggested that the kino veins of *E.camaldulensis* grown in Cyprus resulted from water stress, causing collapse of the cambial tissues, but Skene (1965) points out that there is insufficient evidence to support this view. The actual mechanism by which the lacunae are formed is not completely clear. The walls of the collapsed cells, many of which are broken, remain attached to the living cells around the edges of the lacunae. The kino could accumulate in the lacunae by diffusing through the cell walls after the loss of permeability and death of the protoplasm. However, it appears that the release of kino occurred after the peripheral cambium had started to divide and some of its derivates had grown considerably in size (about four times their original cross-sectional area), and that the unequal distribution of growth among its derivatives caused cell walls to rupture. The continuing growth of these derivatives would create internal pressure inside the lacunae which would force the kino to exude or distort and stretch the surrounding traumatic parenchyma (Jacquiot and Hervet 1954, Skene 1965).

About 49 days after mechanical wounding, a layer of suberized cells began to form on the radial and outer tangential surfaces of the kino vein and about 56 days later, on the inner tangential surface also (Fig. 6.13). The suberized layer was rarely more than six cells thick. The cambium returns to normal xylem formation quite quickly after the formation of the veins except for some minor features (Skene 1965). The traumatic parenchyma become thick-walled and lignified much later than other parenchyma. Thus injury results in an abrupt proliferation of parenchyma with different cell wall characteristics but also very different cell contents. Moreover, as mentioned previously, in some species this takes place on the phloem side and in others on the xylem side of the cambium.

Tippett (1986) found that the development of veins was similar regardless of whether they became included in the xylem or the phloem. Periclinal cell divisions in the cambial zone, and septation of fusiform cells by transverse and pseudo-transverse divisions were observed 7 days after initiation of the changes in *Eucalyptus marginata* (xylem veins) and in *E. wandoo* (phloem veins). Between 7 and 22 days, patches of cells within the broad bands of parenchyma accumulated

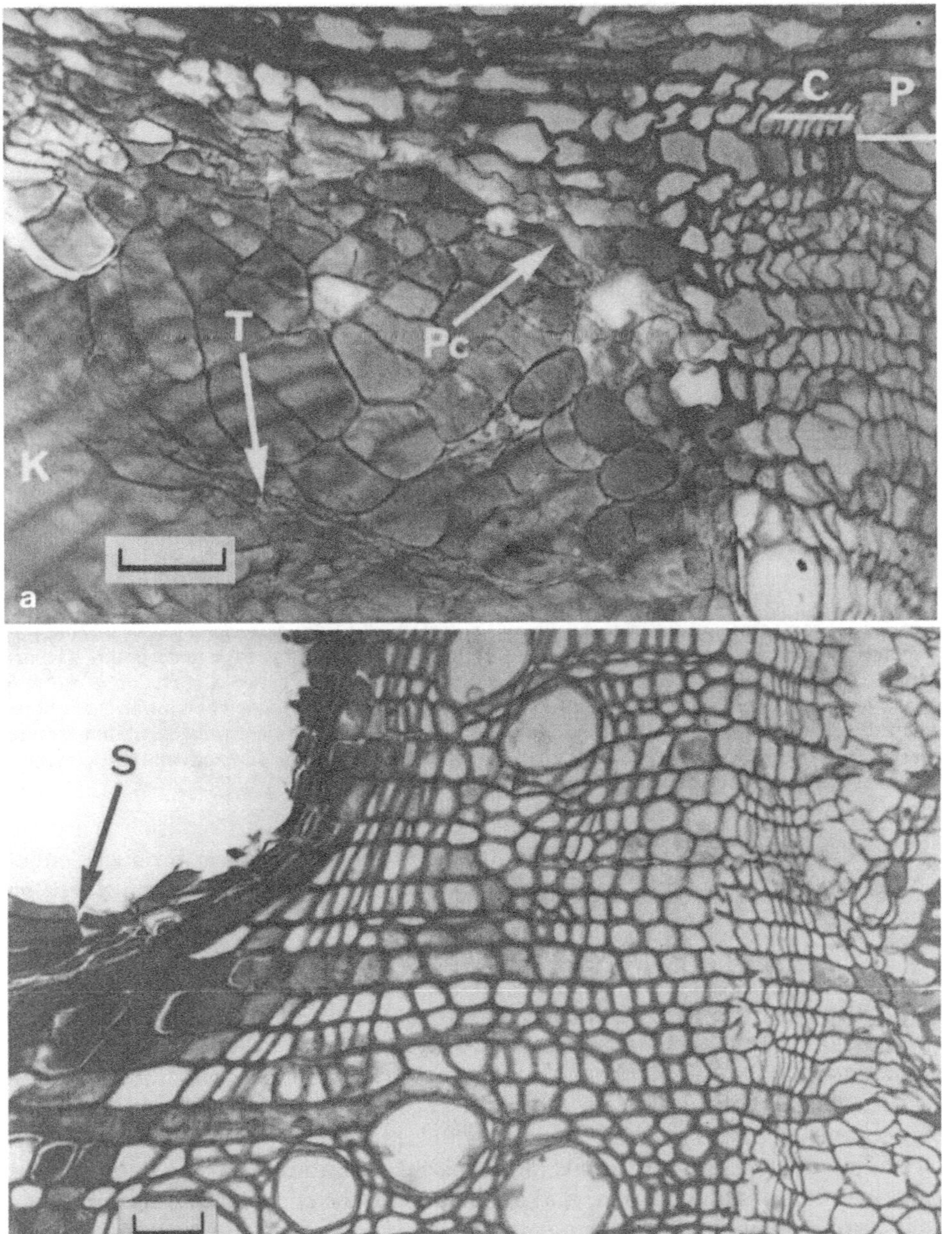

Fig. 6.13. Kino veins in *Eucalyptus obliqua*. **a** Transverse section through a kino vein at an intermediate stage of development, 35 days after wounding, showing cambium (*C*), kino (*K*) and phloem (*P*). The walls of the torn cells (*T*) have been stretched by the growth of the derivatives of the peripheral cambium (*Pc*). **b** Transverse section through a mature kino vein, showing the cambium and xylem produced subsequent to its formation; 92 days after wounding (S suberized cells) (scale 50 μm). (Skene 1965)

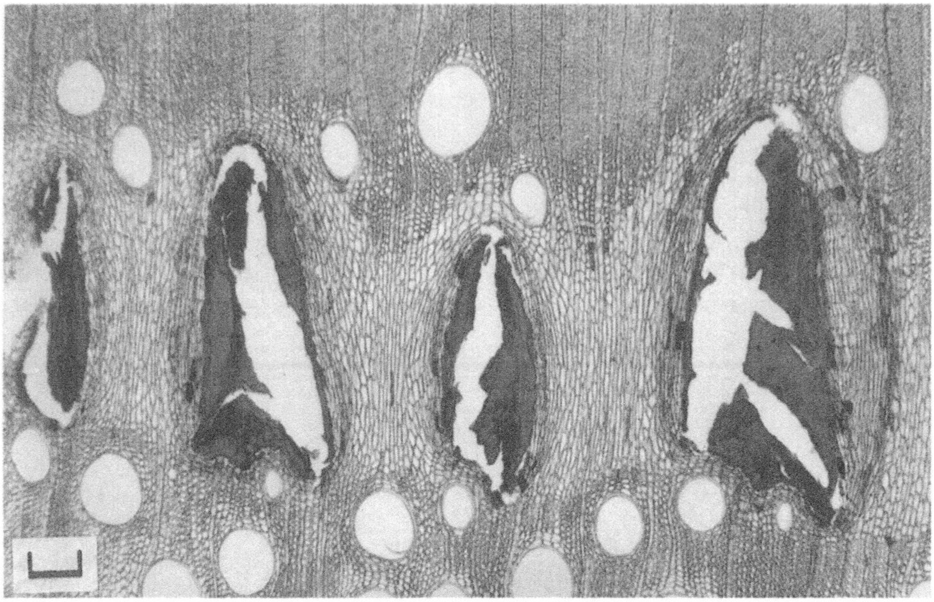

Fig. 6.14. Transverse section from *Eucalyptus maculata* of a fully developed kino vein that has become included in the stem. The center of the tree is to the bottom of the micrograph. The organized structure of kino vein tissue and the abrupt deliniation between cessation of normal wood production and initiation of kino vein development can be seen. The kino in the wedge-shaped lacunae has dried and cracked. Ordered files of cells form a periphial cambium around the edges of these cavities, which are radially separated by traumatic parenchyma. Rays traverse the kino vein, but their shape is deformed (scale 1 mm). (Courtesy G.T. Dale)

polyphenols. The tissue which differentiated into kino veins was produced either to the phloem side or to the xylem side in 7 days. Parenchyma was produced on the phloem side of the cambial zone of *E. wandoo*, while differentiation proceeded on the xylem side and the latter was often stimulated when phloem veins were developing.

At times of the year other than the spring, injury of eucalypt cambium will lead only to traumatic tissues with cells of different wall thicknesses. The cells are sometimes filled with polyphenols. It is in the spring season, or when the cambium is very active, that injury leads to the formation of anomalous parenchyma and the copious synthesis of a class of polyphenols different from those formed in surrounding tissues. The ray initials of the vascular cambium and the initials of the peripheral cambium can grow and divide despite the considerable accumulation of polyphenols in them. Consequently, these polyphenols do not interfere greatly with the metabolism of the cells (Skene 1965).

6.5.4.2 Chemistry of Kino Formation

The wood and bark extractives of *Eucalyptus obliqua* and a number of other eucalypts with pale-colored woods are preponderantly ellagitannins based on gallic acid and its derivatives, which are synthesized by the shikimic acid pathway. The kino of eucalypts is very largely proanthocyanidins derived from the shikimic acid pathway and also acetate units from the TCA cycle. Proanthocyanidins can be detected specifically by their formation of a red color when heated with butanol and hydrochloric acid (in contrast to the brown or no reaction with other compounds) and this reaction has been adapted for anatomical studies (Hillis 1958). Traumatic parenchyma have contents which color red with this reagent; they form a layer of only about 5 cells and in the spring become apparent in less than 20 days after injury (Skene 1965).

Increased amounts of ethylene are present during kino vein formation, and administration of ethylene-releasing compounds to active eucalypt cambium leads to kino formation. A logarithmic dose-response relationship has been found between kino vein size and the amount of ethrel applied to the cambium (Nelson and Hillis 1978a,b). Auxin-like herbicides such as picrolam, 2,4-D, 2,4,5-T and the synthetic auxin α-naphthalene acetic acid stimulate exudation of kino in a number of *Eucalyptus* species. The suggestions that auxin is involved in kino formation, possibly because of its stimulative effect on ethylene production, could not be confirmed (Nelson and Hillis 1978b). Ethylene production may be increased in an auxin-independent manner when many plant tissues are stressed in various ways.

6.5.5 Rubber Tapping

Hevea braziliensis trees repeatedly stimulated with ethrel over a period of 3–9 years showed a reduction in time of the initial flow rates of latex when compared with the unstimulated trees. The severity of reduction of flow rates varies in different clones and with intensity of the original tapping. Large areas of laticiferous tissues below the tapping panel recorded abnormally low turgor pressures and, in addition, there was an increase in the area of stone cells in the soft bark. The levels of nitrogen, phosphorus, potassium, and magnesium were higher in the bark shavings at the tapping cut than elsewhere. The rate of decline of response to stimulation largely depends on the rate of extraction of latex. If the rate of extraction exceeds the rate of refilling of the drainage area and biosynthesis of latex then the yield response declines. Failure to restore the osmotic balance may be due to lack of availability of sugars and other osmotic constituents or hindrance of movement into the latex vessels. The decline is closely associated with the upset in internal water relations of the latex vessels, leading to an instability of the latex which affects the metabolic activity of the organelles and finally causes a loss of nuclear activity. Consequently, the rate of extraction and the limitations of biosynthesis and replacement of the latex from the surrounding regions can explain the observed decline in response to stimulation (Pakianathan et al. 1982).

Chapter 7

Function, Formation and Control of Heartwood and Extractives

The events of plant cell division and cell enlargement have received much attention. On the other hand, the changes during maturation as well as the onset of cellular senescence, which are less easily observed, have received comparatively little attention. When the aging processes in the cells of a living tree are studied, they are found to be much more complex, as these processes occur at different time intervals with different cells and in different dimensional spaces. Moreover, there are severe experimental difficulties in observing the processes and changes in the inner sapwood. Furthermore the environment acting on a tree is complex and constantly changing and the response of the tree to the environment is equally complex and similarly changing.

Most studies on heartwood formation have been undertaken on samples collected at one point in time and conclusions have been drawn from the studies of a single species with the approach of one discipline. This has resulted in some contradictory evidence and conclusions. For the development of an understanding of heartwood formation, greater attention to seasonal and climatic effects is needed, with the realization that adjacent cells can behave differently in different parts of the tree, and that interpretations should be made with a multidisciplinary approach.

7.1 Function and Volume of Sapwood

7.1.1 Function

There is much evidence (Chapter 2.2) to support the view that sapwood has three important functions. These are: to transport large quantities of water, solutes, and gases from the roots to all parts of the tree including the foliage, meristems, and the vascular tissues; to provide a strong structural support for the entire tree so that it can project its foliage to a position where it can compete for the required amounts of light and air; and to serve as an important reservoir for water, energy, minerals and solutes. This description of functions accommodates the variations in width of sapwood in trees of one species growing under different conditions. In this situation a simplistic explanation of the change of sapwood to heartwood appears to be valid, namely that heartwood formation is a physiological aging process taking place over a period of time, and characterized by the death of the storage cells (Bosshard 1965a). This view is supported by the observation that the number of sapwood rings in living stumps of *Pseudotsuga menziesii* is, on average, about the same as in adjacent trees. The evidence indicates that factors controlling the formation of heartwood and the amount and composition of its extractives are incorporated into the ray cells during the early stages of their development (Hemingway and Hillis 1970).

The description of sapwood functions does not explain the wide variations in the number of sapwood rings in different species of some genera (e.g., *Pinus*) and not in others (e.g., *Eucalyptus*). It does not explain the wide differences in the number of sapwood rings in trees of the same size and growth rate, and growing on similar sites but belonging to different genera.

The proposal that the volume of sapwood is a means to conserve the amount of essential nutrients in the tree (Bamber 1976b) does not explain the selective removal of elements when heartwood is formed (Chapter 5.2). It does not explain the existence of large trees of certain species which have large volumes of sapwood or do not form heartwood, or the accumulation of inorganic salts into wetwood (Chapter 2.5.2.4).

In view of the different functions of sapwood it is not surprising that there are variations in the volume present. The rate of maturation and senescense of the living cells of sapwood is not solely determined by the intrinsic characteristics of those cells but is under the control of the whole tree.

7.1.2 Volume

The number of growth rings of sapwood is a more consistent characteristic than that of width. In a number of species this number is positively correlated with the age of the tree (Chapter 2.2.3.3) but there can still be considerable variations in the relationships due to environmental conditions (Chapter 2.2.3.1).

Within the one species the crown volume or leaf biomass has the greatest influence on the width of the growth rings (Chapter 2.2.3) and the larger the crown the greater the width or cross-sectional area of the sapwood. The more rapidly growing trees require a greater uptake of water and hence large volumes of sapwood, and thus less conversion to heartwood, to provide the crown with moisture. As expected, those trees growing closer to water, or with a regular supply of water, have larger sapwood volumes than those which do not. Crown class has a greater association with sapwood width than site index.

There are some contradictory data on the effect of various factors on sapwood width. These may be due to variations between samples of sampling height, crown class, growth rate, site conditions, and age of specimens. Furthermore, the relative proportions of sapwood with a conducting role and with a storage role of water under different conditions requires determination.

7.2 Types and Formation of Heartwood

7.2.1 Introduction

After the formation of the xylem in the cambium, only the radial and other parenchyma cells remain alive for a varying number of years. By definition, heartwood has ceased to contain living cells. It is evident (Chapter 2.5) that several types of heartwood exist and that the biological descriptions (Chapters 6.2 and 6.3) of heartwood formation are inadequate and sometimes contradictory. The complex situation of the living cells at the heartwood periphery, and the experimental difficulties encountered in the study of cell maturation and the onset of cellular senescence at the appropriate seasonal period render the development of an explanation of the death of these cells more difficult than with other organisms. Numerous questions require answer before the formation of various types of heartwood or quasi-heartwood can be explained, and the cause and effect relationships can be clarified. Is the death of undamaged parenchyma cells a random event, the result of an internal specific signal of a genetically inherited programed death, or is it in response to an extracellular stimulus? At what point is an individual cell completely and irreversibly dead, and are all parenchyma cells of equal importance in heartwood formation? Is a gradual rate of death required for heartwood formation with the sequential loss of function of key cells, and must this occur at certain seasonal intervals? Is the collapse of the nucleus of greater importance than the destruction of the cytoplasm, and are the changes interrelated? Is death a failure of a particular organelle with a definite total work capacity, or is it the result of the change of permeability of a membrane with loss of fluid, or the inactivation of a particular set of genes, or the accumulation of an inhibitor, or is it a more complex set of situations? Some of the experimental difficulties encountered in collecting samples to answer these questions may be reduced in studies of the tissues forming exudates or forming lesions in sapwood. But does an injured cell in these regions die by the same or extended physiological events as a cell responding to normal nontraumatic events?

The collection of answers to these complex questions will be aided in the first instance by more precise descriptions of those regions considered as heartwood in its various forms. It is apparent that heartwood formation is a process governed by intricate and delicately balanced physical and biochemical systems. The results of experiments which disturb one of these systems must be considered in relation to the whole system.

7.2.2 Types of Heartwood

The simple description of appearance – despite its limitations – is used in the following discussion. It would be premature to classify heartwoods into a number of classes until there is agreement on the relative importance of more physiological and biochemical characteristics. Those trees which regularly form a sharply defined, symmetrical heartwood can be contrasted with those in which there is a gradual change in appearance or decrease of moisture content in the sapwood, with an eccentric location of heartwood or dark-colored zone which may have an erratic periphery.

7.2.2.1 Regular Heartwoods

The heartwoods in *Robinia pseudoacacia, Quercus robur, Pinus*, and *Eucalyptus* species, and *Cryptomeria japonica* exemplify regular heartwoods with their abrupt formation, concentric position, sharply defined, slightly undulating periphery, and darker color. Although the color intensity usually increases gradually from pith to periphery, the heartwood can contain bands of differing color intensity, presumably due to different concentration of extractives. The extreme case of this situation is the formation of target rings (Chapter 2.5.2.5) which lack or contain low amounts of heartwood extractives. The abnormally low amounts of extractives may be due to inadequate supplies of photosynthate precursor at the time of heartwood formation, but there is no direct evidence to support this view.

The heartwood of conifers contains much less moisture than the sapwood, but this is not always so with the heartwood of angiosperms (Chapter 5.1). One explanation for the formation of heartwood is that it enables, or is the result of, constriction of water-storage tissues. There is some evidence that the cytological behavior of the ray parenchyma of species with wet heartwoods may differ from those where regular heartwood is formed (Chapter 6.1.5).

In most samples the conclusions concerning the relative volume of heartwood reflect those concerning sapwood, although there are some contradictions (Chapter 2.5.2.3). The first appearance of heartwood is usually at breast height (Chapter 2.5.2.2), a region where maximum heartwood area is found throughout the life of most trees. The age at which heartwood formation is initiated in the tree varies widely and is largely characteristic of the species (Chapter 2.2.3.2) and significantly genetically controlled. However, it can be influenced by the rate of growth of the young tree and environmental conditions (Chapter 2.5.2.3). Insect attack, pruning, and cambial damage can increase heartwood formation in some species.

7.2.2.2 Other Types

Trees belonging to a range of genera possess an inner zone of wood which does not contain living cells or reserve materials, and has a variable appearance. In a number of these situations the sapwood gradually transforms with loss of moisture into light-colored wood (sometimes called ripewood), the central portion of which is prone to discoloration. Various causes of discoloration have been postulated, such as the entry of air from branch stubs, wounds in branches or roots (e.g., Paclt 1953b), or by severe cold. These common but external factors result in a variety of reactions, with the darker zone having a series of concentric, but unsymmetrical, dark-colored peripheries sometimes advancing radially over 20 to 30 growth rings (as in black heart in some species, frost heart, crescent rings, or false heartwood). Pathological heartwoods or those regions containing fungal hyphae are likely to be of different origin or due to later changes. The response of cells in ripewood regions to stimulation is different from that of those in the inner sapwood of trees forming regular heartwood. The extractives formed do not penetrate the cell wall to the same extent, and they are apparently more highly polymerized. The large amount of extractives found in lesions in outer sapwood could be due to parenchyma possessing a high vitality and being able to respond at a greater rate to the particular stimulus of the invading organism.

7.2.3 Conclusions

The regularity of initiation of heartwood formation and of the number of growth rings of sapwood in some species indicates a genetic control of the transformation (Chapter 2.5.2.3) and a significant degree of heritability (Chapter 4.1.1). In addition to the age of the parenchyma cells as a controlling factor, a major influence on the life of the cell is the distance from the cambium which results from the need to maintain a balance between crown size, transpirational and storage capacities and diameter of tree stem (Chapter 4.1). The complexity of the situation is further illustrated by the influence of the initial growth rate on the formation of heartwood, at least in some conifers. The variation of appearance and properties of other types of heartwood, and indeed their absence in some species, indicates that specific initiating procedures are involved, and further work is required to define these procedures.

7.3 Features of Heartwood and Woundwood

A major function of heartwood could be a passive one in that its formation enables the maintenance of an optimum volume of sapwood necessary for the efficient development of the tree. Proposals concerning the economy of the tree in the storage and use of water, food reserves (starch and fats), and inorganic constituents have been considered by Rudman (1966b) and Bamber (1976b). Rudman (1966b) has suggested that trees efficient in the use of food reserves do not form heartwood. A passive formation of heartwood could result from accumulation of gas, which results in the blocking of cells from the transpiration stream by aspiration of pits and blockage of vessels with tyloses.

7.3.1 Some Theories of Heartwood Formation

7.3.1.1 Natural Causes

Observance of the formation of regular heartwood in many species after a certain number of years has led to a common view that heartwood formation is simply an aging process (Zycha 1948, Trendelenburg and Mayer-Wegelin 1955). The descriptions in Chapters 2.2 and 2.5 show that the age at which transformation of sapwood to heartwood occurs can be modified. Significant disturbances of hormonal balance, as in the formation of reaction wood, can override the aging of sapwood parenchyma as the basic cause of heartwood formation (Fig. 6.3), if indeed aging is the major parameter.

7.3.1.2 Accumulation of Gas and Control of Water Content

The nature of the water balance between inner and outer sapwoods is an imperfectly understood situation. The differences in moisture content of the sapwood and heartwood regions led Hartig (1888, see Zycha 1948) to propose that heartwood was formed in seedlings beyond a certain age as a result of entry from branch stubs of air of which oxygen was the most important component. Priestley (1932) postulated that heartwood formed because of an accumulation of air in closed vessel systems, with consequent effects on the permanent water content of the wood, which in turn brought about secondary changes in the living parenchyma cells. Zycha (1948) believed that the water requirements of the crown reduced the inner sapwood moisture content, which caused air to be drawn in through dead branch stubs or wounds, thus "triggering" heartwood formation. By drawing air into living segments of *Fagus sylvatica* he was able to produce material resembling heartwood. This view of the involvement of water is consistent with the relatively constant thickness of sapwood within a species and sometimes a genus. It offers an explanation for the existence of a dry wood or transition zone adjacent to the heartwood and for the dry heartwoods of conifers. It does not explain why the life span of ray cells in sapwoods of various trees of the one genus, and even in the same species, should vary so greatly. In addition, the theory implies that all heartwoods are drier than the sapwoods, and this is incorrect for a number of genera.

Frequently the nonaqueous portion of cell lumens has been designated as air but, as recent analyses have shown the major component is carbon dioxide (Chapter 5.1). A gas with a high proportion of carbon dioxide would not facilitate some of the numerous changes that have been postulated to take place.

7.3.1.3 Initiation by Fungi and Hormones

After Münch (1910, quoted by Zycha 1948) succeeded in artificially infecting *Fagus sylvatica* and producing heartwood, there has been an acceptance of the view that fungi initiate heartwood formation in a number of trees. The stimulus was considered to originate from the hyphae of fungi which originally entered from branch stubs or root wounds. Various workers (see Hillis 1962b) have subsequently drawn attention to the weaknesses in the theory. The dark-colored woods containing fungi appear to be the result of secondary changes. Although crystals have been seen in association with fungal hyphae, it is probable that the supersaturated solutions of extractives in heartwood, or other regions, have been nucleated by hyphae after their formation, causing the crystals to appear (e.g., Chattaway 1953).

Translocated toxins from the roots (Raunecker 1956) or hormones from the cambial region (Bamber 1976b) have been proposed as the initiators of heartwood formation. Whatever the cause, it is evident that those factors influencing cellular activity at the heartwood periphery are of key importance in the formation of heartwood.

7.3.2 Anatomical Changes

Aspirated pits in conifers or tyloses in angiosperms are two anatomical features evident between those zones showing differences in water and gas content. Their formation often takes place at the heartwood periphery preceding that of extractives.

Aspiration by membranes of bordered pits in conifers results in an effective control of the water economy of the tree. The tori of the bordered pits of the axial tracheids may become encrusted with phenolic compounds on heartwood formation (Chapter 6.1.3). In some *Pinus* species lignification of the ray parenchyma occurs in the inner sapwood. This lignification can take place without direct contact with a protoplast.

7.3.3 Occurrence of Extractives

If the function of heartwood is to maintain the volume of sapwood in balance with the needs to supply the necessary amounts of water for the growth of the tree, and also to provide a reservoir in equilibrium with the capacity to supply essential metabolic materials, then this function could be controlled in many cases by the barriers provided by aspirated pits or by tyloses. However, a characteristic feature of heartwood, and various other types of colored woods, is the presence of extractives, and sometimes in considerable amounts. Their presence to maintain sapwood in optimum volume by physical separation from other regions is unnecessary in many cases, as aspirated pits and tyloses fill that role. In some cases, some components of the extractives can provide a chemical barrier to penetration by potentially damaging microorganisms. In other cases the presence of extractives serves no known function.

7.4 The Transition Zone and its Formation

The presence of the narrow and sharply defined transition (or white) zone is being recognized in an increasing number of species (Chapter 2.4) surrounding not only heartwood but also some lesions. In view of its varying width throughout the year (Chapter 6.2.1), and its sometimes faint and evanescent appearance, it is probably present in a wider range of species. The sharply defined boundaries of the transition zone distinguish it from the diffuse and wide "intermediate" wood noticed in some species, the function of which in heartwood formation is uncertain.

Part at least of the relatively white or colorless appearance of the transition zone is due to a moisture content that is lower than that of the adjacent sapwood and sometimes the heartwood (Chapter 6.2.3). The sharp boundaries have been associated with a sudden, and sometimes abrupt, aspiration of pits at the outer edge of the transition zone in conifers (and tyloses in some hardwoods). The consequent marked decrease in permeability of these tissues results in their separation from the water stored in the inner sapwood (Fig. 2.4). These changes at the outer periphery of the transition zone have been attributed to water tension in the inner sapwood at times of physiological drought in the rest of the tree so that gas accumulates in the inner layers, the pits aspirate, and tyloses form. This mechanism does not explain the relatively sudden aspiration of pits over a few rows of cells, or why the width of the transition zone does not vary with drought intensity, the contradictory reports of the effect of drought on heartwood formation, or the fact that in some cases the zone enlarges in the dormant season.

Initiation of the formation of the transition zone in those species with regular heartwood formation is more likely to be caused by activation of the parenchyma in the inner sapwood than by water stress. As the parenchyma (and the resin canal epithelium when present) apparently remain alive, the loss of water would take place from the surrounding tissues of the transition zone. Impermeability of the zone is not essential, as is shown by the presence occasionally of permeable parts of heartwood in *Pinus radiata* and *Pseudotsuga menziesii*.

Although starch has been detected in the transition zones of some species, the translocation of a considerable amount of primary metabolites from the sapwood to the transition zone is required to form the high levels of extractives found in some heartwoods. This could take place through the living ray parenchyma over a period of time, and the moisture required by those heartwoods wetter than the transition zones could pass through the same route.

Ethylene, the important initiator or activator of a number of enzymes systems, is liberated in increased amounts by some transition zones, although the precise regions from which it emanates is unknown. It is becoming evident that different processes take place in different layers of the zone with the commencement of polyphenol formation occurring in the inner layers.

7.5 Function of Extractives and Exudates

Extractives are formed in significant and different amounts in different tissues of the tree and the amounts of those present in comparable tissues of trees from various genera can vary widely. Whereas the contribution of the aspiration of pits, or the formation of tyloses, to the efficient operation of the living tree can be appreciated, the contribution of the extractives is less clear, although in some cases extractives supplement or provide this blockage of tissues from the translocation stream. Extractives in all forms of heartwood and woundwood can provide some form of resistance of the living tree to destructive agencies although that resistance in the native habitat varies widely. The amount of extractives present in a tissue bears little relationship to its need for protection. The extractives content of the bark of *Eucalyptus astringens* can be as high as 70 percent whereas that of the heartwood of the same species is about 7 percent (Hillis 1962b) (Fig. 7.1). Furthermore, the total amount in various types of heartwood can vary greatly between species (Chapter 5.7), with no apparent cause for this variation. In addition, the location of extractives can be of major importance. The extractives in the cell wall have a greater effect on most properties of the wood in the living tree or in its utilization than the extractives in the lumen of the cell.

The constant composition of extractives over a wide range of growth conditions of the tree indicates that they must have some specific function in the tree rather than being the result of an erratic disturbance of the metabolism of living cells. Even under abnormal conditions, such as the formation of the contents of

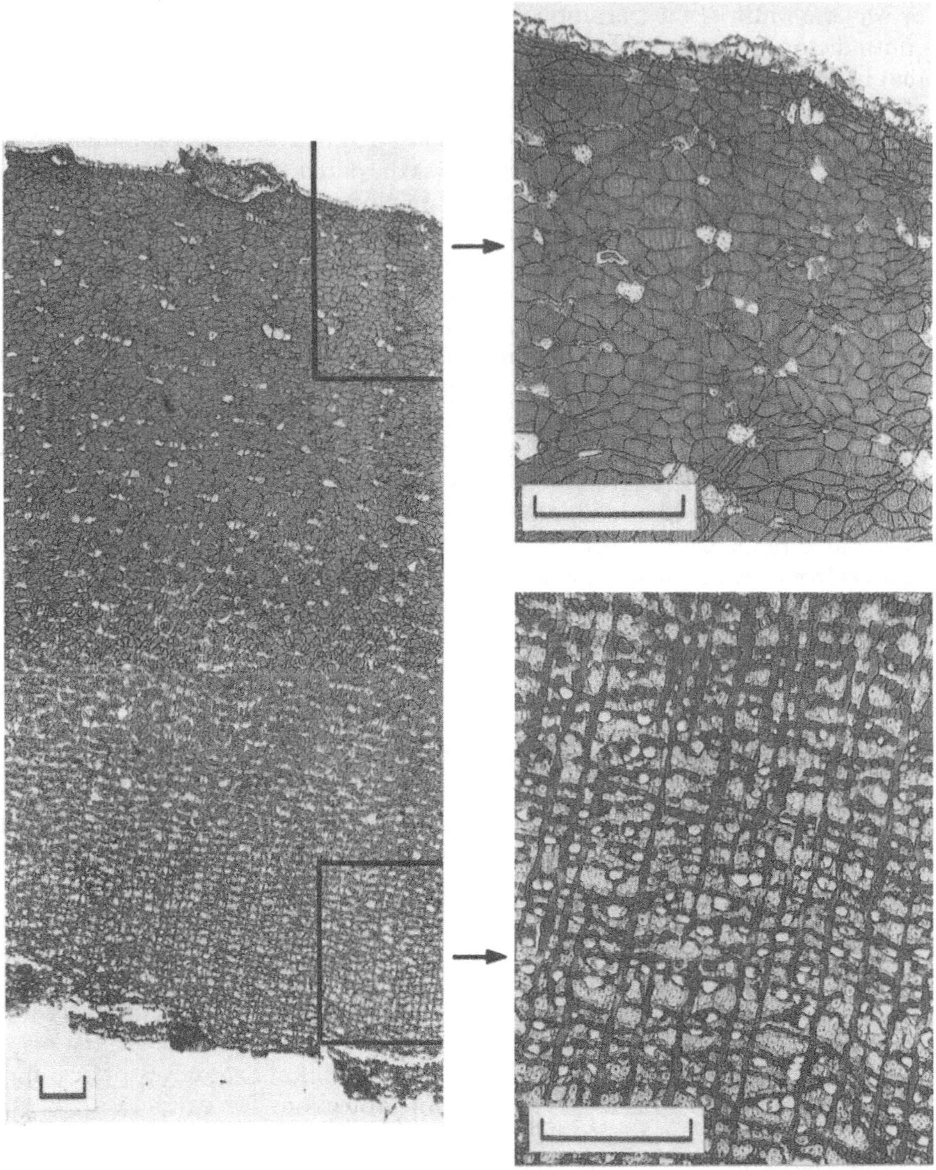

Fig. 7.1. Photomicrograph of a cross-section of the phloem of *Eucalyptus astringens* (scale 500 μm). The cambium is at the bottom and the polyphenols appear black. (Hillis 1960)

shakes, the composition remains the same. The amount formed, however, can vary considerably, particularly with the amount of exudates. The large amounts formed are much greater than any requirement for tree protection, for example the 45 liters of polyphenolic kino from *Eucalyptus calophylla* (Phillips 1931) and the 68 liters of pocket resin from *Pseudotsuga menziesii* (Erdtman et al. 1968).

The copious production of resins by prehistoric and ancient forests (and the availability of amber) may be an evolutionary trend to a more efficient tree physiology (Langenheim 1969). Alternatively, those forests may have contained more large-sized moribund trees, which today would have been harvested. Some investigators have considered the terpenoids in resins to be the end-product in a copiously productive metabolism and a way to dispose of excess acetate. Such a view of the formation of kino would require also a large availability of components of the shikimic acid pathway, and this appears unlikely. Recent studies indicate more active functions for terpenoid extractives and exudates and an important role in the ecological interactions of forest trees and their parasites (Langenheim 1969, Langenheim et al. 1982). The presence of polyphenols, which are far more widespread in heartwoods, possibly has a similar role.

7.6 Formation of Exudates and Extractives

7.6.1 Differences in Composition

Most extractives and exudates contain three major groups of compounds, namely, those derived from pyruvate, from shikimic acid, or from both these compounds (Fig. 7.2). One of these groups of compounds predominate in certain genera, in particular zones of those genera, and even in particular tissues.

Pyruvate from the glycolytic pathway is involved in the tricarboxylic acid cycle, where it is converted to acetyl-CoA, two units of which form mevalonic acid and in turn isopentenylpyrophosphate which is then polymerized to form a variety of terpenoids. For example, it goes through the isoprenoid biosynthetic pathway via geranyl (C_{10}) – or farnesyl (C_{15}) – pyrophosphate to be cyclized into respectively monoterpenes or sesquiterpenes, or with the latter compound, into squalene (C_{30}). Squalene can be folded to make C_{30} terpenoid alcohols (Croteau and Johnson 1985).

Considerable evidence supports the view that gallic acid –3,4,5-trihydroxybenzoic acid – arises from the shikimic acid pathway. In turn, gallic acid forms the dimer hexahydroxydiphenic acid, and both acids esterify with glucose (and other core molecules) to form various gallotannins and ellagitannins (Haslam 1982a, Gross 1985, Hillis 1985b).

The shikimic acid pathway also leads to the phenylpropanoid (C_6 C_3) compounds of which L-phenylalanine is a key compound. A starting point of the biosynthetic routes is the deamination of phenylalanine to cinnamic acid which is further modified by subsequent hydroxylation and methylation. These compounds can esterify with quinic and other acids to form chlorogenic and related acids. Two phenylpropanoid compounds can combine and be modified to form the class of lignans. Cinnamic acid derivatives can be converted to their CoA esters, and then reduced to the corresponding alcohols via the respective hydroxy cinnamaldehydes. [Dehydrogenative polymerization of the alcohols results in the formation of lignin (Higuchi 1985)].

Except where it is incorporated into the pit membranes and cell wall of the inner sapwood of some conifers (Chapter 6.1.5.2), lignin is a structural compo-

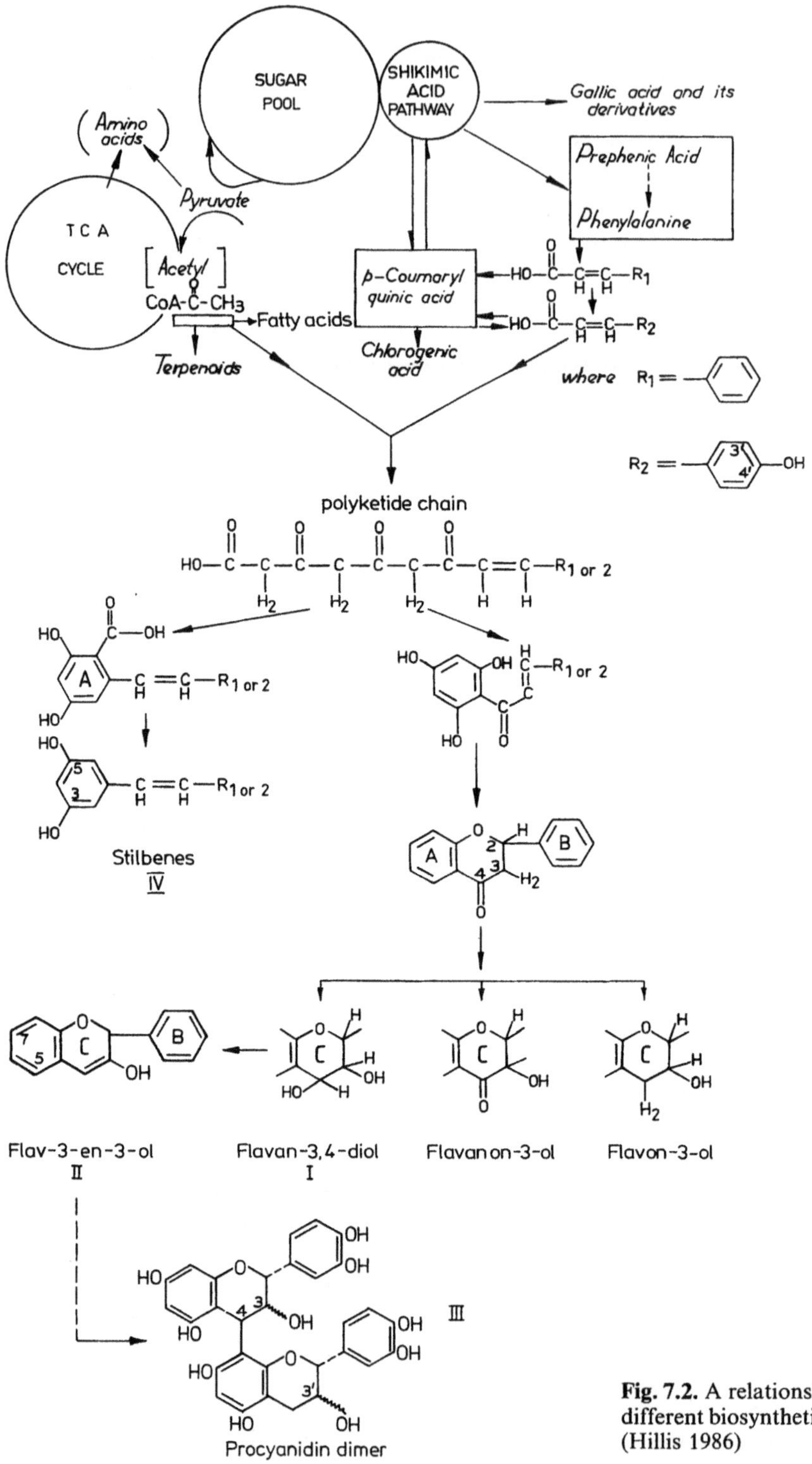

Fig. 7.2. A relationship between different biosynthetic pathways. (Hillis 1986)

nent of the fiber or tracheid cell wall. It is noteworthy that in the cambial region, lignin is the last and only aromatic constituent to be incorporated in the cell wall and that it is formed only via the shikimic acid pathway.

Phenylalanine ammonia lyase catalyzes the elimination of ammonia to yield cinnamic acid in the biosyntheses mentioned above and is in turn hydroxylated to coumaric acid by cinnamate 4-hydroxylase (Fig. 7.2). Although these enzymes have been found in many plant tissues, the involvement of phenylalanine or its ammonia lyase has not been confirmed at the heartwood periphery of *Pinus radiata* and *Eucalyptus polyanthemos* or in eucalypt leaves (Hillis and Ishikura 1970). Phenylalanine and its ammonia lyase may not be obligatory in the biosynthesis of polyphenols in some situations.

Acetyl-CoA carboxylase, regarded as an enzyme of fatty acid biosynthesis (Ebel and Hahlbrock 1982), also catalyzes the carboxylation of acetyl-CoA to malonyl-CoA by an ATP-dependent reaction. Three units condense head-to-tail into a polyketide chain which in turn condenses with the ester group of phenylpropanoid unit derived from the shikimic acid pathway (Fig. 7.2). Stilbene synthase and chalcone synthase fold the intermediary polyketide chain topologically in different ways in different mechanisms of cyclization. Stilbenes result on the one hand and chalcones on the other, which in turn form all classes of flavonoids containing the 6-membered heterocyclic C ring. Thus acetate units result in the A ring, and the phenyl propanoid unit from the shikimic acid pathway results in the B ring for both stilbenes and flavonoids (Grisebach 1985, Kindl 1985).

7.6.2 Site of Formation

7.6.2.1 Exudates

Those exudates, which appear immediately a cellular tissue is ruptured, are invariably acetate-engendered compounds such as the resins from the resin canal systems in pine sapwoods or the latexes in the phloem of rubber trees. In addition to appearing on freshly cut surfaces, this type of exudate can appear in cavities that occur in wood when a canal system is broken in the living tree by wind or some other form of stress.

In addition to these exudates formed by normal metabolism there are exudates resulting from a disturbed metabolism, particularly of the cambium. The composition of these exudates differs from that of the extractives formed under normal conditions in adjacent tissues. Significantly, whereas the first group of exudates are acetate-engendered, these involve as well precursors from the shikimic acid pathway (as with kino) or have a carbohydrate composition (as with gum or manna). It is interesting that the injured cambium of those species with an acetate-oriented metabolism can also produce lignans originating from the shikimic acid pathway (Chapter 5.9.1). Most data indicate that polyphenols in exudates are formed as a result of disturbed metabolism. Kino, for example, results from the changes accompanying the complete cessation of normal wood formation and the complete reorganization of the tissues caused by the cambium differentiating into parenchyma of various types.

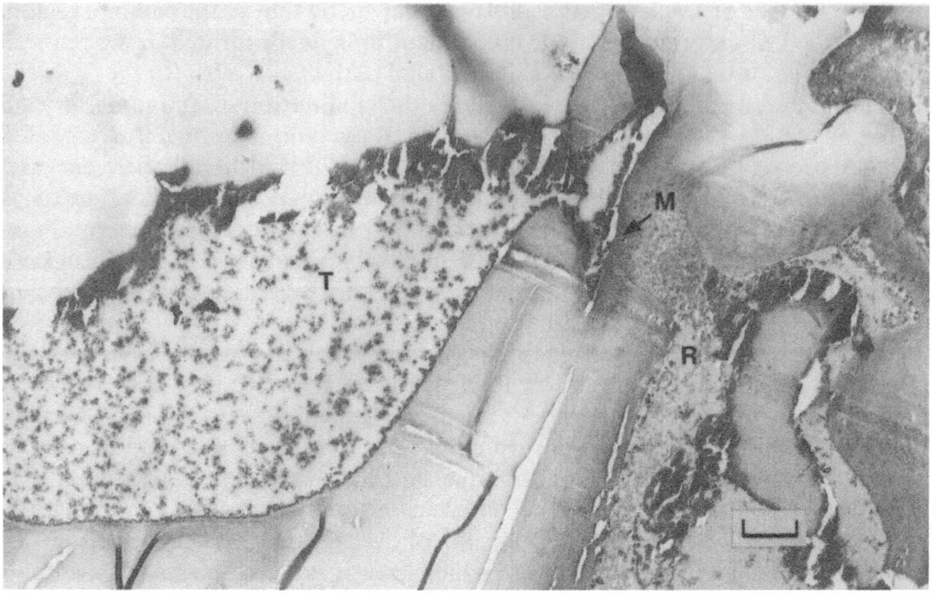

Fig. 7.3. Transverse section through a floccosoid zone in *Tsuga heterophylla*, illustrating occlusions in a tracheid lumen (*T*) and a ray cell (*R*). Dissimilar deposits appear to occlude the pit cavities on either side of the membrane (*M*) (TEM, scale 1 μm). (Krahmer et al. 1970)

The composition of the exudate resulting from a change in metabolism after injury is consistent, and formation takes place over a few layers of cells. Deposits in frost checks have shown that some adjacent parenchyma can behave in a very selective and specific manner and can produce components different from those in normal extractives. Aromatic compounds are absent from the lumens of rays close to shakes in *Tsuga heterophylla*, and the polyphenols or their precursors are formed in the ray cytoplasm adjacent to a pit and then pass through the pit membrane into the tracheids (Krahmer et al. 1970) (Fig. 7.3). This mode of formation differs from that of the terpenoid podocarpic acid, which is formed in the rays of *Dacrydium nausorensis* adjacent to shakes in the heartwood region, and then passes into the lumen and walls of tracheids and the cavity of the shakes where it crystallizes (Hillis et al. 1976). Clearly these components are formed in situ and are not translocated from elsewhere.

7.6.2.2 Extractives

7.6.2.2.1 Translocation Theories

Before the characteristics of the extractives in the different woody tissues of a tree were known in detail, various workers postulated that the extractives in the heartwood accumulated by translocation from other regions. For example, it has been suggested that the precursors of heartwood extractives are formed in the cambium and then diffuse radially to the heartwood periphery (Erdtman and Rennerfelt 1944). A feature which influenced the development of translocation theories was the occurrence of included sapwood on the interior side of a previously killed cambium (Chapter 2.2.1.2). This sapwood retained its storage materials, which were not transformed to extractives.

On the assumption that living processes in plant cells produce byproducts harmful or superfluous to the protoplasts, various proposals have been made of the manner in which they may be isolated within the living plant. These byproducts accumulate in vacuoles in living cells, and in some cases these vacuoles expand to a large size (Fig. 6.9). Suggestions have been made that these "waste" byproducts are translocated, with or without "detoxication", as excreta to regions such as the heartwood so that living tissues are unaffected. When the concentration of "toxic" substances accumulated to lethal levels the innermost living ray cells would become nonfunctional. In addition, the metabolic byproducts could be excreted through the plasmalemma into the nonliving cell wall regions (Stewart 1960, 1966 a, b).

These proposals do not account for a number of characteristics observed with many heartwoods. These include the sudden increase in the amount of extractives sometimes over two or three rows of cells at the periphery. The components of the extractives can have markedly different chemical compositions from those found elsewhere in the tree. Also the composition of heartwood extractives is under strong genetic control which probably evolved under selective pressure over a long period of time. Such a situation does not conform with the view that heartwood extractives are excretory or waste substances from other tissues growing under different environmental conditions. Also waste products are usually less complex products and of variable composition, whereas extractives can be of considerable complexity, requiring significant energy for their formation.

7.6.2.2.2 In situ Formation

Chattaway (1952) considered that death of the living parenchmya cells is preceded by a period of activity that is greater than normal for the species. She suggested that it is this increased rate of metabolism that results in the formation of extractives, which ceases on the death of the cell.

Increasing evidence is accumulating to support the view that extractives, and particularly polyphenols, are formed in situ and are not mobile beyond the intact cell. The similar composition of the extractives in the heartwood of *Pseudotsuga menziesii* and that in the heartwood formed in the living stump after the stem had been removed (Hemingway and Hillis 1970) shows that it is most unlikely that the extractives originated in the foliage. Not only can the extractives in the heartwood differ from those in the sapwood but also they can differ from those in the bark and from the cambial tissues (Chapter 5.6.4.7). Extractives in injured tissues and knot wood can have a further difference in composition (Chapter 6.4). The tissues of mistletoes contain extractives different from those of the adjacent tissues of the eucalypts on which they grow parasitically. Parenchyma resin in conifer can differ markedly in composition from nearby canal resin (Chapter 6.5.2). The composition of polyphenols in cells in close proximity can be significantly different (Chapters 5.7.5 and 6.2.8). The addition of the same labeled precursor to different tissues can result in different extractives (Chapter 6.2.8).

All the above facts agree with the view that extractives are formed in situ from translocated primary metabolites. This requires the presence of cells of sufficient activity and a supply of energy. In view of the small volume of living parenchyma cells at the heartwood periphery they must be capable of considerable activity to synthesize the large amounts of extractives found in some heartwoods. An understanding of the physiology of the transition zone or inner sapwood is fundamental to the future understanding of heartwood formation.

7.6.3 Amounts

While significant amounts of extractives can enter the cell wall of fibers or tracheids (Chapter 6.1.4), these amounts are generally unknown. The contents of cell lumens vary considerably in amount and type and the proportion of them that enter cell walls varies between species forming regular heartwoods. The amounts entering the cell wall of species forming irregular heartwood or quasi-heartwood appear to be lower.

The extractives content (excluding storage materials) of sapwoods is invariably low in quantity. The amounts in heartwoods and bark, on the other hand, can reach high levels (Chapter 5.7). As extractives are formed in situ, the amounts formed must depend on the number and activity of the living parenchyma at those locations, the amount of primary metabolites available at or reaching those sites, and the volumes of the lumens available to store the extractives.

Nelson (1975) found with *Juglans nigra* and *Quercus rubra* that there was no causal relationship between the relative volumes of parenchyma and the quantity of polyphenolic heartwood extractives (Chapter 6.1.5.1). Apparently the rate and duration of aromatic biosynthesis during the formation of heartwood are affected much more by the characteristics of the parenchyma and the physiological conditions near the sapwood-heartwood boundary than by the amounts of parenchyma participating in the biosynthesis. In a study of living stumps of *Pseudotsuga menziesii*, it was concluded that the factors controlling the amount and composition of heartwood polyphenols formed are incorporated into the ray parenchyma during the early stages of their development (Hemingway and Hillis 1970). Further work is required to determine which are the parenchyma cells initiating or controlling biosynthesis of extractives and which cells undertake particular functions.

The size and physiological activity of the sink of primary metabolites in the sapwood could be a controlling factor in the amount of extractives formed at the heartwood periphery. The amount of starch stored in the sapwood of many species is inadequate for extractives formation, for example in *Angophora costata* (Hillis et al. 1962). Also the starch content of the narrow sapwood of *Eucalyptus marginata* rarely exceeds 3 percent, yet the extractives in the outer heartwood of old trees of this species can reach ten times that amount. Most of the substrate for extractives formation must be translocated to the heartwood periphery at the time of formation of the transition zone or heartwood. Some of this substrate will arise from starch hydrolyzed in the middle to inner sapwood of some species; there are reports that starch exists in the transition zone of a few species.

Little information exists concerning the amount of primary metabolites produced by trees at different ages. It can be assumed that after a tree has reached the mature stage the volume of the crown and the amount of photosynthate it produces remains the same. Under these conditions the volume of the wood laid down annually in a mature tree would be the same in average seasons, and consequently the width of a growth ring would steadily decrease with the increasing diameter of the tree, and this is in agreement with observations. Also, the increasing height and spread of the crown would require a greater distance (and energy requirements) for the photosynthates to be translocated to the cambium at the base of the tree. If a constant proportion of the tree's photosynthate is used annually for cell wall formation, then the amount of the remainder, and in turn the heartwood extractives, would be relatively constant. However, this disagrees with frequent observations.

The amount of extractives in heartwood increases with the distance from the pith or with the age of the tree (Hillis 1962b, Posey and Robinson 1969), and age influences the amount of extractives present in *Pinus echinata* more than any other factor. It is uncertain whether the rate of consumption of photosynthate in the cambial region alters, particularly with fast-growing trees, to affect in turn the amount of extractives formed in the heartwood. The increased requirements for the formation of the inner gelatinous layer in tension wood fibers during cambial growth is associated with the smaller amounts of storage carbohydrate in the sapwood (Hillis et al. 1962). The massive S_1 and S_2 layers of the cell wall in compression wood tracheids require increased demands of photosynthate and result in smaller storage volume. As expected, the compression wood of different *Larix* species (Côté and Timell 1967) and *Sequoia sempervirens* (Sherrard and Kurth 1933) contained lesser amounts of extractives than normal wood. Further study is required to determine whether inadequate supplies of photosynthate are responsible for the formation of target ring (Chapter 2.5.2.5.1) in some species. Certainly the formation of extractives in the heartwood of living stumps of *Pseudotsuga menziesii* was not affected by the limited availability of photosynthate (Hemingway and Hillis 1970).

Although the volume available for storage must affect the amount of extractives in a tissue (such as more extractives in earlywood than in latewood), it is overweighed by other factors. The large amounts of polyphenols found in the bark of *Eucalyptus astringens* (Fig. 7.1) are possible because they are contained within the thin walls of apparently expanding cells. Other tissues have little opportunity of enlarging their lumen volumes or altering the porosity of the cell walls. Attempts have been made to study the effects of different lumen volumes on extractives content by comparing sectors with or without reaction wood zones. No close relationship could be found (Hillis 1960) between the amount of extractives and differing lumen volumes in alternating bands of normal and tension wood of an *Acacia* species with different lumen volumes. As extractives are formed in parenchyma, it is interesting that tension wood of *Angophora costata* with a lower volume of axial parenchyma than normal wood in the same cross-section also had a lower polyphenol content (Hillis et al. 1962). The total alcohol-

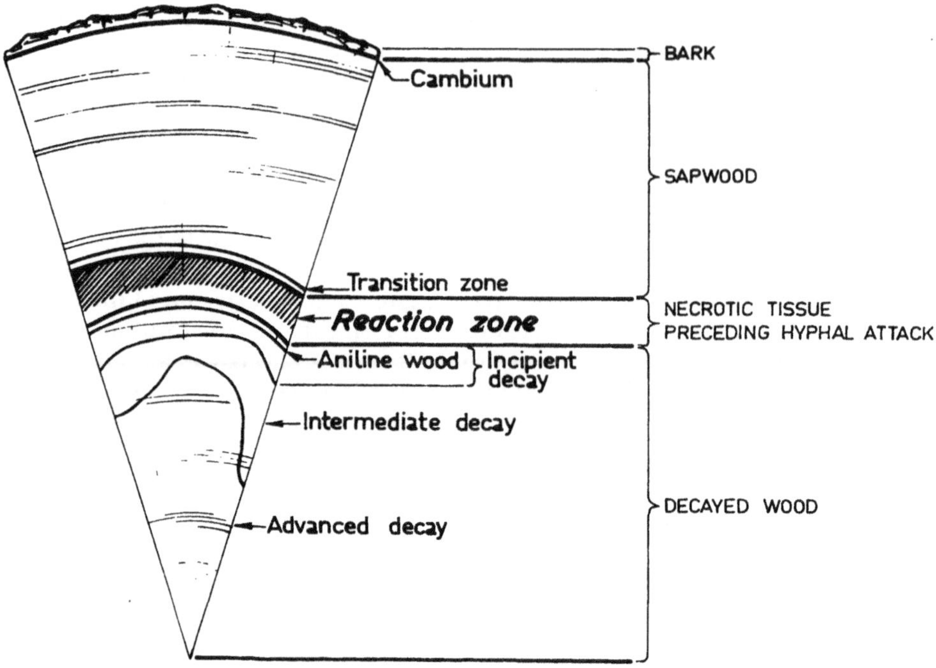

Fig. 7.4. Sketch of zones typically encountered in *Picea abies* attacked by *Fomes annosus*. (Shain 1971)

soluble materials in *Pseudotsuga menziesii* showed a positive relationship with the volume of ray parenchyma (Hemingway and Hillis 1970).

Lesions in the outer and middle sapwood attacked by fungi can have a higher extractives content than the heartwood. Even when a reaction zone resulting from *Fomes annosus* infection (Shain 1971, Chapter 2.5.2.4.4) (Fig. 7.4) and a heartwood periphery in the one cross-section are at about the same distance from the cambium, the reaction zone can have a higher extractives content. This indicates that the amount of extractives is controlled by the activity of the cells in that region. If this is so, then greater amounts of carbohydrate need to be translocated to this region to supply the needs for forming extractives. Consequently the translocation rate will need to be greater for the reaction zone than for the heartwood. Thus some message must be sent to the cambium, or to the leaves, in order to have the greater amount of carbohydrate transported inwards to the reaction zone. Distinctive extractives are then formed in the zone.

No attention has yet been given to the relative amounts of extractives in heartwood and in bark, whether they may change, or how they may be changed.

7.6.4 Type of Extractives in Tissues

There are marked differences between the composition of the resinous extractives in the heartwood and sapwood of those species which contain them. The resin acid content of the heartwood of *Pinus* species is much higher than that of the sapwood, whereas the fatty acid content is correspondingly lower. Heartwood resinous extractives are much closer in composition to canal resins than to sapwood resins.

When heartwood is formed, the amount of polyphenols increases. The small amounts of polyphenols in the sapwood of a number of species involve only the shikimic acid pathway (for example as in *Pinus radiata*). The amount of compounds from this pathway increases at the outer periphery of the transition zone of *Eucalyptus polyanthemos*, but at its boundary with the heartwood the amount of compounds involving both this pathway and the tricarboxylic acid cycle also increases considerably (Hillis and Yazaki 1973b). Polyphenols requiring both major pathways (Chapter 5.6.4) (Fig. 7.2) are probably involved in the heartwood of all species. Assessment of the relative involvement of the two pathways is difficult, as by far the major proportion of extractives is polymeric material of uncertain nature although in a number of cases it is based on proanthocyanidins.

There is evidence in some species that at the heartwood periphery, or in the transition zone, those compounds arising solely from the shikimic acid pathway are formed first (Chapters 5.6.4.1 and 5.6.4.2). An examination of the metabolic pattern of the ray parenchyma in the transition zone is needed to determine the factors which induce the formation of the A ring of both flavonoids and stilbenes as well as the B ring (Chapters 5.6.4.3 and 5.6.4.4). In addition to the different classes of compound found in the heartwood, there can be different substitution patterns, such as the phloroglucinol-pyrogallol pattern of the sapwood becoming the resorcinol-catechol pattern of the heartwood of *Calophospermum mopane*. Furthermore, a marked increase in the amount of proanthocyanidins (with this hydroxyl pattern) was found as the heartwood of this species moved steadily outwards. At the same time a similar decrease occurred with two closely related compounds possessing a methylene bridge and which could have arisen from the same common precursor (Drewes and Roux 1966).

It is evident that the metabolism of the inner sapwood is intensified during some time of the year so that much greater amounts of extractives are formed. In addition, the course of that metabolism is altered so that the extractives found in regular heartwood are different from those of the sapwood. The metabolism of the parenchyma at the periphery of lesions appears to be similarly altered.

7.7 Initiation of Formation of Heartwood, Extractives, and Exudates

Death of the cambium or parenchyma cells does not per se result in the formation of extractives. When living trees of *Populus tremuloides* were subjected to high radiofrequency impulses applied through the bark, the cambium at least was killed, but discoloration of the xylem tissues did not occur until they were exposed to air (Sucoff et al. 1967).

The most common proposals for the initiation of the formation of heartwood have been the loss of water from the tissue and entry of air or infection of the tissues with fungi. Although fungal growth has been associated with discolored wood and woundwood, it is an unlikely initiator of the formation of regular heartwood (Section 7.3.1.3).

7.7.1 Initiation by Displacement of Water

The withdrawal of water from the inner sapwood (often abruptly over one or two growth rings) of conifers (and some angiosperms) precedes the formation of polyphenols in heartwood, woundwood, and lesions. Drywood zones also precede the formation of resin in the sapwood of *Pinus* species after injection with paraquat. Jorgensen (1961) observed that one year and more after branches of *P. resinosa* had been killed by suppression, the stubs within the stem had "shadows", below and for a much greater distance above the stub, of significant amounts of the polyphenols, pinosylvin, and its monomethyl ether. Similar patterns were observed with wounds made by insects or boring instruments, provided that the cells did not die rapidly, desiccation of the wood did not proceed too far, and the temperatures were sufficiently high (with an optimum of 23 °C) to maintain cellular activity (Jorgensen 1961). Neither polyphenols nor dry zones were formed in pines after injury if the temperatures fell below 10 °C (Jorgensen 1961, Coutts 1977) (see also kino formation, Chapter 6.5.4). Controlled desiccation was apparently provided by the resin coating which forms over wounds in a number of conifers (see also Lyr 1967).

Tyloses are also associated with heartwood formation in some species. Using their appearance as an indication of heartwood formation, Zycha (1948) found that the withdrawal of water (and its replacement by air) from a section of *Fagus sylvatica* resulted in the formation of tyloses, and coloration, after a 1-week period. Under normal conditions the water content of this species decreases steadily across the sapwood towards the pith. Zycha (1948) found that the boundary of the red heartwood was at the region where the water content had decreased to 60 percent, which on calculation is the amount at which all cells, except the parenchyma, are free from water. Paclt (1953b) came to a similar conclusion and suggested that aeration occurs during the winter via branch stubs or dead roots leading to the pith.

Coutts (1977) simulated living conditions by injecting poisons (at an unstated time) into the sapwood of *Pinus nigra* var. *maritima*, *P.sylvestris*, and *Abies grandis* to create dry zones by withdrawing water from adjacent living cells. The formation of dry zones can be attributed either to passage of gas through dead or dying ray parenchyma cells, or to some increasingly active processes in those cells. It was noteworthy that injection of 1 percent oxalic acid into *P.radiata* sapwood caused pronounced dehydration, and produced much more polyphenols than did sodium arsenite (Coutts 1976), which is a specific enzyme poison likely to assist formation of polyphenols. It was found (Coutts 1977) that although gas bubbles were present in tracheids throughout the dry zone, they were not present in the ray parenchyma cells in the recently formed portions of that zone. Consequently, it is unlikely that dying ray cells provide a pathway for gas movement between tracheids. Metabolic activity of ray parenchyma and the presence of oxygen were considered to be responsible for the formation of extensive dry zones (Coutts 1977).

As water is not withdrawn from rapidly killed regions, it is apparent that death of ray parenchyma cells does not in itself allow withdrawal of water for the formation of dry zones. As Coutts (1977) points out, the main difficulty in elucidating the mechanism of formation of dry zones is to show how gas can appear in the system so as to allow the release of water from the tracheids of conifers. In the case of infected wood, the force for moving the water appears to be hydrostatic tension (Coutts 1976). Opinions differ as to whether heartwood is most likely to form when hydrostatic tension is reduced (Harris 1954a, b) or increased (Rudman 1966b)in pines. During some times of the year reduced vitality of the ray parenchyma cells can precede heartwood formation in some species (Frey-Wyssling and Bosshard 1959, Nečesaný 1965, Higuchi et al. 1967a), and Stewart (1966a) considered that this reduced vitality renders the cells less able to retain water against tension in the transpiration stream. Even if this were so, there remains an explanation of the replacement of water by gas. Coutts (1977) suggested that cavitation occurs in each conifer tracheid individually, and that the activity of the gradually dying ray parenchyma cells brings about the cavitation. This gradual death was accompanied by the disappearance of starch, the formation of polyphenols in the tracheids and ray parenchyma, and the formation of substances with physical properties capable of causing gas embolism in water columns under tension. Increased polyphenol formation has also been associated with zones containing an abnormally large amount of water. The concentration of phenolic components in wetwood in young *Abies alba* was 3 to 5 times, and in older trees 10 to 20 times, higher than in the sapwood (Bauch et al. 1975). In *Tsuga heterophylla* wetwood a fivefold increase in extractives content in comparison with normal heartwood has been reported, with the extractives having the same composition in both tissues (Schröder and Kozlik 1972).

Carrodus (1971) suggested that the high concentration of carbon dioxide present in the heartwood gases of several species (Chapter 5.1) was conducive to the formation of heartwood polyphenols, but this has not been substantiated (Shain and Hillis 1973, Höll 1974). Certainly Jorgensen (1961) initiated the formation of polyphenols in wounds of *Pinus resinosa* sapwood by wrapping solid carbon dioxide (at −78 °C) around them, but their formation was probably due to other causes. It is known that freezing enhances the production of ethylene (W.C. Cooper et al. 1969).

7.7.2 Initiation by Changes in Ethylene Levels and in Hormonal Balance

Direct injury or stress of the cambium at particular times of the year can result in changes in the anatomy. In the case of the formation of kino veins in several *Eucalyptus* species, the cambial derivatives are first altered from normal parenchyma, vessels and fibers to callus-like, well-organized parenchyma cells frequently iso-diametric in shape and thin-walled. In other situations, cambium under stress produces reaction wood with thick cell walls and modified anatomy. Changes in uninjured cambium have been initiated by deep wounds in the bark of *Abies* species (Mullick 1977). Mechanical injury in the xylem of roots of *Abies balsamea* and *Tsuga canadensis* can result in the uninjured cambium forming resin ducts and increased amount of parenchyma (Tippett et al. 1982). Also the uninjured cambium in the stems of *Acer rubrum*, *A.saccharum*, and *Betula paprifera* form bands of thick-walled cells and wider growth rings when the adjacent xylem is injured (Shigo and Dudzik 1985).

Ethylene is produced in increased amounts by many plant tissues under wound or stress conditions and even by physical bending (Robitaille and Leopold 1974). Ethylene has been implicated in the formation of reaction woods (for example Nelson and Hillis 1978c, Yamanaka 1985). There is increasing evidence that ethylene is a major component of sensitively balanced interactions with auxin and other phytohormones involved in xylem formation. For example, the formation of xylem elements in lettuce pith explants cultured in the presence of auxin was enhanced when a compound releasing ethylene, or ethylene precursors, were added to the culture mediums (Miller and Roberts 1984, Miller et al. 1984).

At most times of the year injury of eucalypt cambium will lead only to traumatic tissues of cells with different wall thicknesses and sometimes the cells are filled with polyphenols. However, in the spring season, or when the cambium is fully active, and after the formation of anomalous parenchyma resulting from injury, a distinctive pathway is initiated for the copious synthesis of a class of polyphenols different from those formed in surrounding tissues. Increased amounts of ethylene are present at the same time, and ethylene-releasing compounds administered to the cambium of *Eucalyptus* or *Prunus* species at this time result in the formation of kino or gum, respectively. These compounds can also enhance the formation of rubber from *Hevea braziliensis* and resin from *Pinus* species. Ethylene is apparently an essential component of the systems controlling the formation of exudates.

Kino rarely forms in eucalypts younger than 4 or 5 years. It is noteworthy that application of an unnatural growth substance, naphthaleneacetic acid, to undamaged bark of juvenile eucalypt seedlings in the spring season can lead to vigorous exudation of kino (R.C. Foster and H. Dowden, 1971 personal communication).

Auxin stimulates ethylene production (Fuchs and Lieberman 1968), but this production may be increased in an auxin-independent manner when many plant tissues are stressed in various ways (such as moisture stress, for example McMichael et al. 1972). This traumatic or stress ethylene is a product of living cells and not simply the result of cell degradation (Abeles 1973, Beyer et al. 1984). Because it can be produced rapidly at the site of trauma, this ethylene, rather than that mediated by auxin, is probably the most likely to be involved in kino formation. Although the involvement of ethylene has been confirmed, a direct and essential role for this hormone in the induction of kino veins has not yet been proven. The formation of lesions ahead of the stimulus and of kino veins away from the immediate region under the influence of wounding may occur through the stepwise propagation of ethylene by autocatalysis. The ability of sapwood parenchyma or of cambial cells to change, over a layer of two or three cells, from one form of metabolism to another reflects the fine balance between the controlling influences of auxins and ethylene that is necessary to maintain the status quo. Living tissues have a very high affinity for ethylene and regardless of whether traumatic or exogenous (from ethrel) ethylene be produced, it can be rapidly sequestered and released at a slow rate (Bengochea et al. 1980) by some unknown control. Apparently the above-normal amounts of ethylene are removed by metabolic conversion to other compounds, which could play a further role in autocatalysis of ethylene production or in stimulating other metabolic pathways. Once these responses are set in motion, they seem to operate independently of ethylene levels and are possibly affected by the different sensitivities of cells to ethylene. Also Paradies et al. (1980) has concluded that ethylene is not an elicitor for the formation of phytoalexins but rather a consequence of elicitation and that it does not affect the amounts which accumulate.

The production of ethylene increases at the heartwood periphery, or in the transition zone, of some species when the cambium is dormant or relatively inactive (Chapter 5.5). This situation implies that ethylene production and changes in hormonal balance are initiated locally and lead to the formation of heartwood and extractives. Similarly, the largest amounts of ethylene are found at the margins of lesions in sapwood and in the dry zone surrounding the biologically damaged regions. The *Sirex* lesions in *Pinus radiata* clearly produce ethylene before polyphenol accumulation commences. Furthermore, when blocks of living sapwood of *P.radiata* were ventilated with air containing ethylene, the production of polyphenols was confined to the slightly desiccated outer 2–3 mm (Shain and Hillis 1973).

Although ethylene production accompanies many changes in wood, and the administration of ethylene produces the same changes, only a correlation can be shown at this stage and not a causal relationship. What is the role of ethylene? It remains to be determined whether ethylene is generated in the dying cells or in the cells adjacent to those dying, or whether ethylene formation is an intracellular response to an extracellular signal. Little is known of why some species appear to produce more ethylene than other species or if some produce ethylene more rapidly on injury than others. Also, is the ethylene-producing capacity of individual parenchyma cells a determining feature in the amount of extractives formed and why is it higher in one situation than in another? The very small volume of parenchyma in *Pinus radiata* is capable of producing substantial volumes of ethylene (Shain and Hillis 1972) and amounts of polyphenols. Other species with larger volumes of parenchyma do not form proportionally larger amounts of polyphenols. However, Nelson (1975) found with *Juglans nigra* and *Quercus rubra* that the rate and duration of aromatic biosynthesis during heartwood formation were affected much more by physiological conditions near the sapwood-heartwood boundary than by the amounts of parenchyma participating in the biosynthesis.

The physiological situation in the inner sapwood, or the transition zone, is likely to be more complex than the one depicted above. Although there is accumulating evidence that increased levels of ethylene are associated with the production of polyphenols and other extractives, the gaseous environment in the inner sapwood may hinder their synthesis. A high proportion of carbon dioxide, and little oxygen, has been found in the gases in this zone (Chapter 5.1). In apple tissues, high concentrations of carbon dioxide, particularly with low concentrations of oxygen, can inhibit the action of ethylene (Mapson and Hulme 1970) and inactivate respiratory enzymes (Chapter 5.1). However, in other tissues that inhibition does not take place (Abeles 1973).

The induction of kino vein formation provides a model system to facilitate the collection of data that will provide a greater understanding of the processes controlling the formation of heartwood polyphenolic extractives.

7.8 Factors Controlling the Nature of Extractives and Exudates

As mentioned above, there is accumulating evidence that increased levels of ethylene are associated with an enhanced formation of extractives and exudates in a number of trees. Ethylene is apparently a factor in the reorganization of cellular enzyme systems, causing the synthesis of polyphenols (and other classes of compounds) to become the dominant activity. There is a ready availability of substrate and an increased channeling of precursors for their formation. But in addition to the increased formation of extractives, a marked change in the nature of them often occurs. There is little evidence of the factors or elicitors responsible for these consistent differences, in the one tree, of the composition of extractives in different tissues grown under various conditions or formed as a result of injury (Chapter 7.6). Are there different pathways to different types of extractives or extraneous components, is there a hierarchy among these components, can the sequence of their formation be changed, and are some pathways blocked so as to lead to the involvement of others?

The deposition of silica, which has no relationship with other extraneous components, occurs abruptly as conglomerates in the mid-sapwood of *Syncarpia glomulifera*. This species contains varying amounts of silica according to the age and growth rate of the tree (Hillis and de Silva 1979) and it subsequently synthesizes large amounts of extractives at the heartwood periphery (Fig. 7.5). Extractives are formed after tyloses in hardwoods or after the aspiration of pits. There are indications of an order in the formation of polyphenols which may be controlled within the one cell or by the enhancement or suppression of the activities of specialized cells. This order of formation may be associated with the availability of energy at that time. Polyphenols involving only the shikimic acid pathway appear to be formed before those involving the tricarboxylic acid cycle. In situations where components of this cycle could be directed away from energy-demanding roles, such as cell-wall formation, flavonoids and terpenoids containing acetate-engendered components can be formed. The structures of extractives in different tissues, and of exudates such as kino, may provide valuable chemical counterparts for the morphological theory of Haeckel that ontogeny recapitulates phylogeny.

In addition to different classes of compounds being formed in different tissues, different types of substitution within those classes can exist (Section 7.6.4). Some differences may have significance in tree protection. Pinosylvin is the major polyphenol in the lesions formed in *Sirex*-affected sapwood of *Pinus radiata* (Hillis and Inoue 1968), whereas in heartwood the less fungi-toxic monomethyl ether (Rennerfelt and Nacht 1955, Hart and Shrimpton 1979) is the major stilbene in addition to nontoxic flavonoids.

The extractives in sapwood induced by different forms of damage can be chemically different from those in normal heartwood (Chapter 6.4.1). Con-

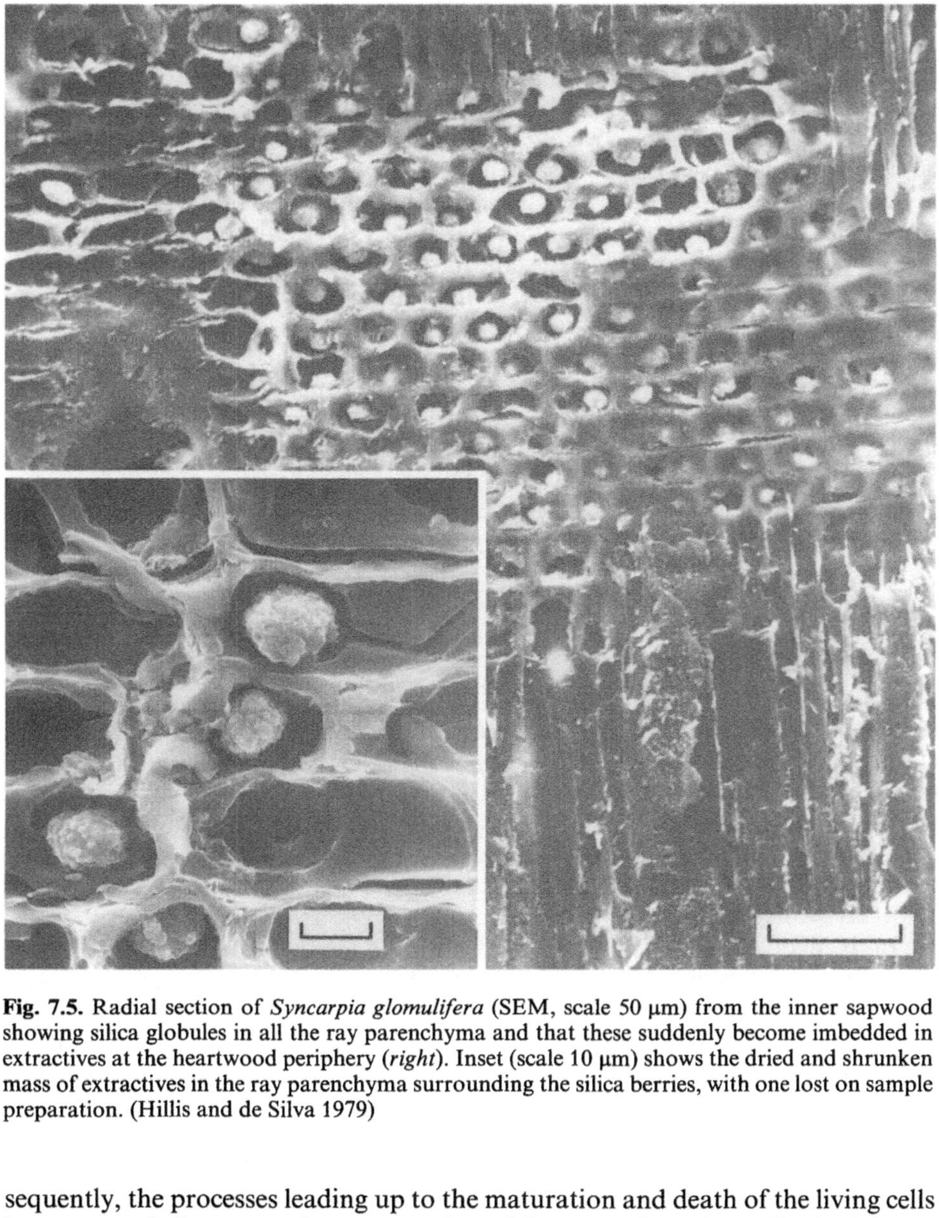

Fig. 7.5. Radial section of *Syncarpia glomulifera* (SEM, scale 50 μm) from the inner sapwood showing silica globules in all the ray parenchyma and that these suddenly become imbedded in extractives at the heartwood periphery (*right*). Inset (scale 10 μm) shows the dried and shrunken mass of extractives in the ray parenchyma surrounding the silica berries, with one lost on sample preparation. (Hillis and de Silva 1979)

sequently, the processes leading up to the maturation and death of the living cells have a great influence on the composition of the extractives. Accordingly, it is not only the constant composition of normal heartwood extractives that is under genetic control, but also the processes which initiate and elicit their formation.

Conclusions similar to those concerning the formation of heartwood extractives and of exudates are embodied in Müller and Börger's theory of phytoalexin formation [reviewed by Bailey and Mansfield (1982)]. This theory concerned the antifungal compounds formed by plants in response to fungal infection and which restricted the growth of the invading fungi in resistant plants. It was later found these compounds (extractives) were formed in response to other inducers or elicitors even including abiotic ones (Bailey 1982). The role of phytoalexins is associated with the accumulation of compounds often absent from healthy, living tissues (particularly of herbaceous plants). Isoflavonoids are prominent among phytoalexins and these compounds have so far not been found in uninjured sapwood but their role may be assumed by stilbenes and other compounds (Chapter 6.4.1). Tissues of the same species but infected by different fungi generally contain a similar range of compounds which can differ in their ratio of amounts to each other. Infection of different species by the same microorganisms, on the other hand, can lead to the formation of different compounds sometimes utilizing different biosynthetic pathways.

The distinction between antimicrobial phytoalexins, stress metabolites formed after some type of injury, and heartwood extractives is unclear. Much of the literature that is devoted to the perceived exclusiveness of phytoalexins and stress metabolites is directly relevant to the sudden formation of heartwood extractives and exudates – and vice versa. These studies share a number of aspects in common and consideration of their interrelationship would be mutually beneficial.

7.9 Activities at Cellular Levels

Increased amounts of a range of compounds are formed in heartwood, lesions, and exudates. The most commonly encountered classes of compounds are the polyphenols, terpenoids and fatty acid derivatives to which most attention is usually given. Increased amounts of other classes, such as nitrogen-containing compounds (Chapter 6.4.1), are occasionally reported in situations that are less well-documented than those concerning the above compounds.

Acetyl-CoA (malonyl-CoA) formed from pyruvic acid or fatty acids by β-oxidation is involved in the biosynthesis of both the terpenoids via mevalonic acid and as well the A ring of flavonoids and stilbenes (Fig. 7.2). Separately, glucose-6-phosphate is metabolized to the B ring of both stilbenes and flavonoids through the shikimic acid or the cinnamic acid pathways. It is widely believed that phenylalanine is an obligatory intermediate in the biosynthesis of phenylpropanoid compounds and the B ring. However, phenylalanine ammonia lyase has not been detected in the inner sapwood adjacent to the heartwood of several species (Higuchi and Fukazawa 1966, Higuchi et al. 1967a, Hillis and Ishikura 1970). Nevertheless, phenylalanine and glucose have been rapidly converted to polyphenols in eucalypt leaves with low phenylalanine ammonia lyase activity (Hillis and Ishikura 1970). Possibly other pathways than the accepted one also operate (Swain 1962, Ebel and Hahlbrock 1982). With the development of cytological studies, the various organelles and cytological conditions governing the separate stages in the biosynthesis of extractives and exudates will be identified. Characterization of the enzyme systems responsible for these biosyntheses will assist cytological studies and determination of the regulating systems. Quantitative studies are hindered by the small proportion of parenchyma in the woody tissues of the conifers, the different functions and activities of different cells among the parenchyma generally (Chapters 6.1.5 and 6.2.6), and their action individually or as a group and at different periods.

Higuchi et al. (1967b) found that the respiratory patterns differed considerably across cross-sections of *Cryptomeria japonica, Chamaecyparis obtusa, Thujopsis dolabrata*, and *Magnolia obovata* collected in the late spring season. In the cambial zone glycolysis and the tricarboxylic acid cycle were predominant in respiration. However, the relative contribution of the pentose phosphate pathway increased in the inner sapwood and this favored the formation of cinnamic acid as a precursor of heartwood extractives. They confirmed the earlier work of Hillis and Inoue (1966), who showed that arsenite, an inhibitor of the utilization of acetate in the tricarboxylic acid cycle, increased the formation of flavonoids by providing unused acetate for incorporation in the A ring of the flavonoids (administration of iodoacetic acid also had this effect). In addition they (Higuchi et al. 1967b) showed that arsenite (and sodium fluoride) also increased the activity in the pentose phosphate pathway and thus created conditions favorable for the biosynthesis of the B ring of flavonoids and stilbenes and for the formation of $NADPH_2$ and erythrose-4-phosphate and more shikimic acid.

Other studies (Chapters 6.2 and 6.3) have shown an increase in oxygen uptake and enzyme activity in the transition zone in the dormant season, but the contribution of different respiratory pathways is less clear. Although there is little information on the composition of gases in the transition zone, it would be expected that anaerobic conditions exist before the polyphenols are formed and when the mitochondria and tricarboxylic acid cycle would not operate.

The study of vegetative tissues is revealing the enzymes and co-factors required for the biosynthesis of the aromatic ring A of flavonoids by stepwise addition of three acetate units from malonyl-CoA to the phenylpropanoid moiety of 4-coumaryl-CoA to the chalcone which is isomerized to naringenin (Rolfs and Kindl 1984). The enzymes for hydroxylation to dihydroflavonols have been isolated and in flowers found to be located in the microsomal fraction and to require NADPH as a co-factor. The enzymes synthesizing flavonoids occur in the endoplasmic reticulum (Wagner et al. 1984), but the causes of successive or selective development of flavonoids are unknown (Ebel and Hahlbrock 1982). Further studies on these simpler systems will assist the understanding of flavonoid synthesis at the heartwood periphery.

Hydroxylation of the B ring of flavonoids takes place by a different pathway from that involved in the formation of the galloyl moieties of hydrolyzable tannins. This is shown by the flavonoids with mono- or di-hydroxy B-ring substitutions in the heartwood of *Rhus succedanea* in contrast with the gallotannins of the sapwood (Hillis and Inoue 1966).

The increased activities of peroxidases and phenol oxidases (Chapter 6.1.6, 6.2.6, and 6.3.3) at the periphery of the heartwood and disease-injured tissues would result in the high degree of polymerization which is characteristic of the extractives of these tissues. The complexity of these polymers prevents an elucidation of their precursors, which will only be obtained by a study of their formation before oxidation-polymerization reactions commence. In addition, there are suggestions that these reactions can occur at different stages and to different extents. With regular heartwoods the lower molecular size of the polymers formed enables their penetration into the cell wall. With irregular (facultative) heartwoods the large size of the polyphenolic polymers hinders their penetration into the cell wall so that shrinkage and some other properties show little change from the sapwood (Frey-Wyssling and Bosshard 1959, Bosshard 1968). Greater attention to the various stages of development within transition zones is likely to yield information on the factors controlling the formation of extractives and of heartwood and the amount of energy required for that formation. A study of the more accessible wound tissues, such as kino veins and fungal-induced lesions, should elucidate the factors eliciting the selective development of components of extractives so that their production may be controlled.

7.10 Conclusions

During the evolution of the plant kingdom, many woody species developed a heartwood in the core of their stems. Today the heartwoods contain extractives, the amounts of which vary between genera and species, and in some specimens the amounts are large. Furthermore, the composition and properties of these extractives differ between heartwoods of different genera and species and often other tissues of the same tree. In addition, the cambium of some species exude, after injury, compounds differing in composition from those in surrounding tissues. Lesions or defects in trees can develop extractives different in composition from those elsewhere in the tree. These extractives and exudates have consistent compositions which are sometimes highly specific. Presumably the compounds have been selected over time for particular functions, and the amounts formed have been required to meet those functions.

The genetic program of the species largely controls the composition of the extractives or exudates and to a lesser extent the amount. Heartwood extractives and tree exudates are formed in situ from translocated carbohydrates originating from the leaves or from stored metabolites. Some of the stages of the mechanisms controlling the amounts formed are being established.

Mechanical injury alone leads to only a very small increase in the amount of extractives in living sapwood; a very small zone of diminished water content can be formed and sometimes trace amounts of ethylene. Injury of the cambium during most times of the year also has little effect, but at other times copious amounts of exudate may form. Injection into the sapwood of chemicals which affect some enzyme systems can diminish the water content of surrounding tissues and result in ethylene formation, followed by the production of extractives. Administration of these chemicals to active cambium can initiate exudate formation. Fungal infection of sapwood, and in one case at least the injection of a fungal metabolite, can also result in dehydration of tissues, formation of ethylene, and then extractives. A similar sequence occurs with the formation of a transition or white zone and then a heartwood, with a low moisture content. Dehydration occurs before disintegration of the nucleus of the parenchyma. The sequence of events in those species with wet heartwood is less understood.

The events leading to heartwood formation occur after a certain number of annual growth cycles are completed, and these can be characteristic of a genus or species. At that time the metabolism of the parenchyma in the inner sapwood intensifies and alters. These changes are most evident in the narrow transition zone at the heartwood periphery of a number of species. The volume of the sapwood when these changes occur is associated with the relationship of the leaf biomass to the stem volume in an apparent conservation of water in the stem. The centrifugal movement of water is under active control, as shown by the selective transfer of certain inorganic nutrients into the sapwood from the heartwood periphery and particularly from the transition zone. Similar movements of water and selected inorganic nutrients into the sapwood occur when lesions develop increased amounts of extractives. Once extractives formation commences in the transition zone, or at the heartwood periphery, or in the lesion, information is relayed to the cambium and the sapwood for the supply of primary metabolites. The formation of heartwood extractives appears to commence in the dormant season. The amount of metabolites provided depends on the health of the tree, and the amount of extractives formed depends also on the physiological conditions at the periphery of the heartwood or lesion.

The composition of the extractives of various tissues or of exudates is genetically controlled. Parenchyma cells can form different suites of chemicals according to the location in which these are formed. Heartwood extractives usually involve acetate units from the tricarboxylic acid energy cycle. Apparently at the heartwood periphery a significant increase occurs in the activity of the basic enzymic reactions, leading particularly to the formation of acetate units. Ethylene is a prominent component of the systems which initiate the enhancement of these basic activities. The next stage involves the formation of increased amounts of extractives, often with a composition different from that in the sapwood or adjacent tissue. In addition some parenchyma have the ability for the synthesis of particular types of extractives, and other parenchyma can form selectively just one compound. There appear to be elicitors for a particular suite of compounds or for one compound, once basic enzymic reactions have been initiated. The elicitors for extractives formation associated with fungi can influence the composition of extractives involving stilbenes and flavonoids, and even the same fungal species infecting the same xylem substrate can result in extractives with slightly different composition. Presumably the elicitors differ for those situations where extractives result from activities other than microbiological invasion. These elicitors are apparently genetically selected, and they choose between balanced pathways.

The importance of the transition zone in all these transformations has been shown with its increased levels of respiration, and of ethylene, particularly during the dormant period of the tree. The incorporation of early precursors, such as shikimic acid, into polyphenols has also been demonstrated, and these products can be formed sequentially in the zone. It is not certain how the primary metabolites for these syntheses enter the impermeable transition zone.

With the development of sensitive and delicate techniques and the increased knowledge of basic aspects of cytology and biochemistry, an understanding of the mechanisms of formation of secondary components of the tree will be obtained. It will require data concerning the three-dimensional movement of fluids within the tree, the time of year in which movement and cell activities occur, and the influence of environment on those situations. The gradual accumulation of this knowledge will assist our control of the growth of trees for the production of wood to meet our needs. Furthermore, with this knowledge we can begin to match the finesse of the tree. The tree using solar energy and a common and low-cost substrate can biosynthesize selectively without byproducts a range of secondary products of considerable complexity for its needs. Eventually we should be able to biosynthesize, at low energy cost and from renewable resources, complex chemicals to meet the needs of mankind.

References

Abe M, Yokota T, Yasue M, Hosaka H 1953 On the turpentining from Jezomatsu (Picea jezoensis). I. Bull Gov For Exp Stn (Tokyo) 63: 118–145

Abeles F B 1973 Ethylene in plant biology. Academic Press, London, New York, 302 pp

Abercrombie T J 1985 Arabia's frankincense trail. Nat Geogr 168: 475–513

Abraham P D, P'ng T C, Lee C K, Sivakumaran S, Manikam B, Yeoh C P 1975 Ethrel stimulation of Hevea. Int Rubber Conf, Kuala Lumpur, Preprint, 37 pp

Abraham P D, Wycherley PR, Pakianathan S W 1968 Stimulation of latex flow in Hevea brasiliensis by 4-amino-3,5,6-trichloropicolinic acid and 2-chloroethanephosphonic acid. J Rubber Res Inst Malaya 20(5): 291–305

Albrektson A 1984 Softwood basal area and needle mass of Scots pine (Pinus sylvestris) trees in central Sweden. Forestry 57: 35–43

Amos G L 1952 Silica in timber. CSIRO Aust Bull 267, 58 pp

Anderegg R J, Rowe J W 1974 Lignans, the major component of resin from Araucaria angustifolia knots. Holzforschung 28: 172–175

Anderson A B 1946 Chemistry of western pines. Ind Eng Chem 38: 450–459

Anderson A B 1952 Pinitol from sugar pine. Tappi 35: 198–202

Anderson A B 1953 Pinitol from sugar pine stump wood. Ind Eng Chem 45: 593–596

Anderson A B 1955 Extract stimulation for wood rosin production – increasing the extractive content in ponderosa pine. For Prod J 5: 417–420

Anderson A B, Riffer R, Wong A 1968 Chemistry of the genus Sequoia. V. Phytochemistry 7: 1367–1371

Anderson D M W 1985 Gums and resins, and factors influencing their economic development. In: Wickens G E, Goodwin J R, Field D V (eds) Plants for arid lands. Proc Kew Int Conf Economic plants for arid lands, 23–27 July 1984. Allen Unwin, London, 343–356

Anderson D M W, McNab C G A, Anderson C G, Brown P M, Pringuer M A 1983 Studies of uronic acid materials. Part 58. Int Tree Crops J 22: 147–154

Anderson R, Popoff T, Theander O 1975 A new lignan from Norway spruce. Acta Chem Scan B 29:835

Anon 1957 International glossary of terms used in wood anatomy : Prepared by the Int Assoc Wood Anatomists. Trop Woods 107: 1–36

Anon 1976 Brazil – a geography. Brazilian Embassy, London, 55 pp

Anon 1978 Almaciga resin. Forpride Digest 7 (No 2,3): 2–5

Anon 1981a Comparative properties of old – and young – growth giant sequoia of potential significance to wood utilisation. Div Agric Sci, Univ Calif USA Bull 1901, 19 pp

Anon 1981b Second annual report on carcinogens. US Dept Health Human Serv, Publ Health Ser NTP 81–43 Washington DC, Dec, p.51

Archer B L, Audley B G 1973 Rubber, gutta percha, and chicle. In : Miller L P (ed) Phytochemistry. 2. Van Nostrand Reinhold, New York, 310–343

Archer S A, Hislop E C 1975 Ethylene in host-pathogen relationships. Ann Appl Biol 81: 121–126

Arni P C, Cochrane G C, Gray J D 1965a The emission of corrosive vapours by wood. I. Survey of the acid-release properties of certain freshly felled hardwoods and softwoods. J Appl Chem 15: 305–313

Arni P C, Cochrane G C, Gray J D 1965b The emission of corrosive vapours by wood II. The analysis of the vapours emitted. J Appl Chem 15: 463–468

Aspinall G O, McKenna J P 1968 Araucaria bidwillii gum. II. Carbohyd Res 7: 244–254

Attiwill P M 1979 Nutrient cycling in a Eucalyptus obliqua forest. III. Growth, biomass, and net primary production. Aust J Bot 27: 439–458

Attiwill P M 1980 Nutrient cycling in a Eucalyptus obliqua forest. IV. Nutrient uptake and nutrient return. Aust J Bot 78: 199–222

von Aufsess H, Schulz H, Mössnang M 1985 Ein ungewöhnlicher Spritzkern an der Buche (Fagus sylvatica L.) und seine Auswirkung auf die Druckfestigkeit. Holz Roh-Werkst 43:350

Back E 1960a On the relative composition of canal resin and ray parenchyma resin in Picea abies stemwood. Svensk Papperstidn 63: 647–651

Back E 1960b Tracheidal and parenchymatous cells in Picea abies and their behaviour in sulphite pulping. Svensk Papperstidn 63: 695–698

Back E L 1969 Intercellular spaces along the ray parenchyma-the gas canal system of living wood. Wood Sci 2: 31–34

Bailey J A 1982 Mechanisms of phytoalexin accumulation. In: Bailey and Mansfield J W (eds) Phytoalexins. Blackie, Glasgow, London, 289–318

Bailey J A, Mansfield J W (eds) 1982 Phytoalexins. Blackie, Glasgow, London, 334 pp

Bailey P J, Preston R D 1969 Some aspects of softwood permeability. I. Structural studies with Douglas-fir sapwood and heartwood. Holzforschung 23: 113–125

Baillie M G L, Hillam J, Briffa K R, Brown D M 1985 Re-dating the English art-historical tree-ring chronologies. Nature (London) 315: 317–319

Baker R T, Smith H G 1910 A research on the pines of Australia. Technol Museum Sydney (Australia) Tech Educ Ser 16, 458 pp

Bakshi B K, Singh S 1964 Mortality of Eucalyptus citriodora. Indian For 90: 15–18

Balatinecz J J, Kennedy R W 1967 Maturation in ray parenchyma cells in pine. For Prod J 17(10): 57–64

Ballou C E, Anderson A B 1953 On the cyclitols present in sugar pine (Pinus lambertiana). J Am Chem Soc 75: 648–650

Bamber R K 1972 Properties of the cell wall of the resin canal tissue of Pinus lambertiana and P.radiata. J Inst Wood Sci 6: 32–35

Bamber R K 1976a The occurrence of secondary walls in the resin canal tissue in the genus Pinus. J Inst Wood Sci 7: 15–17

Bamber R K 1976b Heartwood, its function and formation. Wood Sci Tech 10: 1–8

Bamber R K, Colley R L 1983 A histochemical test for sapwood and heartwood in Pinus radiata. J Inst Wood Sci 9: 228

Bamber R K, Davies G W 1969 Lignification of ray parenchyma cell walls in the wood of Pinus radiata D Don. Holzforschung 23: 83–84

Bamber R K, Floyd A G, Humphreys F R 1969 Wood properties of flooded gum. Aust For 33: 3–12

Bamber R K, Fukazawa K 1985 Sapwood and heartwood: a review. For Prod Abstr 8 (9): 265–278

Bamber R K, Humphreys F R 1965 Variation in sapwood starch levels in some Australian forest species. Aust For 29: 15–23

Bannan M W 1936 Vertical resin ducts in the secondary wood in the Abietineae. New Phytol 35: 11–46

Baqui S A, Shah J J 1985 Histoenzymatic studies in wood of Acacia auriculiformis during heartwood formation. Holzforschung 39: 311–320

Baqui S, Shah J J, Pandalai R C, Kothari I L 1979 Histochemical changes during transition from sapwood to heartwood in Melia azedarach. Indian J Exp Biol 17: 1032–1037

Barker E J 1979 Growth and wood properties of Pinus radiata in relation to applied ethylene. N Z J For Sci 9: 15–19

Barker J A, Schmid J J 1976 Paraquat induced lightering in slash pine. Lightwood Res Coor Proc Ann Meet Jacksonville, FL, 88–101

Barnacle J E, Ampong F F K 1974 Refractory intermediate wood in round teak fence posts. Ghana J Sci 14: 193–198

Barton G M 1963 Conidendrin in floccosoids of western hemlock. For Prod J 13: 304

Barton G M 1967 Differences in phenolic extracts from healthy Douglas–fir roots and those infected with Poria weirrii. Can J Bot 45: 1545–1552

Barton G M, Gardner J A F 1962 The occurrence of matairesinol in mountain hemlock, western hemlock and balsam. J Org Chem 27: 322–323

Barton G M, Gardner J A F 1966 Brown stain formation and the phenolic extractives of western hemlock (Tsuga heterophylla). Can Dep For Publ 1147, 20 pp
Barton G M, Smith R S 1971 Brown stain in kiln-dried Abies amabilis lumber. Can For Serv Bi-m Res Notes 27(3): 21–23
Basham H G, Cowling E M 1975 Distribution of essential elements in forest trees and their role in wood deteriortion. In: Becker G, Liese W (eds) Material und Organismen, Beih 3, 155–165
Basset C, Sherwood R T, Kepler J A, Hamilton P B 1967 Production and biological activity of fomannosin, a toxic sesquiterpene metabolite of Fomes annosus. Phytopathology 57: 1046–1052
Bate-Smith E C, Swain T 1953 Identification of leuco-anthocyanins as "tannins" in food. Chem Ind 1953: 377
Bauch J 1980 Variation of wood structure due to secondary changes. In: Bauch J (ed) Natural variations of wood properties. Mitt Bundesforschungsanst Forst-Holzwirtsch 131, 69–97
Bauch J, Baas P (eds) 1984 Development and characteristics of discoloured wood. IAWA Bull N S 5(2): 91–154
Bauch J, Berndt H 1973 Variability of the chemical composition of pit membranes in bordered pits of gymnosperms. Wood Sci Technol 7: 6–19
Bauch J, Höll W, Endeward R 1975 Some aspects of wetwood formation in fir. Holzforschung 29: 198–205
Bauch J, Liese W, Scholz F 1968 Über die Entwicklung und stoffliche Zusammensetzung der Hoftüpfelmembranen von Längstracheiden in Coniferen. Holzforschung 22: 144–153
Bauch J, Schweers W, Berndt H 1974 Lignification during heartwood formation: comparative study of rays and bordered pit membranes in coniferous woods. Holzforschung 28: 86–91
Bauch J, Shigo A L, Starck M 1980 Wound effects in the xylem of Acer and Betula species. Holzforschung 34: 153–160
Baur P S, Walkinshaw C H 1974 Fine structure of tannin accumulations in callus cultures of Pinus elliottii. Can J Bot 52: 615–619
Becker G 1962 Über den Eiweissgehalt von Nadelhölzern. Holz Roh-Werkst 20: 368–375
Beevers H 1960 Respiratory Metabolism in Plants. pp 51–57 et seq. Row, Peterson Co New York, 232 pp
Belluschi P G, Johnson N E, Heikkenen H J 1965 Douglas-fir defects caused by the Douglas-fir beetle. J For 63: 252–256
Bengochea T, Dodds J H, Evans D E, Jerie P H, Niepel B, Shaari A R, Hall M A 1980 Studies on ethylene binding by cell-free preparation from cotyledons of Phaseolus vulgaris. Planta 148: 397–406
Benic R 1956 (Investigations on the proportion, and some physical properties of the sapwood and heartwood of Fraxinus angustifolia). Glas Sumske Pok 12: 13–104 (from For Abstr 1957, 18: 400)
Benson-Cooper D M, Knowles R L, Thomson F J, Cown D J 1982 Computed tomographic scanning for the detection of defects within logs. For Res Inst N Z Bull 8, 8 pp
Berlyn G P 1969 Microspectrophotometric investigations of free space in plant cell walls. Am J Bot 56: 498–506
Berlyn G P 1970 Ultrastructural and molecular concepts of cell-wall formation. Wood Fiber 2: 196–227
Beyer E M, Morgan P W, Yang S F 1984 Ethylene. In: Wilkins M B (ed) Advanced plant physiology. Pitman, London, 111–127
Birch A J, Hughes G K, Smith E 1954 Constituent of gmelinol. III. Final elucidation. Aust J Chem 7: 83–86
Birchem R, Brown C L 1979 Ultrastructure of paraquat-treated slash pine (Pinus elliottii). Am J Bot 66: 1208–1218
Bisset N G, Chavanel V, Lantz J-P, Wolff R E 1971 Constituants sesquiterpéniques et triterpéniques des resines du genre Shorea. Phytochemistry 10: 2451–2463
Bland D E 1971 The relation of lignin and polyphenol content of Tasmanian alpine ash to shrinkage and recovery. Wood Sci Technol 5: 117–26
Bland D E, Hillis W E 1969 Microspectrophotometric investigation of lignin and polyphenol distribution in wood sections. Appita 23: 204–210

Bolza E 1980 Some health hazards in the forest products industries. CSIRO, Div Building Res Reprint 966, 9 pp
Bordasch R P, Berryman A A 1977 Host resistance to the fir engraver beetle, Scolytus ventralis (Coleoptera: Scolytidae) 2. Repellency of Abies grandis resins and some monoterpenes. Can Entomol 109: 95–100
Bosshard H H 1953 Der braune Kern der Asche. Holz Roh-Werkst 11 : 349–353
Bosshard H H 1955 The physiology of brown heart of ash. Schweiz Z Forstwes 106: 592–612
Bosshard H H 1965a Aspects of the ageing process in cambium and xylem. Holzforschung 19: 65–69
Bosshard H H 1965b Mosaikfarbkernholz in Fagus silvatica. Schweiz Z Forstwes 116/1: 1–11
Bosshard H H 1967 Über die fakultative Farbkernbildung. Holz Roh-Werkst 11: 409–416
Bosshard H H 1968 On the formation of facultatively colored heartwood in Beilschmiedia tawa. Wood Sci Technol 2: 1–12
Böttcher P, Liese W 1975 Zur Verkernung des Wurzelholzes von Fichte und Lärche. Forstwiss Cbl 94: 152–160
Bourdeau P F, Schopmeyer C S 1958 Oleoresin exudation pressure in slash pine: its measurement, heritability and relation to oleoresin yield. In: Thimann K V (ed) The physiology of forest trees. Ronald, New York, 313–319
Brack C L, Dawson M P, Gill A M 1985 Bark, leaf and sapwood dimensions in Eucalyptus. Aust For Res 15: 1–7
Bray M W, Curran C E 1937 Sulfate pulping of southern yellow pines. Paper Trade V 105(20): 39–47
Bray M W, Paul B H 1934 Evaluation of southern pines for pulp production. Paper Trade J 99(5): 38–43
Briggs L H, Cain B F 1959 Chemistry of the Podocarpaceae. Constituents of the heartwood of Podocarpus spicatus. Tetrahedron 6: 143–147
Briggs L H, Frieberg A G 1937 Resinol of Oleo cunninghamii. J Chem Soc 1937: 271–273
Brix H, Mitchell A K 1983 Thinning and nitrogen fertilisation effects on sapwood development and relationships of foliage quantity to sapwood area and basal area in Douglas-fir. Can J For Res 13: 384–389
Brockway G E, Hillis W E 1955 Tan bark eucalypts of the semi-arid regions of south western Australia. Emp For Rev 34: 31–41
Brooks F T, Moore W C 1925 Silver-leaf disease. V. J Pomol Hort Sci 5: 61–73
Brown C L, Nix L E 1975 Uptake and transport of paraquat in slash pine. For Sci 21: 359–364
Brown C L, Clason T R, Michael J L 1976 Paraquat induced changes in reserve carbohydrates, fatty acid and oleoresin content of young slash pines. Lightwood Res Coor Coun Proc Ann Meet Jacksonville, FL, 8–19
Brown C L, Sommer H E, Birchem R 1979 Additional observations on the utilization of reserve carbohydrates in paraquat induced resinosis. In: Esser M H (ed) 6th Lightwood Res Coor Coun, Proc Ann Meet Atlanta, GE, 4–11
Brown H P, Panshin A J, Forsaith C C 1949 Textbook of wood technology. Vol 1, McGraw-Hill New York, 622 pp
Brown K M, Leopold A C 1973 Ethylene and the regulation of growth in pine. Can J For Res 3: 143–145
Browne F G 1955 Forest trees of Sarawak and Brunei and their products. Gov Print Off Kuching, Sarawak, 395 pp
Buckland N J, Dalley O T, Matheson C 1954 The distribution and nature of resinous material in Pinus radiata. Aust Pulp Paper Ind Tech Assoc Proc 8: 165–182
van Buijtenen J P, Santamour F S 1972 Resin crystallisation related to weevil resistance in white pine (Pinus strobus). Can Entomol 104: 215–219
Bulgrin E H, Ward J C 1968 Factors contributing to heartwood – boundary stain in living oak. Wood Sci 1: 58–64
Burden R S, Kemp M S 1983 (–)-7-Hydroxycalamenene, a phytoalexin from Tilia europea. Phytochemistry 22: 1039–1040
Burden R S, Kemp M S 1984 Sesquiterpene phytoalexins from Ulmus glabra. Phytochemistry 23: 383–385
Burg S P 1962 The physiology of ethylene formation. Ann Rev Plant Physiol 13: 265–302

Burgess P F 1966 Timbers of Sabah. Publ For Dep Sabah Malaysia, 501 pp
Burkhill I H 1935 A dictionary of the economic products of the Malay Peninsula. Crown agents for the colonies, 2 Vols. London, 2402 pp
Burmester A 1983 Veränderung von Holzfeuchtigkeit, Dichte und Schwindung bei Laubhölzern durch jahreszeitlich bedingte Einflüsse. Holz Roh-Werkst 41: 493–498
Büsgen M, Münch E 1929 The structure and life of forest trees, 3rd revised edn by E Münch. (Engl transl by Thomson T) Chapman Hall London, 436 pp
Cain W F 1980 Carbon-14 in modern American trees. Chem Abstr 92: 220169
Cain W F, Suess H E 1976 Carbon-14 in tree rings. J Geophys Res 81: 3688–3694
Calvin M 1983 New sources for fuel and materials. Science 219: 24–26
Campbell J R, Swan E P, Wilson J W 1965 Comparison of wood and growth zone resinous extracts in Douglas fir. Pulp Paper Mag Can 66: T-248, T-252
Campbell W G, McGowan J C 1939 The composition and origin of a stony deposit found in galleries of the bee-hole borer in teakwood. Emp For J 18: 91–94
Carnmalm B 1954 Constitution of resin phenols and their biogenetic relations. XVIII. Acta Chem Scand 8: 1827–1829
Carrodus B B 1971 Carbon dioxide and the formation of heartwood. New Phytol 70: 939–943
Carrodus B B 1972 Variability in the proportion of heartwood formed in woody stems. New Phytol 71: 713–718
Carrodus B B, Triffett A C K 1975 Analysis of composition of respiratory gases in woody stems by mass spectroscopy. New Phytol 74: 243–246
Cartwright K T St G, Findlay W P K 1936 Decay of timber and its prevention. His Majesty's Stn Off London, p 267
Chafe S C, Durzan D J 1973 Tannin inclusions in cell suspension cultures of white spruce. Planta 113: 251–262
Chalk L 1951 Water and growth of Douglas-fir. Q J For 45(4): 237–242
Chalk L, Bigg J M 1956 The distribution of moisture in the living stem in sitka spruce and Douglas fir. Forestry 29: 5–21
Chalutz E, de Vay J E, Maxie E C 1969 Ethylene-induced isocoumarin formation in carrot root tissue. Plant Physiol 44: 235–241
Chase W W 1934 The composition, quantity and physiological significance of gases in tree stems. Univ Minnesota USA Tech Bull 99
Chattaway M M 1949 The development of tyloses and secretion of gum in heartwood formation. Aust J Sci Res B 2: 227–240
Chattaway M M 1951 Morphological and functional variations in the rays of pored timbers. Aust J Sci Res B 4(1): 12–27
Chattaway M M 1952 The sapwood-heartwood tranformation. Aust For 16 : 25–34
Chattaway M M 1953 The occurrence of heartwood crystals in certain timbers. Aust J Bot 1: 27–38
Chattaway M M 1955a Crystals in woody tissues. I. Trop Woods 102: 55–74
Chattaway M M 1955b The anatomy of bark. VI. Peppermints, boxes, ironbarks, and other eucalypts with cracked and furrowed barks. Aust J Bot 3: 170–176
Chattaway M M 1956 Crystals in woody tissues. II. Trop Woods 104: 99–124
Chen C L, Chang H M, Cowling E B, Huang Hsu C Y, Gates R 1976 Aporphine alkaloids and lignans formed in response to injury of sapwood in Liriodendron tulipifera. Phytochemistry 15: 1161–1167
Chidester G H, Bray M W, Curran C E 1939 Growth factors influencing the value of jack pine for kraft and sulfate pulps. Pap Trade J 109(13): 36–45
Chochieva M M, Tsvetaeva I P, Yur'eva M K, Zaitseva A F, Petropavlovskii G A, Nikitin N I 1959 Distribution of arabogalactan in Larix dahurica. Chem Abstr 53: 4439
Choong E T 1969 Effect of extractives on shrinkage and other hydroscopic properties of ten southern pine woods. Wood Fiber 1: 124–133
Chow S-Z 1972 Hydroxyl accessibility, moisture content, and biochemical activity in cell walls of Douglas-fir trees. Tappi 55: 539–544
Chudnoff M 1961 The physical and mechanical properties of Eucalyptus camaldulensis. Bull Natl Univ Inst Agric Bet Dangan 66
Clark J, Gibbs R D 1957 Studies in tree physiology. IV. Can J Bot 35: 219–253

Clarke A E, Anderson R L, Stone B A 1979 Form and function of arabino-galactans and arabinogalacton-proteins. Phytochemistry 18: 521–540

Clarke I D, Steiner E T, Frey R W 1942 Effect of age on the size and tannin content of Chinese chestnut trees grown in Maryland. J Am Leather Chem Assoc 37: 591–598

Clark-Lewis J W, Dainis I 1967 Flavan derivatives. XIX. Phenolic hydroxylation patterns of heartwood flavonoids characteristic of sections and sub-sections of the genus Acacia. Aust J Chem 20: 2191–2198

Clifton N C 1969 Resin pockets in Canterbury radiata pine. N Z J For 14: 38–49

Conn E E 1964 Enzymology of phenolic biosynthesis. In: Harborne J B (ed) Biochemistry of phenolic compounds. Academic Press, London, New York, 399–435

Conner A H, Diehl M A, Rowe J W 1980a Tall oil precursors and turpentine in black and white spruce. Wood Sci 13: 111–116

Conner A H, Diehl M A, Rowe J W 1980b Tall oil precursors in three western pines: ponderosa, lodgepole and limber pine. Wood Sci 12: 183–191

Conner A H, Diehl M A, Rowe J W 1980c Tall oil precursors and turpentine in jack and eastern white pine. Wood Sci 12: 194–200

Conner A H, Diehl M A, Wroblewska H, Rowe J W 1977 Effect of paraquat treatment of Northern and Western Conifers. Lightwood Res Coor Coun, Proc Ann Meet Atlanta, Fl 34–56

Cooper G A 1974 The effect of black walnut extractives on sorption, shrinkage and swelling. Wood Sci 6: 380–388

Cooper W C, Rasmussen G K, Waldon E S 1969 Ethylene production in freeze-injured citrus trees. J Rio Grande Valley Hort Soc 23: 29–37

Coppen J J W, Greenhalgh P, Smith A E 1984 Gum naval stores: an industrial profile of turpentine and rosin production from pine resin. Dep Trop Dev Res Inst G187, 40 pp

Côté W A 1963 Structural factors affecting the permeability of wood. J Polym Sci Part C (2): 231–242

Côté W A, Day A C, Simson B W, Timell T E 1966 Studies on larch arabinogalactan. I. The distribution of arabinogalactan in larch wood. Holzforschung 20: 178–192

Côté W A, Simson B W, Timell T E 1967 Studies on larch arabinogalactan II. Degradation of arabinogalactan within the living tree. Holzforschung 21: 85–88

Côté W A, Timell T E 1967 Studies on larch arabinogalactan. III. Distribution of arabinogalactan in tamarack. Tappi 50: 285–289

Coutts M P 1969 The formation of polyphenols in small blocks of Pinus radiata sapwood with and without the fungal symbiont of Sirex. Aust For Res 4(1): 29–34

Coutts M P 1970 The influence of phenolic compounds in Pinus radiata on the growth of Amylostereum areolatum. Aust For Res 4(4): 15–18

Coutts M P 1976 The formation of dry zones in the sapwood of conifers. I. Eur J For Pathol 6: 372–381

Coutts M P 1977 The formation of dry zones in the sapwood of conifers. II. Eur J For Pathol 7: 6–12

Coutts M P, Dolezal J E 1966 Polyphenols and resin in the resistance mechanism of Pinus radiata attacked by the wood wasp Sirex noctilio and its associated fungus. For Res Inst (Canberra Aust) Leaflet 101, 19 pp

Cowling E B, Merrill W 1966 Nitrogen in wood and its role in wood deterioration. Can J Bot 44: 1539–1554

Cown D J 1973 Resin pockets: their occurrence and formation in New Zealand forests. N Z J For 18: 233–251

Cown D J, McConchie D L 1980 Wood property variations in an old-crop stand of radiata pine. N Z J For Sci 10: 508–520

Craib W G 1923 Regional spread of moisture in the wood of trees. III. Notes R Bot Garden, Edinburgh, XIV(LXVI) (Jan), 1–11

Craker L E, Wetherbee P J 1973 Ethylene, carbon dioxide and anthocyanin synthesis. Plant Physiol 52: 177–179

Croteau R, Johnson M A 1985 Biosynthesis of terpenoid wood extractives. In: Higuchi T (ed) Biosynthesis and biodegradation of wood components. Academic Press, London, New York, 379–439

Cummins N H O 1972 Heartwood differentiation in Pinus species – a modified azo-dye test. N Z J For Sci 2: 188–191

Cutter B E, McGinnes E A, McKown D H 1980 Inorganic concentrations in selected woods and charcoals. Wood Fiber 12: 72–79
Da Costa E W B, Rudman P, Gay F J 1958 Investigations on the durability of Tectona grandis. Emp For Rev 37: 291–298
Da Costa E W B, Rudman P, Gay F J 1961 Relationships of growth rate and related factors to durability in Tectona grandis. Emp For Rev 40: 308–319
Dadswell H E, Hillis W E 1962 Wood. In: Hillis W E (ed) Wood extractives. Academic Press, London, New York, 3–55
Daube W 1883 Chemische Analysen des Kern- und Splintholzes wichtiger Waldbäume. Forstl Blätter 20: 177–192 (Original not seen)
Davidson R L 1980 Handbook of water-soluble gums and resins. McGraw-Hill, New York, 657 pp
Davies G W, Carrodus B B 1978 Anatomical studies on paraquat treated Pinus radiata. Appita 31: 359–364
Day W R 1959 Observations on eucalypts in Cyprus. 1. The character of gum veins and anatomical indications for their origin. Emp For Rev 38: 35–44
Dell B, McComb A S 1978 Plant resins – their formation, secretion and possible functions. Adv Bot Res 6: 275–316
Desch H E 1968 Timber; its structure and properties. Macmillan, London, 399 pp
Dickenson P B, Sivakumaran S, Abraham P D 1975 Ethad (R) and other new stimulants for Hevea brasiliensis. Preprint Int Rubber Conf, Kuala Lumpur, 28 pp
Dietrich G 1973 Untersuchungen über die Astbildung und natürliche Astreinigung der Weisstanne. Forstwiss Forsch Beih Forstwiss Cbl 34, 95pp
Dietrichs H H 1964a Das Verhalten von Kohlenhydraten bei der Holzverkernung. Holzforschung 18: 14–24
Dietrichs H H 1964b Chemisch – physiologische untersuchungen über die Splint-Kern-Umwandlung der Rotbuche (Fagus sylvatica). Mitt Bundesforschungsanst Forst Holzwirtsch 58, 149 pp
Dietrichs H H, Hausen B M 1971 (Dalbergiones-active substances from Dalbergia species). Holzforschung 25: 182–187
Dixon A F G 1971 The role of aphids in wood formation. I. Appl Ecol 8 : 165–179
Dobie J, Wright D M 1978 Metric conversion factors for forest products in western Canada. Forintek West For Prod Lab, Tech Rep No 1, 60 pp
Doran J C 1975 Occurrence of kino veins in two provenance trials of Eucalyptus regnans. Aust For Res 7: 21–27
Drew J, Roberts D R 1978 Extractive enhancement in pines by paraquat treatment. For Prod J 28: 21–26
Drewes S E, Roux D G 1966 Stereochemistry and biogenesis of mopanols and peltogynols and associated flavonoids from Colophospermum mopane. J Chem Soc: 1644–1653
Dujesiefkern D, Liese W, Bauch J 1984 Discolouration in the heartwood of oak trees. IAWA Bull N S 5(2): 128–132
Eades H W 1958 Differentiation of sapwood and heartwood in western hemlock by color tests. For Prod J 8(3): 104–106
Ebel J, Hahlbrock K 1982 Biosynthesis. In: Harborne J B, Mabry T J (eds) The flavonoids, Advances in research. Chapman, London, 641–679
Ebermann R, Stich K 1982 Peroxidase and amylase isoenzymes in the sapwood and heartwood of trees. Phytochemistry 21: 2401–2402
Ebermann R, Stich K 1985 Distribution and seasonal variation of wood peroxidase activity in oak (Quercus robur). Wood Fiber Sci 17(3): 391–396
Eckstein D, Wazny T, Bauch J, Klein P 1986 New evidence for the dendrochronological dating of Netherlandish paintings. Nature (London) 320: 465–466
Ehlers R C, Sandermann W 1961 Untersuchungen über harzhaltige Tropenhölzer. Holz Roh-Werkst 19: 187–195
Einspahr D W 1972 Forest tree improvement and the pulp and paper industry. Tappi 55: 1660–1669
Einspahr D W, Goddard D E, Gardner H S 1964 Slash pine, wood and fiber property heritability study. Silvae Genet 13 (4): 103–109

Ekman R 1979 Distribution of lignans in Norway spruce. Acta Acad Abo 39 B (3), 6 pp
Ellis E L 1965 Inorganic elements in wood. In: Côté W A (ed) Cellular structure of woody plants. Syracuse Univ Press, New York, 181–189
Englerth G H, Scheffer T C 1955 Tests of decay resistance of four western pole species. J For 53(8): 556–561
Engström P, Back E 1959 The proportion of canal resin in Picea abies pulpwood. The resin in parenchymatous cells and resin canals of conifers. VIII. Svensk Papperstidn 62: 545–553
Erdtman H 1943 The phenolic constituents of pine heartwood. IV. Membrane forming substances in pine heartwood. Svensk Papperstidn 46: 226–228
Erdtman H 1958 (Heartwood and heartwood chemistry). Svensk Papperstidn 61(18B): 625–632
Erdtman H 1959 Conifer chemistry and taxonomy of conifers. In Proc 4th Int Congr Biochem, Vienna 1958, vol II, Symp II: Biochem Wood. Pergamon, London, 1–17
Erdtman H 1963 Some aspects of chemotaxonomy. In: Swain T (ed) Chemical plant taxonomy. Academic Press, London, New York, 89–125
Erdtman H, Frank A, Lindstedt G 1951 Constituents of pine heartwood. XXVII. Svensk Papperstidn 54: 275–279
Erdtman H, Kimland B, Norin T, Daniels P J L 1968 The constituents of the 'pocket resin' from Douglas fir Pseudotsuga menziesii. Acta Chem Scan 22: 938–942
Erdtman H, Rennerfelt E 1944 Der Gehalt des Kiefernholzes an Pinosylvin-Phenolen. Svensk Papperstidn 47: 45–56
Erteld W, Mette H-J, Archterberg W 1964 Defects in wood. Leonard Hill London, 76 pp (and plates)
Esau K 1965 Plant anatomy. 2nd edn John Wiley New York, 735 pp
Esser M H 1976 (ed) Proc Lightwood Res Coord Counc, Chevron Chem, FL, USA, 154 pp
Esser M H 1977 (ed) Proc Lightwood Res Coord Counc, Southeast For Exp Stn, Asheville NC 28802 USA, 193 pp
Esser M H 1978 (ed) Proc 5th Lightwood Res Coord Counc, Southeast For Exp Stn, Asheville NC 28802 USA, 211 pp
Esser M H 1979 (ed) Proc 6th Lightwood Res Coord Counc, Southeast For Exp Stn, Asheville NC 28802 USA, 151 pp
Fahn A 1974 Plant anatomy, 2nd edn. Pergamon, Oxford, 66 pp
Fahn A 1979 Secretory tissues in plants. Academic Press, London, New York, 302 pp
Fahn A, Arnon N 1963 The living wood fibres of Tamarix aphylla and the changes occurring in them in transition from sapwood to heartwood. New Phytol 62: 99–104
Fahn, A, Zamski E 1970 The influence of pressure, wind, wounding and growth substances on the rate of resin duct formation in Pinus halepensis wood. Israel J Bot 19: 429–446
Fairhall A W, Young J A 1970 Radiocarbon in the environment. Adv Chem 93: 401–418
Farkas G L, Kiraly Z 1962 Phenolic compounds in the physiology of plant diseases and disease resistance. Phytopathology Z 44: 105–150
Farmer R H, Campbell W G 1951 The composition and origin of a crystalline deposit in iroko wood Chlorophora excelsa. Emp For Rev 30: 64–65
Fengel D 1965 Electron-microscopic contributions to the fine structure of beechwood. Holz Roh-Werkst 23: 257–263
Fengel D 1970 Ultrastructural changes during aging of wood cells. Wood Sci Technol 4: 176–188
Fengel D, Wegener G 1984 Wood: Chemistry, ultrastructure, reactions. De Gruyter, Berlin, New York, 613 pp
Fielding J M 1952 The moisture content of the trunks of Monterey pine trees. Aust For 16: 3–21
Fink S 1982 Histochemische Untersuchungen über Stärkeverteilung und Phosphataseaktivität im Holz einiger tropischer Baumarten. Holzforschung 36: 295–302
Fisher E E 1945 Manna-formation in Myoporum platycarpum. J Counc Sci Ind Res (Aust) 18: 159–166
Fletcher J M 1986 Dating of art-historical artefacts. Nature (London) 320: 466 pp
Flückiger F A 1894 Australische manna. Arch Pharm 232: 311–314
Forster M O, Rao K A N 1925 d-Mannitol from Gardenia turgida. J Chem Soc 125: 2176–2180

Fossum T, Hartler N, Libert J 1972 The inorganic content of wood. Svensk Papperstidn 75: 305–309

Foster R C 1964 Fine structure of tyloses. Nature (London) 204:494–495

Franke G 1983 Bermerkungen zur Produktion von Guttapercha. Beitr Trop Landwirtsch Veterinärmed 21: 431–436

Franklin E C 1976 Within-tree variation of monoterpene composition and yield in slash pine clones and families. For Sci 22: 185–191

Franklin E C, Squillace A E 1973 Short-term progeny tests and second generation breeding in slash pine. Can J For Res 3: 165–169

Franklin E C, Taras M A, Volkman D A 1970 Genetic gains in yields of oleoresin wood extractives and tall oil. Tappi 53: 2302–2304

Fraser H F, Swan E P 1972 A chemical test to differentiate Abies amabilis from A.lasiocarpa wood. Can For Ser Bi-Mon Res Notes 28(5): 32

Frey-Wyssling A 1964 Cytology of ageing ray cells. In: Zimmermann M H (ed) Formation of wood in forest trees. Academic Press, London, New York, 457 pp

Frey-Wyssling A, Bosshard H H 1959 Cytology of the ray cells in sapwood and heartwood. Holzforschung 13: 129–137

Frühwald A, Bauch J, Göttsche-Kühn H 1984 Holzeigenschaften von Fichten aus Waldschadensgebieten. Holz Roh-Werkst 42: 441–449

Fuchs Y, Lieberman M 1968 Effects of kinetin IAA and gibberellin on ethylene production and their interactions in growth of seedlings. Plant Physiol 43: 2029–2036

Fujikawa S, Ishida S 1973 Study on the pit of wood cells using scanning electron microscopy. IV. The membrane structure of bordered pit in Coniferae. Mokuzai Gakkaishi 19: 413–420

Fujikawa S, Ishida S 1974 Membrane structure of bordered pit on the tangential wall of tracheid in Coniferae species. Mokuzai Gakkaishi 20: 103–110

Fujita M, Nakagawa K, Mori N, Harada H 1978 The season of tylosis development and changes in parenchyma cell structure in Robinia pseudoacacia. Bull Kyoto Univ For 5: 183–190

Fukada T 1963 Studies on the chemical composition of woods. 1. On the amino acids. J Jpn Wood Res Soc 9: 166–170

Fukazawa K, Higuchi T 1965 Studies on the mechanisms of heartwood formation. Mokuzai Gakkaishi 11: 196–201

Fukazawa K, Imagawa H 1970 Some preliminary observations of ray parenchyma cell by the scanning electron microscopy. Res Bull Coll Exp For Hokkaido Univ 27: 79–90

Fukazawa K, Yamamoto K, Ishida S 1980 The season of heartwood formation in genus Pinus. In: Bauch J (ed) Natural variations of wood properties. Mitt Bundesforschungsanst Forst-Holzwirtsch 131: 113–130

Gahan P B 1981 Cell senescence and death in plants. In: Bowen I D, Lockshin R A (eds) Cell death in biology and pathology. Chapman Hall London, 145–164

Gansel C R, Squillace A E 1976 Geographic variation of monoterpene in cortical oleoresin of slash pine. Silvae Genet 25: 150–154

Gardner J A F 1962 The tropolones. In: Hillis W E (ed) Wood extractives. Academic Press, London, New York, 317–330

Gardner J A F, Barton G M 1960 The distribution of dihydroquercetin in Douglas fir and western larch. For Prod J 10: 171–173

Gener P, Du Plessix C J 1975 Effective use of tapping system and stimulation methods towards rational exploitation of Hevea. Proc Int Rubber Conf, Kuala Lumpur, Malaysia, 384–395

Gerry E 1921 American storax production: results of different methods of tapping red gum trees. J For 19(1): 1–10

Gerry E 1943 Western hemlock floccosoids. US For Ser Rep FPL – 1392

Ghosh S S, Purkayastha S K 1962 Anatomical studies of wood and bark of Acacia senegal trees with special reference to gum exudation. Indian For 88: 92–99

Gibbs J N 1968 Resins and the resistance of conifers to Fomes annosus. Ann Bot (London) 32: 649–665

Gibbs R D 1958 Water content in trees. In: Thimann K V, Critchfield W B, Zimmermann M H (eds) The physiology of forest trees. Ronald, New York, 43–87

Glock W S 1951 Cambial frost injuries and multiple growth layers at Lubbock, Texas. Ecology 32: 28–36

Gollob L 1980 Monterpene composition in bark beetle-resistant loblolly pine. Naturwissenschaften 67: 409–410

Gonzales E V 1980 Identification of white crystals from narra (Pterocarpus indicus) wood. Chem Abstr 92: 90897

Gonzales E V, Abejo F G 1978 Properties of Manila copal (Almaciga resin) from 15 different localities in The Philippines. Forpride Digest 7 (2,3): 10–15

Good H M, Nelson C D 1951 A histological study of sugar maple decayed by Polyporus glomeratus. Can J Bot 29: 215–222

Good H M, Murray P M, Dale H M 1955 Studies on heartwood formation and staining in sugar maple, Acer saccharum. Can J Bot 33: 31–43

Goodwin R H, Goddard D R 1940 Oxygen consumption in isolated woody tissues. Am J Bot 27: 234–237

Goodwin T W, Mercer E I 1983 Introduction to plant biochemistry. Pergamon Oxford, 400–454

Gorham J 1980 The stilbenoids. Progr Phytochem 6: 203–252

Gottwald H, Willeitner H 1976 Fleckenbildungen im Kernholz von Entandrophragma utile. Holz Roh-Werkst 34: 433–438

Gref R, Ericsson A 1985 Wound-induced changes of resin acid concentrations in living bark of Scots pine seedlings. Can J For Res 15: 92–96

Gregory R A, Walters R S, Gabriel W J, Carl C M, Yawney H W, Donnelly J R, Hogler N K, Garrett L D, Sendak P E, Jenkins W L 1982 Sugar maple research: sap production, processing, and marketing of maple syrup. US For Ser Gen Tech Rep NE–72, 109 pp

Grier C C, Waring R H 1974 Conifer foliage mass related to sapwood area. For Sci 20: 205–206

Grisebach H 1985 Biosynthesis of flavonoids. In: Higuchi T (ed) Biosynthesis and biodegradation of wood components. Academic Press, London, New York, 291–324

Gross G G 1985 Biosynthesis and metabolism of phenolic acids and monolignols. In: Highuchi T (ed) Biosynthesis and biodegradation of wood components. Academic Press, London, New York, 228–271

Grozdits G A, Ifju G 1969 Development of tensile strength and related properties in differentiating coniferous xylem. Wood Sci 1: 137–147

Hägglund E 1951 The chemistry of wood. Academic Press, London, New York, 15–16

Hale J D, Clermont L P 1963 Influence of prosenchyma cell-wall morphology on basic physical and chemical characteristics of wood. J Polym Sci C 2: 253–261

Hancock M V 1957 The distribution of dihydroquercetin and leucocyanidin in a Douglas-fir tree. For Prod J 7: 335–338

Harkin J M, Obst J R 1973 Lignification in trees: indication of exclusive peroxidase participation. Science 180: 296–298

Harris J M 1953 Heartwood formation in Pinus radiata. Nature (London) 172: 552

Harris J M 1954a Heartwood formation in Pinus radiata D Don. New Phytol 53: 517–524

Harris J M 1954b Heartwood formation in Pinus radiata. N Z For Serv For Res Inst Tech Pap 1, 31 pp

Harris J M 1961 The resinification of knots in radiata pine. N Z For Serv For Res Inst Res Note 28, 11pp

Harris J M 1963 The effect of pruning on the resinification of knots in radiata pine. N Z For Serv For Inst Res Note 34, 7 pp

Harris J M 1965 Enrichment of radiata pine heartwood with extractives. Proc IUFRO Sect 41 Mtg Melbourne, Australia vol I, 17 pp

Harris J M, Kripas S 1959 Notes on the physical properties of ponderosa pine, monticola pine, western red cedar and Lawson cypress grown in New Zealand. N Z For Serv For Res Inst Res Note 16, 24 pp

Harris J M, Orman H R 1958 The physical and mechanical properties of New Zealand grown Douglas fir. N Z For Serv For Res Inst Tech Pap 24, 86 pp

Harrison-Smith J L 1940–41 Kauri gum. N Z J For 4(5): 284–292

Harsch W 1912 Die Kiefer des Württembergischen Schwarzwaldes. Laupp, Tübingen, 51–57

Hart J H 1968 Morphological and chemical differences between sapwood, discolored sapwood and heartwood in black locust and osage orange. For Sci 24: 334–338

Hart J H 1981 Role of phytostilbenes in decay and disease resistance. Ann Rev Phytopathology 19: 437–458
Hart J H, McNabb H S 1963 Development of wound heartwood in Iowa hardwoods. Phytopathology 53: 87–881
Hart J H, Shrimpton D M 1979 Role of stilbenes in resistance of wood to decay. Phytopathology 69: 1138–1143
Hart J H, Wardell J F, Hemingway R W 1975 Formation of oleoresin and lignans in sapwood in white spruce in response to wounding. Phytopathology 65: 412–417
Hartig R 1894 Forst Naturwiss Z 3: 1–13 (quoted by Yazawa et al. 1965)
Hasegawa M 1958 Flavonoids contained in Prunus woods. J Jpn For Soc 40: 111–121
Hasegawa M, Shirato T 1959 Abnormal constituents of Prunus wood : Isoolivil from the wood of Prunus jamasakura. J Jpn For Soc 41: 1–3
Hasegawa M, Shiroya M 1965 The formation of phenolic compounds at the sapwood – heartwood boundary. Proc Mtg Sect 41 IUFRO vol 1, 10 pp
Hasegawa M, Shiroya M 1966 Translocation and transformation of sucrose in the wood of Prunus yedoensis. Bot Mag (Tokyo) 79: 595–601
Hasegawa M, Shiroya M 1968 Translocation and transformation of sucrose in the wood of Prunus jedoensis. II. Bot Mag (Tokyo) 81: 141–144
Haslam E 1982a The metabolism of gallic acid and hexahydroxydiphenic acid in higher plants. Prog Chem Org Nat Prod 41:1–46
Haslam E 1982b Proanthocyanidins. In: Harborne J B, Malory T J (eds) The flavonoids. Advances in research. Chapman Hall, London, 417–448
Hathway D E 1959 Experiments on the origin of oak-bark tannin. Biochem J 71: 533–537
Hathway D E 1962 The use of hydroxystilbene compounds as taxonomic traces in the genus Eucalyptus. Biochem J 83: 80–84
Hathway D E, Seakins J W T 1959 Hydroxystilbenes of Eucalyptus wandoo. Biochem J 72: 369–374
Hatt H H, Hillis W E 1947 The manna of Myoporum platycarpum R. Br. as a possible commercial source of mannitol. J Counc Sci Ind Res (Aust) 20: 207–224
Hausen B M 1981 Woods injurious to human health. De Gruyter Berlin New York, 190 pp
Haworth R D, Kelly W 1937 Constituents of natural phenolic resins. VIII. J Chem Soc 1937: 384–391
Hayashi Y, Kitao K, Sato A 1968 Chemical studies on tropical woods. I. A novel flavone from Haplormosia monophylla wood. Wood Research No 44: 68–76
Hayashi Y, Sakurai K, Takahashi T, Kitao K 1974 The heartwood constituents of Haplormosia monophylla. Mokuzai Gakkaishi 20: 595–599
Hedley B T 1936 The future of natural resins. J Oil Colour Chem Assoc 19: 75–95
Hemingway R W, Hillis W E 1970 Heartwood formation in living stumps of Douglas fir. Wood Sci Technol 4: 246–254
Hemingway R W, Hillis W E 1971 Changes in fats and resins of Pinus radiata associated with heartwood formation. Appita 24: 439–443
Hemingway R W, McGraw G W, Barras S J 1977 Polyphenols in Ceratocystis minor infected Pinus taeda. J Agric Food Chem 25: 717–722
Hergert H L, Goldschmid O 1958 Biogenesis of heartwood and bark constituents. I. Taxifolin glucoside. J Org Chem 23: 700–704
Higuchi T 1976 Biochemical aspects of lignification and heartwood formation. Wood Res 59/60: 180–199
Higuchi T 1985 Biosynthesis of lignin. In: Higuchi T (ed) Biosynthesis and biodegradation of wood components. Academic Press, London, New York, 141–160
Higuchi T, Fukazawa K 1966 Studies on the mechanism of heartwood formation. III. Mokuzai Gakkaishi 12: 135–139
Higuchi T, Fukazawa K, Nakashima S 1964 Study on the mechanism of heartwood formation. 1. Histochemistry of the wood tissue. Mokuzai Gakkaishi 10: 235–241
Higuchi T, Fukazawa, Shimada M 1967a Biochemical studies on the heartwood formation. Res Bull Coll Exp For Hokkaido Univ 25: 167–194
Higuchi T, Onda Y, Fujimoto Y 1969 Biochemical aspects of heartwood formation with special reference to the site of biogenesis of heartwood compounds. Wood Res 48: 15–30

Higuchi T, Shimada M, Watanabe K 1967b Studies on the mechanism of heartwood formation V. Changes in the pattern of glucose metabolism in heartwood formation. Mokuzai Gakkaishi 13: 269–273
Higuchi T, Shimada M, Watanabe K 1967c Studies on the mechanism of heartwood formation. VI. On the artificial heartwood of Cryptomeria japonica and Pinus densiflora. Mokuzai Gakkaishi 13: 274–279
Hillis W E 1952 The chemistry of eucalypt kinos. II. Aromadendrin, kaempferol and ellagic acid. Aust J Sci Res 5A: 379–386
Hillis W E 1954 The precursors of the red colour in leather. J Soc Leather Tra Chem 38: 91–102
Hillis W E 1956 Leucoanthocyanins as the possible precursors of extractives in woody tissues. Aust J Biol Sci 9: 263–280
Hillis W E 1958 Formation of condensed tannins in plants. Nature (London) 182 : 1371
Hillis W E 1960 Factors influencing the formation of phloem and heartwood polyphenols. Holzforschung 14: 105–110
Hillis W E (ed) 1962a Wood extractives and their significance to the pulp and paper industries. Academic Press, London, New York, 513 pp
Hillis W E 1962b The distribution and formation of polyphenols within the tree. In: Hillis W E (ed) Wood extractives and their significance to the pulp and paper industries. Academic Press, London, New York, 59–131
Hillis W E 1964 The formation of polyphenols in trees. 2. The polyphenols of Eucalyptus sieberiana kino. Biochem J 92: 516–521
Hillis W E 1966 Variation in polyphenol composition within species of Eucalyptus. Phytochemistry 5: 541–556
Hillis W E 1968a Chemical aspects of heartwood formation. Wood Sci Technol 2: 241–259
Hillis W E 1968b Heartwood formation and its influence on utilisation. Wood Sci Technol 2: 260–267
Hillis W E 1971 Distribution, properties and formation of some wood extractives. Wood Sci Technol 5: 272–289
Hillis W E 1972a Properties of eucalypt woods of importance to the pulp and paper industry. Appita 26: 113–122
Hillis W E 1972b Distribution, properties and formation of some wood extractives. Wood Sci Technol 5: 272–289
Hillis W E 1975 Ethylene and extraneous material formation in woody tissues. Phytochemistry 14: 2559–2562
Hillis W E 1977 Secondary changes in wood. In: Loewus F A, Runeckles V C (eds) Recent advances in phytochemistry 11: 247–309. Plenum, New York, 527 pp
Hillis W E 1978 Wood quality and utilization. In: Hillis W E, Brown A G (eds) Eucalypts for wood production. CSIRO Australia, 259–289
Hillis W E 1985a Occurrence of extractives in wood tissue. In: Higuchi T (ed) Biosynthesis and biodegradation of wood components. Academic Press, London, New York, 209–228
Hillis W E 1985b Biosynthesis of tannins. In: Higuchi T (ed) Biosynthesis and biodegradation of wood components. Academic Press, London, New York, 325–347
Hillis W E 1986 Forever Amber – a story of the secondary wood components. Wood Sci Technol 20: 203–227
Hillis W E, Carle A 1958a Extractives from klinki pine Araucaria klinkii. Nature (London) 1982: 1593–1594
Hillis W E, Carle A 1958b The polyphenols and shikimic acid of eucalypt cambium and wood. Holzforschung 12: 136–141
Hillis W E, Carle A 1959 The influence of extractives on eucalypt pulping and paper making. Appita 13: 74–83
Hillis W E, Carle A 1962 The origin of the wood and bark polyphenols of Eucalyptus species. Biochem J 82: 435–439
Hillis W E, Carle A 1963 The chemistry of eucalypt kinos. IV. Eucalyptus hemiphloia kino. Aust J Chem 16: 147–159
Hillis W E, de Silva D 1979 Inorganic extraneous constituents of wood. Holzforschung 33: 47–53

Hillis W E, Ditchburne N 1974 The prediction of heartwood diameter in radiata pine trees. Can J For Res 4: 524–529; 1975, 5: 743
Hillis W E, Hasegawa M 1963 The formation of polyphenols in trees. 1. Administration of 14C glucose and subsequent distribution of radioactivity. Phytochemistry 2: 195–199
Hillis W E, Humphreys F R, Bamber R K, Carle A 1962 Factors influencing the formation of phloem and heartwood polyphenols. Holzforschung 16: 114–121
Hillis W E, Inoue T 1966 The formation of polyphenols in trees. III. The effect of enzyme inhibitors. Phytochemistry 5: 483–490
Hillis W E, Inoue T 1968 The formation of polyphenols in trees. IV. The polyphenols formed in Pinus radiata after Sirex attack. Phytochemistry 7: 13–22
Hillis W E, Ishikura N 1970 The biosynthesis of polyphenols in tissues with low phenylalanine ammonia–lyase activity. Phytochemistry 9: 1517–1528
Hillis W E, Ishikura N, Foster R C, Marks G C 1968 The role of extractives in the formation of ectotrophic mycorrhizae. Phytochemistry 7: 409–410
Hillis W E, Orman H R 1962 The extractives of New Zealand Nothofagus species. J Linnean Soc (Bot) 58: 175–184
Hillis W E, Swain T 1959 Phenolic constituents of Prunus domestica. III. J Sci Food Agric 10: 533–537
Hillis W E, Yazaki Y 1973a Polyphenols of Intsia heartwoods. Phytochemistry 12: 2491–2495
Hillis W E, Yazaki Y 1973b Wood polyphenols of Eucalyptus polyanthemos. Phytochemistry 12: 2969–2977
Hillis W E, Yazaki Y 1974 Kinos of Eucalyptus species and their acid degradation products. Phytochemistry 13: 495–498
Hillis W E, Yazaki Y, Bauch J 1976 The significance of anomalous extractives in the heart-shakes in Dacrydium species. Wood Sci Technol 10: 79–95
Hirai S 1951 Study on the process of heartwood growth in the Japanese larch stem. Trans 59th Meet Jpn For Soc, 231–234
Hirai S 1952 The early stage of the transformation of sapwood of Japanese larch into heartwood. Res Bull Coll Exp For Hokkaido Univ 15: 239–253
Hirakawa Y, Ishida S, Ohtani J 1979 A SEM observation of organelles in the cambial and living xylem cells in Todomatsu (Abies sachalinensis). Res Bull Exp For Hokkaido Univ 36: 459–468
Hirata T, Saiki H, Harada H 1972 Observations of crystals and silica inclusions in parenchyma cells of certain tropical woods by scanning electron microscope. Bull Kyoto Univ For 44: 194–199
Höll W 1972 Stärke und Stärkeenzyme in Holz von Robinia pseudoacacia. Holzforschung 26: 41–45
Höll W 1973 Physiological and biochemical gradients in trunks of hardwoods with special reference to heartwood formation. Proc Meet IUFRO Division 5, Pretoria S Afr, 494–502
Höll W 1974 Dark CO_2 fixation by cell-free preparations of the wood of Robinia pseudoacacia. Can J Bot 52: 727–734
Höll W 1975 Radial transport in rays. In: Zimmermann M H, Milburn J A (eds) Transport in plants. I. Phloem transport. Springer, Berlin, Heidelberg, New York, 432–450
Höll W, Lendzian K 1973 Respiration in the sapwood and heartwood of Robinia pseudoacacia. Phytochemistry 12: 975–977
Höll W, Meyer H 1977 Enzyme catalyzed CO_2 fixation by wood preparations of Tilia cordata and Fraxinus excelsior. Holzforschung 31: 184–187
Höll W, Poschenrieder G 1975 Radial distribution and partial characterization of lipids in the trunk of three hardwoods. Holzforschung 29: 118–123
Höll W, Priebe S 1985 Storage lipids in the trunk and rootwood of Tilia cordata from the dormant to the growing period. Holzforschung 39: 7–10
Holton R W, Winston R D 1976. Turpentine and tall oil analysis of 8% paraquat-treated Virginia loblolly and short-leaf pines in the Tennessee Valley. Lightwood Res Coor Counc Proc Ann Meet, Jacksonville, FL, 77–80
Hook D D, Brown C L, Wetmore R H 1972 Aeration in trees. Bot Gaz 133(4): 443–454
Hoque E 1982 Biochemical aspects of stress physiology of plants and some considerations of defence mechanisms. Eur J For Pathol 12: 280–296

Hosli J P 1982 Wood preservation in the pre-industrial period. Int J Wood Pres 2: 29–36
Howes F N 1949 Vegetable gums and resins. Chronica Botanica, Waltham MA, USA. 180 pp
Howes F N 1950 Age-old resins of the Mediterranean region and their uses. Econ Bot 4(4): 307–316
Hyodo H, Yang S F 1971 Ethylene–enhanced synthesis of phenylanine ammonia-lyase in pea seedlings, Plant Physiol 47: 765–770
Huber G 1956 Die Saftströme der Pflanzen. Springer, Berlin, Göttingen, Heidelberg, 126pp
Hugentobler U H 1965 Zur Cytologie der Kernholzbildung. Vierteljahrsch Naturforsch Ges Zürich 110(2): 321–342
Hughes J C, Swain T 1960 Scopolin production in potato tubers infected with Phytophthora infestans. Phytopathöology 50: 398–400
Humphreys F R, Martin P J 1956 A survey of the tannin content of the bark of plantation-grown Pinus radiata. Aust For 20: 96–101
Hunter A G, Goggans J F 1968 Variation in specific gravity, diameter growth and colored heartwood of sweetgum in Alabama. Tappi 51: 76–79
Hyodo H, Yang S F 1971 Ethylene enhanced synthesis of phenylalanine ammonia-lyase in pea seedlings. Plant Physiol 47: 765–770
Ihara N 1972 A dendrometrical study on heartwood formation of forest trees. Bull Kyushu Univ For 46 (March): 123–129
Imamura H, Fushiki H, Ishihara S, Oshashi H 1972 Phenolic constituents of merbau wood and the specks on board. Res Bull Fac Agric Gifu Univ 33: 39–46
Imamura H, Nomura K, Hibino Y, Ohashi H 1974 A new flavonol in the shake of merbau wood (Intsia sp). Res Bull Fac Agric Gifu Univ 36: 93–101
Irvine C J 1936 A note on gum vein formation in saplings of mountain ash (E.regnans). J Counc Sci Ind Res (Aust) 9: 91–94
Isenberg I H 1936 Age and the chemical composition of white fir wood. J Am Chem Soc 58: 2231–2233
Ishida S, Ohtani J, Kawarada T 1976 Study of tyloses by the scanning electron microscopy. Yearly and seasonal development of tyloses in Harienju Robinia pseudo-acacia. Trans 8th Meet Hokkaido Branch Jpn Wood Res Soc, 6–9
Ito M 1953 Studies on the heartwood and its quantity in trunk. Science Rept No. 1 Fac Liberal Arts Educ. Gifu Univ Jpn, 63–67
Jacobs M R 1937 Field studies on the gum veins of the eucalypts. Commonwealth For Bur Bull 20. Gov Printer, Canberra, Australia, 36 pp
Jacobs M R 1955 Growth habits of the eucalypts. For Timb Bur Canberra, Australia, 262 pp
Jacquinot C, Hervet N 1954 Contribution à l'étude des "poches à kino" des Eucalyptus. Rev Pathol Vég Ent Agric Fr 33: 199–207
Jandrier M 1893 C R Acad Sci 117: 498
Jane F W 1970 The structure of wood. Adam Charles Black, London, 478 pp
Janin G, Clément A 1972 (Evidence for the presence of calcium carbonate crystals in Poplar wood attributable to the distribution of minerals during the formation of heartwood). Ann Sci For 29: 67–105
Jensen K F 1967 Measuring oxygen and carbon dioxide in red oak trees. US For Serv Res Note NE–74, 4 pp
Jensen K F 1969 Oxygen and carbon dioxide concentrations in sound and decaying red oak trees. For Sci 15: 246–251
Jerie P H, Shaari A R, Hall M A 1979 The compartmentation of ethylene in developing cotyledons of Phaseolus vulgaris. Planta 144: 503–507
Joel D M, Fahn A 1980 Ultrastructure of resin ducts of Mangifera indica. Part I. Ann Bot (London) 46: 225–233; Part II. Ann Bot (London) 46: 779–783
Johansson M, Theander O 1974 Changes in sapwood of roots of Norway spruce attacked by Fomes annosus. I. Physiol Plant 30: 218–225
Jones R L 1968 Ethylene enhanced release of α-amylase from barley aleurone cells. Plant Physiol 43: 442–444
Jorgensen E 1961 The formation of pinosylvin and its monomethyl ether in the sapwood of Pinus resinosa. Can J Bot 39: 1765–1772
Jorgensen E 1962 Observations on the formation of protection wood. For Chron 38: 292–294

Jorgensen E, Balsillie D 1969 Formation of heartwood phenols in callus tissue of red pine (Pinus resinosa). Can J Bot 47: 1015–1016
Joyce E R, Drew J 1979 SCM field experiments with paraquat substitutes. In: Esser M H (ed) 6th Ann Lightwood Res Conf Proc, Atlanta GA, 82–103
Joyce E R, Peters W J, Roberts D R 1977 Final report on Cooperative Orange Park Florida paraquat study. Lightwood Res Coor Counc Proc Ann Meet, Atlantic Beach, FL, 65–70
Jurášek L 1956 Vznik thyl v bukoven dreve. Drev Výsk 1:7–15
Jurášek 1958 The influence of wood temperature and moisture content on the formation of tyloses in beech. Drev Výsk 3: 5–13
Kai Y, Teratani F 1977 Studies on the colour of the heartwood of sugi Cryptomeria japonica. I. Mokuzai Gakkaishi 23: 499–503
Kandeel S A, McGinnes E A 1970 Ultrastructure of ring shake in scarlet oak. Wood Sci 2: 171–178
Kärkkäinen M 1972 On the proportion of heartwood in Norway spruce and Scots pine. Silva Fennica 6(3): 193–208
Kaufmann M R, Troendle C A 1981 The relationship of leaf area and foliage biomass to sapwood conducting area in four subalpine forest tree species. For Sci 27: 477–482
Kemp M S, Burden R S 1984 Isolation and structure determination of γ–pyrufuran from the wood of Pyrus communis. J Chem Soc Perkin Trans 1: 1441–1443
Kemp M S, Burden R S, Loeffler R S T 1983 Isolation, structure determination, and total synthesis of dibenzofurans from the wood of Pyrus communis. J Chem Soc Perkin Transl: 2267–2272
Kemp M S, Holloway P J, Burden R S 1985 3β, 19α-dihydroxy-2-oxours-12-en-28-oic-acid: a pentacyclic triperpene induced in the wood of Malus pumila infected with Chondostereum purpureum. J Chem Soc M: 1848–1876
Kennedy R W, Wilson J W 1956 Variation in taxifolin content of a Douglas-fir stem exhibiting target ring. For Prod J 6: 230–231
Kessell S L 1921 The occurrence of "gum veins" and "pockets" in marri (E.calophylla). Western Aust For Dep Rep, 9 pp
Kessel S L, Stoate T N 1938 Nutrition of radiata pine. W Aust For Dep Bull 50
Kiatgrajai P, Conner A H, Rowe J W, Peters W, Roberts 1976a Attempt to induce lightwood in balsam fir and tamarack by treating with paraquat. Wood Sci 9: 31–36
Kiatgrajai P, Rowe J W, Conner A H, Peters W, Roberts D R 1976b Attempt to induce lightwood in eastern hemlock by treating with paraquat. Wood Sci 8: 170–173
Kile G A, Turnbull C R A 1974a The effect of radiata pine resin and resin components on the growth of the Sirex symbiont Amylostereum areolatum. Aust For Res 6(4): 27–34
Kile G A, Turnbull C R A 1974b Drying in the sapwood of radiata pine after inoculation with Amylostereum areolatum and Sirex mucus. Aust For Res 6(4): 35–40
Kile G A, Turnbull C R A, Podger F D 1981 Effect of regrowth dieback on some properties of Eucalyptus obliqua trees. Aust For Res 11: 55–62
Kile G A, Wade G C 1975 Trametes versicolor in apple. II. Phytopathology Z 82: 1–24
Kimland B, Norin T 1972 Wood extractives of common spruce, Picea abies. Svensk Papperstidn 75: 403–409
Kindl H 1985 Biosynthesis of stilbenes. In: Higuchi T (ed) Biosynthesis and biodegradation of wood components. Academic Press, London, New York, 349–377
King F E, Clarke-Lewis J W, Forbes W F 1955 Extractives from hardwoods, epiafzelechin a member of the catechin series. J Chem Soc: 2948–2953
King H G C, White T 1957 Tannins and polyphenols of Schinopsis (quebracho) species: their genesis and interrelation. J Soc Leather Trades Chem 41: 368–384
Kininmonth J A 1972 Permeability and fine structure of certain hardwoods and effects on drying. II. Holzforschung 26: 32–38
Knowles E, White T 1958 The protection of metals with tannins. J Oil Color Chem Assoc 41: 10–23
Knuchel H 1947 Holzfehler. 3rd edn. Classen, Stuttgart, Zurich, pp 30,35; 119 pp
Koch P 1972a Utilisation of the southern pines. US Dep Agric For Serv Agric Hand 420, vol I, 267–275

Koch P 1972b Utilisation of the southern pines. US Dep Agric For Serv Agric Hand 420, vol I, 267–275

Koch P 1972c Utilisation of the southern pines. US Dep Agric For Serv Agric Hand 420, vol II, 1476–1493

Kohara J 1972 Ki No Bunka. Kashima, Tokyo, 217

Kondo T 1964 On the wood enzyme. J Jpn Wood Res Soc 10: 43–48

König E 1957 Fehler des Holzes. Classen, Stuttgart, Zurick, 119 pp

Koran Z, Côté W A 1965 The ultrastructure of tyloses. In: Côté W A (ed) The cellular ultrastructure of woody plants. Univ Press Syracuse, N Y, 319–333

Kozlik C J 1982 Kiln-drying Douglas-fir pole sections: effects on strength and checking. For Prod J 32(6): 24–30

Kozlowski T T, Keller T 1966 Food relations of woody plants. Bot Rev 32: 293–382

Krahmer R L, Côté W A 1963 Changes in coniferous wood cells associated with heartwood formation. Tappi 46(1): 42–49

Krahmer R L, Hemingway R W, Hillis W E 1970 The cellular distribution of lignans in Tsuga heterophylla wood. Wood Sci Technol 4: 122–139

Krahmer R L, Wellons J D 1973 Some anatomical and chemical characteristics of Douglas-fir bark. Wood Sci 6(2): 97–105

Kramer P J, Kozlowski T T 1979 Physiology of trees. McGraw-Hill, New York, 811 pp

Kuo M L, Arganbright D G 1980a Cellular distribution of extractives in redwood and incense cedar. I. Holzforschung 34: 17–22

Kuo M L, Arganbright D G 1980b Cellular distribution of extractives in redwoods and incense cedar. II. Holzforschung 34: 41–47

Kuroda H, Shimaji K 1983 Distribution of coloring substances in sugi heartwood. Holzforschung 37: 225–230

Kurth E F 1950 Methyl ester of dehydroperillic acid, an odoriferous constituent of western red cedar. J Am Chem Soc 72: 5778–5779

Kurth E F 1952 The volatile oils. In: Wise L E, Jahn E C (eds) Wood chemistry, 2nd edn. Reinhold, New York, 548–589

Kutscha N P, Lomerson J T, Dyer M V 1978 Separation of eastern spruce and balsam fir by chemical methods. Wood Sci Technol 12: 293–308

Kutscha N P, Sachs I B 1962 Color tests for differentiating heartwood and sapwood in certain softwood tree species. US For Ser Rep FPL 2246, 16 pp

Lai Y Z 1971 Proposed biogenetic pathway of tropolones in the family Cupressaceae. Cellul Chem Technol 5: 621–623

Laidlaw R A, Smith G A 1965 The proteins of the timber of scots pine (Pinus sylvestris). Holzforschung 19: 129–134

Lairand D B 1963 Contribution to the cytochemistry of wood. Drev Výsk 1: 1–10 (CSIRO Trans 7409)

Lambert M J 1981 Inorganic constituents in wood and bark of New South Wales forest tree species. For Comm NSW Res Note 45, 43 pp

Lambert M J, Turner J, Kelly J 1983 Nutrient relationships of tree species in a New South Wales subtropical rainforest. Aust For Res 13: 91–102

Langenheim J H 1969 Amber: a botanical inquiry. Science 163: 1157–1169

Langenheim J H, Lincoln D E, Stubblebine W H, Gabrielli A C 1982 Evolutionary implications of leaf resin pocket patterns in the tropical tree Hymenaea (Caesalpinioideae legumonosae). Am J Bot 69: 595–607

Lappi-Seppälä M 1952 Männyn sydänpunsta ja runkomuodosta. (Über Verkernung und Stammform der Kiefer.) Communicationes Inst For Fenniae Helsinki 40 (25): 1–26

Lee K Y, Fukazawa K, Ishida S 1985 Distribution, morphology and chemical composition of the crystals in the wood and bark of Populus maximowiczii. Res Bull Coll Exp For Fac Agric Hokkaido Univ Jpn 42: 595–608

Leibundgut H 1983 (Investigations on Larix decidua of various provenances. 3). Schweiz Z Forstwes 134 (1): 61–62

Leistner E 1985 Occurrence and biosynthesis of quinones in woody plants. In: Higuchi T (ed) Biosynthesis and biodegradation of wood components. Academic Press, London, New York, 273–290

Liese J 1930 Deut Forstwirt 12: 812–818

Liese W, Bauch J 1967 On the closure of bordered pits in conifers. Wood Sci Technol 1: 1–13

Lindgren B, Norin T 1970 (The chemistry of natural resins). Svensk Papperstidn 72: 143–153 (Can Trans 3182)

Lloyd J A 1978 Distribution of extractives in Pinus radiata earlywood and latewood. N Z J For Sci 8: 288–294

Lohwag K 1968 Eichenholz wertlos durch Mondringe? Holzforsch Holzverwert 20: 61–62

Long J N, Smith F W, Scott D R M 1981 The role of Douglas-fir stem sapwood and heartwood in the mechanical and physiological support of crowns and development of stem form. Can J For Res 11: 459–464

Lucas A 1962 Ancient Egyptian materials and industries. 4th edn. Arnold, London, 323–324

Lyr H 1967 Über den jahreszeitlichen Verlauf der Schutzkernbildung bei Pinus silvestris nach Verwundungen. Arch Forstwes 16: 51–57

MacDonald I R C, Porter L J 1969 The resin and fatty acid composition of Pinus radiata whole wood. N Z J Sci 12: 352–362

MacDougal D T 1936 The communication of the pneumatic systems of trees with the atmosphere. Proc Am Philos Soc 76: 823–845

MacDougal D T, Working E B 1933 The pneumatic system of plants, especially trees. Carnegie Inst, Washington USA Pub 441, 87 pp

MacDougal D T, Overton J B, Gilbert M S 1929 The hydrostatic – pneumatic system of certain trees. Carnegie Inst, Washington USA Pub 397

MacKinney A L, Chaiken L E 1935 Heartwood in second growth loblolly pine. US For Serv Appalachian For Exp Sta Tech Note 18, 13 pp

MacLean H, Gardner J A F 1958 Distribution of fungicidal extractives in target pattern heartwood of western red cedar. For Prod J 8: 107–108

Mahood S A, Gerry E 1921 The production of American storax. Druggists Circular, New York, January, 1–3

Maiden J H 1889a The examination of kinos as an aid in the diagnosis of eucalypts. I. Proc Linnean Soc N S W 4(3): 605–618

Maiden J H 1889b The examination of kinos as an aid in the diagnosis of eucalypts. II. Proc Linnean Soc N S W 4(3): 1277–1287

Maiden J H 1889c Botany Bay or Eucalyptus kino. Pharmaceutical Soc 20 (N S): 221–222, 321–323

Maiden J H 1891 The examination of kinos as an aid in the diagnosis of eucalypts. III. Proc Linnean Soc N S W 6: 389–426

Maiden J H 1894 Australian sandarac. Agric Gaz N S W 5: 301–305

Maiden J H 1901 The gums, resins and vegetable exudations of Australia. J Proc Roy Soc N S W 35: 161–212

Maiden J H, Smith H G 1895 Contributions to a knowledge of Australian vegetable exudations. J Proc Roy Soc N S W 29: 393–404

Malan F S, de Villiers A M 1976 Die invloed van radiale harsgein-filtreerde kernskeure op die gesaagde opbrengs vit Pinus elliottii. S Afr For J 97: 47–57

Mantell C L, Kopf C W, Curtis J L, Roger E M 1942 The technology of natural resins. John Wiley, New York, 506 pp

Mapson L W, Hulme A C 1970 The biosynthesis, physiological effects and mode of action of ethylene. In: Reinhold L, Liwschitz Y (eds) Progress in phytochemistry. vol II. Wiley-Interscience, New York, 343–384

Marchand P J 1984 Sapwood area as an estimator of foliage biomass and projected leaf area for Abies balsamea and Picea rubens. Can J For Res 14: 85–87

Matsukuma N, Kawano H, Shibata Y, Kondo T 1965 Studies on the intermediate zone of sugi wood. Some physiological activities of the xylem tissues. Mokusai Gakkaishi 11: 227–231

McCaw 1983 Wood defect associated with fire scars on jarrah (Eucalyptus marginata). Aust For Res 13: 261–266

McGinnes E A Jr 1968 Extent of shake in black walnut. For Prod J 18(5) : 80–82

McGinnes E A, Chang C I-J, Wu K Y-T 1971 Ring shake in some hardwood species: the individual tree approach. J Polym Sci (C) No 36: 153–176

McGinnes E A, Kandeel S A, Szopa P S 1969 Frequency and selected anatomical features of included sapwood in eastern red cedar. Wood Sci 2: 100–106
McGinnes E A, Shigo A L 1975 A note on effects of wounds on heartwood formation in white oak (Quercus alba). Wood Fiber 6: 327–331
McGookin A, Heilbron I M 1926 Tannin occurring in the kino of Eucalyptus calophylla. J Pharmacol Exp Therap 26: 421–446
McMichael B L, Jordan W R, Powell R D 1972 An effect of water stress on ethylene production by intact cotton petioles. Plant Physiol 49: 658–660
Meiggs R 1982 Trees and timber in the ancient Mediterranean world. Clarendon, Oxford, 553 pp
Mergen F, Hoekstra P E, Echols R M 1955 Genetic control of oleoresin yield and viscosity in slash pine. For Sci 1 (1): 19–30
Merrill W E, Cowling E B 1966 Role of nitrogen in wood deterioration : amounts and distribution of nitrogen in tree stems. Can J Bot 44: 1555–1580
Meyer B S 1956 The hydrodynamic system. In: Ruhland W (ed) Handbuch der Pflanzenphysiologie. III. Springer, Berlin, Göttingen, 596–614
Meyer R W, Barton G M 1971 A relationship between collapse and extractives in western red cedar. For Prod J 21(4): 58–60
Meyer R W, Leney L 1968 Shake in coniferous wood – an anatomical study. For Prod J 18(2): 51–56
Meyer V, Jacobsen P 1924 Lehrbuch der organischen Chemie. II, Part V. De Gruyter, Berlin, 166
Michels P 1943 Der Naßkern der Weißtanne. Holz Roh-Werkst 6: 87–99
Miller A R, Pengelly W L, Roberts L W 1984 Introduction of xylem differentation in Lactuca by ethylene. Plant Physiol 75: 1165–1166
Miller A R, Roberts L W 1984 Ethylene biosynthesis and xylogenesis in Lactuca pith explants cultured in vitro in the presence of auxin and cytokinin: The affect of ethylene precursors and inhibitor. J Exp Bot 35: 691–698
Miniutti V P 1977 Microscopic observations of paraquat-induced lightwood in slash pine. Wood Sci 9: 113–117
Möller C M, Müller D 1938 (Respiration in old tree stems.) For Forsgovaes 15: 113–138 (CSIRO Trans 9944)
Morgan J W W, Orsler R J 1969 The interaction of wood with organic solvents. II. Holzforschung 23: 48–51
Morrison J C, Polito V S 1985 Gum duct development in almond fruit Prunus dulcis. Bot Gaz 146(1): 15–25
Mullick D B 1977 The non-specific nature of defense in bark and wood during wounding, insect and pathogen attack. In: Loewus F A, Runeckles V C (eds) Recent advances in phytochemistry II. Plenum, New York, 395–441
Mutton D B 1962 Wood resins. In: Hillis W E (ed) Wood extractives. Academic Press, London, New York, 331–363
Nair M N B, Chavan R R 1983 Nuclear changes in the ageing ray parenchyma cells in relation to heartwood formation. IAWA Bull N S 4: 265–270
Nair M N B, Shah J J 1983 Histochemistry of paraquat treated wood in Azadirachta indica. IAWA Bull N S 4: 249–254
Nair M N B, Shah J J, Pandalai R C 1981 Wood anatomy and histochemical changes of sapwood during heartwood formation in Bridelia retusa. Proc Indian Acad Sci (Plant Sci) 90: 425–433
Nair M N B, Shah J J, Subramanyam S V 1983 Ultrastructure and histochemistry of traumatic gum ducts in the wood of Azadirachta indica. IAWA Bull N S 4: 103–112
Nečesaný 1956 (Classification of beech heart). Drev Výsk 11: 93–98 (For Abstr 18: 2251)
Nečesaný 1958a Jàdro Buku, Struktura, Vznik a Vyvoj. Vydavatelstvo Slov Akad Vied, Bratislava, 231 pp (22 plates)
Nečesaný 1958b The change of the parenchymatic cells vitality and the physiological base for the formation of beech heart. Drev Výsk 3: 15–26
Nečesaný 1965 Heartwood formation as a physiological aging process. Proc Meet Sect 41 IUFRO Melbourne 1: 8 pp

Nečesaný 1966 (Changes in the vitality of parenchyma cells as the phyiological bases of heartwood formation.) Holzforsch Holzverwert 18: 61–65
Nečesaný 1968 The biophysical characteristics of two types of heartwood formation in Quercus cerris. Holzforsch Holzverwert 20: 49–52
Nečesaný 1973a Kinetics of secondary changes in living xylem. I. Time dependent formation of tyloses and polyphenolic substances. Holzforschung 27: 73–76
Nečesaný 1973b Kinetics of secondary changes in living xylem. II. Some biophysical aspects. Holzforschung 27: 77–79
Nekrasova G N 1973 (Oxygen absorption by aspen wood in relation to heartwood formation.) Biol Nauk 1: 67–70 (For Abstr 34:4762)
Nelson N D 1975 Extractives produced during heartwood formation in relation to amounts of parenchyma in Juglans nigra and Quercus rubra. Can J For Res 5: 291–301
Nelson N D 1976 Gross influences on heartwood formation in black walnut and black cherry trees. USDA For Ser FPL Res Pap 268: 11 pp
Nelson N D 1978 Xylem ethylene, phenol-oxidising enzymes and nitrogen and heartwood formation in walnut and cherry. Can J Bot 56: 626–634
Nelson N D, Hillis W E 1976 Resinosis induced in pine trees with ethrel. Australian patent application
Nelson N D, Hillis W E 1978a Genetic and biochemical aspects of kino vein formation in Eucalyptus. I. Genetic variation. Aust For Res 8: 75–81
Nelson N D, Hillis W E 1978b Genetic and biochemical aspects of kino vein formation in Eucalyptus. II. Hormonal influence. Aust For Res 8: 83–91
Nelson N D, Hillis W E 1978c Ethylene and tension wood formation in Eucalyptus gomphocephala. Wood Sci Technol 12: 309–315
Nelson N D, Maeglin R R, Wahlgren H E 1969 Relationship of black walnut wood color to soil properties and site. Wood Fiber 1(1): 29–37
Nelson N D, Rietveld W J, Isebrands J G 1981 Xylem ethylene production in five black walnut families in the early stages of heartwood formation. For Sci 27: 537–543
Nicholls J W P 1965 Heritability of heartwood formation in Pinus radiata. Nature (London) 207: 320
Nicholls J W P, Brown A G 1974 The inheritance of heartwood formation in Pinus radiata D Don. Silvae Genet 23: 138–141
Nicholls J W P, Dadswell H E 1965 Assessment of wood qualities for tree breeding. 3. In Pinus radiata. CSIRO (Aust) Div For Prod Tech Pap 37: 34 pp
Nicholls J W P, Griffin A R 1978 Variation in wood characteristics in a field trial of Eucalyptus obliqua, E. regnans and some intermediate forms. Aust For Res 8: 93–102
Nicholls J W P, Matheson A C 1980 Variation in wood characteristics in thinnings from a field trial of Eucalyptus obliqua. Aust For Res 10: 239–247
Nicholls J W P, Phillips F H 1970 Preliminary study of coppice-grown Eucalyptus viminalis as a source of chip material. CSIRO (Aust) Div For Prod Tech Pap 58: 28 pp
Nickel S 1960 Beitrag zur quantitativen Proteinstickstoff-Bestimmung in Holz. Holzforschung 14: 150–152
Nix L E 1976 Paraquat induction of resin soaking in pines in the South Carolina Piedmont. In: Esser M H (ed) Lightwood Res Coor Counc Proc Ann Meet, Jacksonville FL, 102–108
Nobuchi T, Harada H 1968 Electron microscopy of the cytological structure of the ray parenchyma cells associated with heartwood formation of sugi (Cryptomeria japonica). Mokuzai Gakkaishi 14: 197–202
Nobuchi T, Harada H 1983 Physiological features of the "white zone" of sugi (Cryptomeria japonica D Don) – cytological structure and moisture content. Mokuzai Gakkaishi 29: 824–832
Nobuchi T, Harada H 1985 Ultrastructural changes in parenchyma cells of sugi (Cryptomeria japonica) associated with heartwood formation. Mokuzai Gakkaishi 31: 965–973
Nobuchi T, Kamizono Y, Harada H 1976 Cytological changes of parenchyma cells associated with heartwood formation – in three softwood species, Sugi, Momi and Akamatsu. Bull Kyoto Univ For 48: 178–186
Nobuchi T, Kuroda K, Iwata R, Harada H 1982 Cytological study of the seasonal features of heartwood formation of sugi (Cryptomeria japonica). Mokuzai Gakkaishi 28: 669–676

Nobuchi T, Matsuno H, Harada H 1984a Relationship between heartwood phenols and cytological structure in the transition zone from sapwood to heartwood of sugi (Cryptomeria japonica). Pacific Regional Wood Anatomy Conf (IAWA/IUFRO) Tsukuba, 132–134

Nobuchi T, Matsuno H, Harada H 1985 Radial distribution of heartwood phenols and the related cytological structure in the fresh wood of sugi (Cryptomeria japonica). Mokuzai Gakkaishi 31: 711–718

Nobuchi T, Sato T, Iwata R, Harada H 1984b Season of heartwood formation and the related cytological structure of ray parenchyma cells in Robinia pseudoacacia. Mokuzai Gakkaishi 30: 628–636

Nobuchi T, Takahara S, Harada H 1979 Studies on the survival rate of ray parenchyma cells with ageing process in coniferous secondary xylem. Bull Kyoto Univ For No 51: 239–246

Nördlinger H 1860 Die technischen Eigenschaften der Hölzer. Cotta'scher Verlag (see Yazawa and Ishida 1965)

Nutsubidze N N, Abashidze O V 1980 (Formation of ellagitannins in aging parenchymal cells of oak tree woods.) Bull Acad Sci Georg SSR 100 2: 473–476

Nyrén V, Back E 1960 Characteristics of parenchymatous cells and tracheidal ray cells in Picea abies. Svensk Papperstidn 63: 501–509

Ohashi H, Imamura H 1973 Resin exudates of the diseased Japanese cypress (Chamaecyparis obtusa). II. Mokuzai Gakkaishi 19: 599–603

Orman H R, Will G M 1960 The nutrient content of Pinus radiata trees. N Z J Sci 3: 510–522

Packman D F 1960 The acidity of wood. Holzforschung 14: 178–183

Paclt J 1953a Über die Entstehung der Mondringigkeit der Eiche. Phytopathol Z 20: 210–213

Paclt J 1953b Kernbildung der Buche (Fagus sylvatica). Phytopathol Z 20: 255–259

Pakianathan S W 1970 The search for new stimulants. Planters' Bull Rubber Res Inst Malaya 111: 351–365

Pakianathan S W, Hamzah S B, Sivakumaran S, Gomez J B 1982 Physiological and anatomical investigations on long-term ethephon stimulated trees. J Rubber Res Int Malaysia 30 (2): 63–79

Pandalai R C, Nair G M, Shah J J 1985 Ultrastructure of ray parenchyma cells in the wood of Melia azedarach. Wood Sci Technol 19: 201–209

Panshin A J, de Zeeuw C 1970 Textbook of wood technology. 1. McGraw-Hill, New York, 706 pp

Paradies I, Konze J R, Elstner E F, Paxton J 1980 Ethylene: indicator but not inducer of phytoalexin synthesis in soybean. Plant Physiol 66: 1106–1109

Parameswaran N, Bauch J 1975 On the origin of phenolic compounds in the wood rays of Abies alba. Wood Sci Technol 9: 165–173

Parker J 1954 Available water in stems of some Rocky Mountain conifers. Bot Gaz 115: 380–385

Passialis C, Tsoumis G 1984 Characteristics of dicoloured and wetwood in fir. IAWA Bull N S 5: 111–120

Patel J D, Bhat K V 1984 Induction of discoloured wood in Samanea saman. IAWA Bull N S 5: 152–154

Paul B H 1930 Heartwood in second-growth southern pine. South Lumberman Oct.1

Paul B H 1932 The relation of certain forest conditions to the quality and value of second-growth loblolly pine lumber. J For 30: 4–21

Paul B H 1952 Variability in wood of southern pines as influenced by silvicultural practices. US For Ser Rep FPL – R1923: 10 pp

von Pechmann H, von Aufsess H 1973 Struktur und Farbe als Qualitäts merkmale von Eichenfurnierholz. Holzforschung 27: 51–59

Peck E C 1953 Moisture content of wood. US For Ser Rep FPL D768

Pensar G 1967 (The distribution and composition of extractives in wood. I. Earlywood and latewood in spruce.) Inst Träkemi Åbo Akad Åbo, Finl, 31 pp

Pensar G 1968 (The distribution and composition of extractives in wood. 2. Earlywood and latewood in pine.) Inst Träkemi Åbo Akad Åbo, Finl, 33 pp

Pensar G 1969 (Studies on the radial distribution and composition of wood rosin in cross sections of Scandinavian spruce and pine.) Inst Träkemi Åbo Akad Åbo, Finl, 35 pp

Perilä O 1954 (The location of the resins fats and waxes in the stem cells of birch.) Pap Puu 36: 213–216

Perkin A G, Everest A E 1918 The natural organic coloring matters. Longmans Green, London, 655 pp
Perry D H, Lenz M, Watson J A L 1985 Relationships between fire, fungal rots and termite damage in Australian forest trees. Aust For 48(1): 46–53
Peters W J, Roberts D R 1976 Paraquat induced lightwood develops in all Southern pines. Lightwood Res Coor Counc Proc Ann Meet, Jacksonville FL, 55–65
Peters W J, Roberts D R 1977 Ethrel-bipyridylium synergism in slash pine. Lightwood Res Coor Counc Proc Ann Meet, Atlantic Beach FL, 78–83
Peters W J, Roberts D R, Munson J W 1978 Ethrel diquat-paraquat inter-action in lightwood formation. Lightwood Res Coor Counc Proc Ann Meet, Atlanta, GE, 31–39
Petterson R W, Bodig J 1983 Prediction of fracture toughness of conifers. Wood Fiber Sci 15: 302–316
Pew J C 1948 Flavanone from Douglas-fir heartwood. J Am Chem Soc 70: 3031–3034
Pew J C 1949 Membraneous substances in common heartwoods. J For 47: 196–199
Phelps J E, McGinnes E A 1980a Discolorations and streaks in black walnut – a microscopic evaluation. In: Bauch J (ed) Natural variations of wood properties. Mitt Bundesforschungsanst Forst-Holzwirtsch 131: 99–112
Phelps J E, McGinnes E A 1980b Color of black walnut – macroscopic and microscopic studies. In: Bauch J (ed) Natural variations of wood properties. Mitt Bundesforschungsanst Forst-Holzwirtsch 131: 169–192
Phelps J E, Nelson N D, Rietveld W J, McGinnes E A 1983 Rates of ethylene production by parenchyma cells in black walnut sapwood. Wood Fiber Sci 15: 23–27
Phillips L W 1931 Exudation from marri. J R Soc W Aust 18: viii
Pilz 1907 The sapwood/heartwood transition in Scots pine. Allg Forst-Jagd Z 83: 265–272
Polge H 1982 Influence de la compétition et de la disponsibilité en eau sur l'importance de l'aubier du douglas. Ann Sci For 39 (4): 379–397
Poluboyarinov O I, Nekrasova G N 1977 Effect of pruning aspen on increment, structure and wood properties. Lesn Zhur 3: 13–17
Porter L J 1969 The resin and fatty acid content of living Pinus radiata wood. N Z J Sci 12: 687–693
Posey C E, Robinson D W 1969 Extractives of shortleaf pine: analyses of contributing factors and relations. Tappi 52: 110–115
Preusser H J, Dietrichs H H, Gottwald H 1961 Electron-microscope investigations on ultra-thin sections of ray parenchyma. Holzforschung 15: 65–75
Priestley J H 1932 The growing tree. Forestry 6: 105–112
Prior C 1976 Resistance by Corsican pine to attack by Heterobasidion annosum. Ann Bot (London) 40: 261–279
Puritch G S 1977 Distribution and phenolic composition of sapwood and heartwood in Abies grandis and effects of the balsam woolly aphid. Can J For Res 7: 54–62
Purkayastha S K, Agrawal S P, Tandon R D, Chauhan L 1980 Variation in the proportion of heartwood in Eucalyptus tereticornis. Indian For 106: 466–473
Raffa K F, Berryman A A 1982 Physiological differences between lodgepole pines resistant and susceptible to the mountain pine beetle. Environ Ent 11: 486–492
Rahe J E, Arnold R M 1975 Injury-related phaseollin accumulation in Phaseolus vulgaris and its implications with regard to specificity of host-parasite interaction. Can J Bot 53: 921–928
Rao P R, Seshadri T R 1948 Epicatechol from Acacia catechu. J Sci Ind Res (India) 7B: 59
Rao R K 1983 Role of wood and plant anatomy in archeological studies in prehistoric India. J Indian Acad Wood Sci 14: 1–8
Raunecker H 1956 (Is red heart of beech merely a result of ageing.) Allg Forst-Jagd Z 127: 16–31; For Abstr 1957, 18: 2252
Record S J, Hess R W 1943 Timbers of the new world. Yale Univ Press New Haven, 640 pp
Record S J, Mell C D 1924 Timbers of tropical America. Yale Univ Press, New Haven, 610 pp
Reed A H 1954 The story of kauri. Reed, Auckland, N Z, 439 pp
Reid J S 1956 The properties and uses of rimu. N Z For Serv Inform Ser 2: 7 pp
Reid R W 1961 Moisture changes in lodgepole pine before and after attack by the mountain pine beetle. For Chron 37: 368–375

Rennerfelt E, Nacht G 1955 The fungicidal activity of some constituents from heartwood of conifers. Svensk Bot Tidskr 49: 419–432

Resch H, Ecklund B A 1964 Permeability of wood exemplified by measurements on redwood (Sequoia sempervirens). For Prod J 14: 199–206

Rhodes M J C, Wooltorton L S C 1971 The effect of ethylene on the respiration and on the activity of phenylalanine ammonia lyase in swede and parsnip root tissue. Phytochemistry 10: 1989–1997

Rhodes M J C, Wooltorton L S C 1973 Stimulation of phenolic acid and lignin biosynthesis in swede root tissue by ethylene. Phytochemistry 12: 107–118

Roberts D R 1973 Inducing lightwood in pine trees by paraquat treatment. US For Serv Res Note SE-191: 4 pp

Roberts D R 1976 Treatment of slash pine with 12- and 24- percent paraquat. In; Esser M H (ed) Lightwood Res Coor Counc Proc Ann Meet, Jacksonville FL, 81–87

Roberts D R, Mason J N, Proveaux A T, Peters W J, Lawrence R V 1973 A new and more efficient method of naval stores production. Naval Stores Rev 83(5): 4–5

Robinson G M 1937 Leucoanthocyanins. III. Formation of cyanidin chloride from a constituent of Butea frondosa. J Chem Soc 1937: 1157–1160

Robinson G M, Robinson R 1933 A survey of anthocyanins. III. Notes on the distribution of leucoanthocyanins. Biochem J 27: 206–212

Robitaille H A, Leopold A C 1974 Ethylene and the regulation of apple stem growth under stress. Plant Physiol 32: 301–304

Rogers J S, Calderwood H N, Beebe C W 1950 Study of the tannin content of barks from the Florida scrub oaks. J Am Leather Chem Assoc 45: 733–751

Rogers R, Hinckley T M 1979 Foliar weight and area related to current sapwood in oak. For Sci 25: 298–303

Rolfs C H, Kindl H 1984 Stilbene synthase and chalcone synthase. Plant Physiol 75: 489–492

Roux D G, Evelyn S R 1958 Condensed tannins. 3. Biogenesis of condensed tannins based on leucoanthocyanins. Biochem J 70: 344–349

Roux D G, Evelyn S R 1960 Condensed tannins. 4. The distribution and deposition of tannins in heartwoods of Acacia mollissima and Schinopsis spp. Biochem J 76: 17–23

Rowan S J 1970 Fusiform rust gall and canker formation and phenols of lobolly pine. Phytopathology 60: 1221–1226

Rowe J W, Conner A H 1979 Extractives in eastern hardwoods – a review. US For Serv Rep FPL 18: 20–21

Rowe J W, Conner A H, Diehl M A, Wroblewska H 1976 Effect of treating northern and western conifers with paraquat. In: Esser M H (ed) Lightwood Res Coor Coun Proc Ann Meet Jacksonville Fla, 66–76

Rowe J W (ed). In preparation Natural products extraneous to the lignocellulosic cell wall of woody plants. Springer, Berlin, Heidelberg, New York, Tokyo

Roy D N, Konan S K, Purdy J R 1984 Sublimation of occidentalol, a sesquiterpene alcohol, from eastern white cedar (Thuja occidentalis) in a drying kiln. Can J For Res 14: 401– 403

Rudinsky S A 1966 Host selection and invasion by the Douglas fir beetle (Dendroctonus pseudotsugae) in coastal Douglas fir forests. Can Ent 98: 98–111

Rudman P 1966a Studies in wood preservation. III. The penetration of the fine structure of wood by inorganic solutions including preservatives. Holzforschung 20: 60–67

Rudman P 1966b Heartwood formation in trees. Nature (London) 210: 608–610

Rudman P, Da Costa E W B 1959 Variation in extractive content and decay resistance in the heartwood of Tectona grandis. J Inst Wood Sci 3: 33–42

Russell C E, Berryman A A 1976 Host resistance to the fir engraver beetle. 1. Monoterpene composition of Abies grandis pitch blisters and fungus-infected wounds. Can J Bot 54: 14–18

Ryan F J, Schwarz O J 1979 Carbon fixation and enzyme activities in Virginia pine after paraquat treatment. In: Esser M H (ed) Lightwood Res Coor Counc Proc Ann Meet, Atlanta GE, 12–20

Sachs I B 1978 Paraquat-induced lightwood in red pine observed by scanning electron microscopy. In: Becker R P, Johri Om (eds) Scanning Electron Microscopy. SEM Inc AMF O'Hare, Ill, 11: 217–222

Sachs I B, Ward J C, Bulgrin E H 1966 Heartwood stain in red oak. Holz Roh-Werkst 24: 489–497

Sachsse H 1965 (Investigations of the influence of pruning on the formation of colored heartwood and tension wood in some poplar species.) Holz Roh-Werkst 3: 425–434. CSIRO Transl 7929

Sachsse H 1967 On the water/gas – proportion within the pore-capacity of the stemwood of living trees with regard to the heartwood formation. Holz Roh-Werkst 25: 291–303

Saiki H 1973 Electron microscopy of wood cell walls impregnated with aqueous solution of silver nitrate. J Jpn Wood Res Soc 19: 367–372

Saka S, Goring D A I 1983 The distribution of inorganic constituents in black spruce wood as determined by TEM-EDXA. Mokuzai Gakkaishi 29: 648–656

Salamon M, Kozak A 1968 Shrinkage of sapwood and heartwood of young Douglas-fir trees. For Prod J 18: 90–94

Sameshima K, Okumura D, Kawano H, Kondo T 1967 Studies on the intermediate zone of sugi wood. II. J Jpn Wood Res Soc 13: 349–354

Sandberg K R, Herrick F W, Selders A A 1977 Paraquat treatment of north-west conifers for lightwood induction. Wood Sci 10: 28–30

Sandermann W 1966 (Wood constituents, their chemistry and biochemistry.) Naturwissenschaften 53: 513–525

Sandermann W, Braun D, Angustin H 1965 Über ungewöhnliche mineralische Einlagerungen in tropischen Baumarten. Holz Roh-Werkst 23: 87–96

Sandermann W, Dietrichs H H 1957 (Investigation on termite-proof wood species.) Holz Roh-Werkst 15: 281–297

Sandermann W, Dietrichs H H 1959 Chemische Untersuchungen an Teakholz. Holzforschung 13: 137–148

Sandermann W, Dietrichs H H, Simatupang M H, Puth M 1963 Untersuchungen über kautschukhaltige Hölzer. Holzforschung 17: 161–168

Sandermann W, Hausen B, Simatupang M 1967 Orientierende Versuche zur Differenzierung von Splint und Kern sowie zum Sichtbarmachen der Übergangszone von Fichte und anderen Nadelhölzern. Papier 21: 349–354

Sandermann W, Rothkamm M 1959 (The determination of pH values of woods, and their practical importance.) Holz Roh-Werkst 17: 433–441

Sandermann W, Simatupang M 1966 Zur Chemie und Biochemie des Teakholzes (Tectona grandis). Holz Roh-Werkst 24: 190–204

Sands R 1975 Radiata pine bark-aspects of morphology, anatomy and chemistry. N Z J For Sci 5: 74–86

Sargent C S 1926 Manual of the trees of North America. Houghton, Mifflin, Boston, MA

Sarkar S K, Phan C T 1974 Effect of ethylene on the phenylalanine ammonia-lyase activity of carrot tissues. Physiol Plant 32: 318–321

Sauter J J 1966a Über die jahresperiodischen Wassergehaltsänderungen und Wasserverschiebungen im Kern und Splintholz von Populus. Holzforschung 20: 137–142

Sauter J J 1966b Untersuchungen zur Physiologie der Pappelholzstrahlen. I. Z Pflanzenphysiol 55: 246–258

Sauter J J 1966c Untersuchungen zur Physiologie der Pappelholzstrahlen. II. Z Pflanzenphysiol 55: 349–362

Sauter J J 1967 Der Einfluß verschiedener Temperaturen auf die Reserverstärke in parenchymatischen Geweben von Baumsprossachsen. Z Pflanzenphysiol 56: 340–352

Sauter J J 1972 Respiratory and phosphatase activities in contact cells of wood rays and their possible role in sugar secretion. Z Pflanzenphysiol 67: 135–145

Sauter J J 1983 Efflux and reabsorption of sugars in the xylem. II. Seasonal changes in sucrose uptake in Salix. Z Pflanzenphysiol 111: 429–440

Sauter J J, Iten W, Zimmermann M H 1973 Studies on the release of sugar into the vessels of sugar maple. Can J Bot 51: 1–8

Sawabe O, Mori K, Takenchi T 1973 Micro-pore structure in cell wall of wood. J Jpn Wood Res Soc 19: 55–62

Scheffer T C, Cowling E B 1966 Natural resistance of wood to microbial deterioration. Ann Rev Phytopathol 4: 147–160

Schink B, Ward J C 1984 Microaerobic and anaerobic bacterial activities involved in formation of wetwood and discoloured wood. IAWA Bull N S 5: 105–109

Schnell R L 1977 Paraquat and pine trees in East Tennessee, Progress report. In: Esser M H (ed) Lightwood Res Coor Counc Proc Ann Meet, Atlantic Beach, FL, 62–64
Schnell R L, Toennisson R L 1978 Paraquat and pine trees in East Tennessee. In: Esser M H (ed)Lightwood Res Coor Counc Proc Ann Meet, Atlanta, GE, 4–6
Schober R 1953 (Japanese larch: an investigation of growth, yield and biology.) Schriftenr Forst Fak Univ Göttingen 7/8: 258 pp
Schrödinger E 1945 What is life? Cambridge Univ Press, London (reprinted 1967)
Schroeder H A 1968 The p-hydroxycinnamyl compounds of Siam benzoin gum. Phytochemistry 7: 57–61
Schroeder H A, Kozlik C J 1972 The characterization of wetwood in western hemlock. Wood Sci Technol 6: 85–94
Schuck H J 1982 The chemical composition of the monoterpene fraction in wounded wood of Picea abies. Eur J For Pathol 12: 175–181
Schwarz O J, Ling C, Crouch S 1977 Transport and cytochemical action of paraquat in loblolly pine. In: Esser M H (ed) Lightwood Res Coord Counc Proc Ann Meet, Atlantic Beach, FL, 106–111
Schwerin G 1958 The chemistry of reaction wood. II. Holzforschung 12: 43–48
Scott N C 1964 Moisture patterns in Douglas – fir and Tanoak slash. US For Serv Berkeley Res Note PSW 55: 5 pp
Scurfield G, Anderson C A, Segnit E R 1974 Silica in woody stems. Aust J Bot 22: 211–229
Sequin A 1796 Rapport au comité de salut public sur les nouveaux moyens de tanner les cuirs proposés par le cit. Ann Chim Paris 20(15) (see White 1957)
Shah J J, Baqui S, Pandalai R C, Patel K R 1981a Histochemical changes in Acacia nilotica during transition from sapwood to heartwood. IAWA Bull N S 2: 31–36
Shah J J, Nair M N B, Pandalai R C 1981b Effect of ethephon (2 – chloroethyl phosphonic acid) on heartwood formation in Azadirachta indica. Indian J Exp Biol 19: 216–219
Shah J J, Nair G M, Kothari I L 1982 Ultrastructural changes in the gum-resin ducts of the bark of Commiphora wightii induced by mechanical injury. IAWA Bull N S 3: 185–192
Shain L 1967 Resistance of sapwood in stems of loblolly pine to infection by Fomes annosus. Phytopathology 57: 1034–1045
Shain L 1971 The response of sapwood of Norway spruce to infection by Fomes annosus. Phytopathology 61: 301–307
Shain L 1979 Dynamic responses of differentiated sapwood to injury and infection. Phytophathology 69: 1143–1147
Shain L, Hillis W E 1971 Phenolic extractives in Norway spruce and their effects on Fomes annosus. Phytopathology 61: 841–845
Shain L, Hillis W E 1972 Ethylene production in Pinus radiata in response to Sirex-Amylostereum attack. Phytopathology 62: 1407–1409
Shain L, Hillis W E 1973 Ethylene production in xylem of Pinus radiata in relation to heartwood formation. Can J Bot 51: 1331–1333
Shain J , Mackay J F G 1973a Seasonal fluctuations in respiration of aging xylem in relation to heartwood formation in Pinus radiata. Can J Bot 51: 737–741
Shain L, Mackay J F G 1973b Phenol-oxidizing enzymes in the heartwood of Pinus radiata. For Sci 19: 153–155
Shannon L M, Uritani I, Imazecki H 1971 De novo synthesis of peroxidase isozymes in sweet potato slices. Plant Physiol 47: 493–498
Shaw N 1914 Chinese forest trees and timber supply. T. Fisher Unwin London, 351 pp
Sherrard E C, Kurth E F 1933 Distribution of extractive in redwood. Ind Eng Chem 25: 300–302
Sherry S D 1971 The black wattle (Acacia mearnsii). Univ Natal, Press Pietermaritzburg, S Afr, 402 pp
Shibata N, Harada H, Saika H 1980 Formation and wall structure of tyloses in Castanea crenata. Bull Kyoto Univ For 52: 206–215
Shigo A L 1972 Ring and ray shakes associated with wounds in trees. Holzforschung 26: 60–62
Shigo A L 1983 Tree defects: a photo guide. US For Ser Gen Tech Rep NE-82: 167 pp
Shigo A L 1984 Compartmentalization: a conceptual framework for understanding how trees grow and defend themselves. Ann Rev Phytopathol 22: 189–214

Shigo A L, Dudzik K R 1985 Response of uninjured cambium to xylem injury. Wood Sci Technol 19: 195–200
Shigo A L, Hillis W E 1973 Heartwood, discolored wood and microorganisms in living trees. Ann Rev Phytopathol 11: 197–222
Shigo A L, Larson E 1969 A photo guide to the patterns of discoloröation and decay in living northern hardwood trees. US For Serv Res Pap NE-127: 100 pp
Shrimpton D M 1973 Extractives associated with wound response of lodgepole pine attacked by the mountain pine beetle and associated microorganisms. Can J Bot 51: 527–534
Shrimpton D M, Watson J A 1971 Response of lodgepole pine seedlings to inoculation with Europhium clavigerum, a blue stain fungus. Can J Bot 49: 373–375
Shrimpton D M, Whitney H S 1968 Inhibition of growth of blue-stain fungi by wood extractives. Can J Bot 46: 757–761
Sifton H B 1957 Air-space tissue in plants. Bot Rev 23: 303–312
van der Sijde H A 1976 Resin-infiltrated heartshakes in Pinus elliottii grown in the Transvaal. S Afr For J 97: 29–35
Sinnott E W 1918 Factors determining character and distribution of food reserves in woody plants. Bot Gaz 66: 162–175
Sioumis A A, Lau L S 1976 Paraquat-induced resinosis in Pinus radiata. I. Appita 29: 272–275
Sioumis A A, Smelstorius J A, Lau L S 1979 Paraquat-induced resinosis in Pinus radiata. III. Appita 30: 189–194
Sioumis A A, Tham S H, Lau L S 1977 Paraquat-induced resinosis in Pinus radiata. II. Appita 30: 401–404
Sivakumaran S 1982 Ethephon stimulation – minimum time after application for response. Planters Bull Rubber Res Inst Malaysia 172: 85–88
Sivakumaran S, Yeang H Y, Yong W M 1983 Endogenous ethylene in Hevea bark tissues. J Rubber Res Inst Malaysia 32 (2): 155–163
Skene D S 1965 The development of kino veins in Eucalyptus obliqua. Aust J Bot 13: 367–378
Smedman L A, Snajberk K, Zavarin E, Mon T R 1969 Oxygenated monoterperiods and sesquiterpenoid hydrocarbons of the cortical turpentine from different Abies species. Phytochemistry 8: 1471–1479
Smith F, Montgomery R 1959 The chemistry of plant gums and mucilages. Rheinhold, New York, 627 pp
Smith H G 1903 Aluminium the chief inorganic element in a proteaceous tree, and the occurrence of aluminium succinate in trees of this species. J Proc R Soc N S W, 37: 107–120
Smith H G 1912 On the crystalline deposit occurring in the timber of the 'Colonial Beech', Gmelina leichhardtii. J Proc R Soc N S W, 46: 187–198
Smith H G 1913 On the kinos or astringent exudations of one hundred species of Eucalyptus. Rep Australasian Assoc Advancement Sci 14, Gov Printer, Melbourne, Australia, 83–95
Smith J H G, Walters J, Wellwood R W 1966 Variation in sapwood thickness of Douglas-fir in relation to tree and section characteristics. For Sci 12: 97–103
Smith R H 1961 Techniques for determing the toxicity of pine resin vapours to Dendroctonus breviconis and D. jeffreyi. J Econ Ent 54: 359–365
Smith R H 1966 The monoterpene composition of Pinus ponderosa xylem resin and of Dendroctonus brevicomis pitch tubes. For Sci 12: 63–68
Smith W R, Goebel N B 1952 The moisture content of green hickory. J For 50: 616–618
Snell J A, Brown J K 1978 Comparison of tree biomass estimators – DBH and sapwood area. For Sci 24: 455–457
Somerville A 1980 Resin pockets and related defects of Pinus radiata grown in New Zealand. N Z J For Sci 10: 439–444
Squillace A E 1971 Inheritance of monoterpene composition in cortical oleoresin of slash pine. For Sci 17: 381–387
Squillace A E, Gansel C R 1974 Juvenile: mature correlations in slash pine. For Sci 20: 225–229
Squillace A E, Moyer E 1976 Genetic and environmental variation of induced resin-soaking in slash pine. In: Esser M H (ed) Proc Ann Meet, Lightwood Res Coord Counc Proc, Jacksonville, FL, 93–101

Squire G B, Swan E P, Wilson J W 1967 Intra–increment variation in Douglas fir flavonoids by new technique. Pulp Pap Mag Can 68: T-431-T-437
Stalker I V 1971 A safer test for distinguishing heartwood and sapwood in pines. J Inst Wood Sci 5(6): 21–24
Stewart C M 1960 Detoxication during secondary growth in plants. Nature (London) 186: 374–375
Stewart C M 1966a Excretion and heartwood formation in living trees. Science 153: 1068–1074
Stewart C M 1966b The chemistry of secondary growth in trees. CSIRO Div For Prod Melbourne Aust Technol Pap 43: 107 pp
Stewart C M 1967 Moisture content of living trees. Nature (London) 214: 138–140
Stewart C M, Kottek J, Dadswell H E, Watson A J 1961 The process of fibre separation. III. Tappi 44: 798–813
Stich K, Ebermann R 1984 Peroxidase and Polyphenoloxidase Isoenzyme in Splint und Kernholz der Eiche. Holzforschung 38: 239–242
Stone J E 1973 The fibre saturation point, wall density and other characteristics of wood. Proc Meet, IUFRO Div 5, S Afr: 988–999
Stone R N 1975 (Ed) Proc Lightwood Research Coord Counc. Chevron Chem, FL, 145 pp
Strauss S H, Critchfield W B 1982 Inheritance of β–pinene in xylem oleoresin of knobcone x monterey pine hybrids. For Sci 28: 687–696
Stubbs J, Roberts D R, Outcalt K W 1984 Chemical stimulation of lightwood in southern pines. Gen Tech Rep SE-25 Asheville NC. US Dep Agric For Ser S E For Exp Stn: 51 pp
Stutz R E 1959 Control of brown stain in sugar pine with sodium azide. For Prod J 9: 459–463
Sucoff E, Ratsch H, Hook D D 1967 Early development of wound-initiated discoloration in Populus tremuloides. Can J Bot 45: 649–656
Sullivan J D 1967 Color characterisation of wood: color parameters of individual species. For Prod J 17: 25–29
Swain T 1962 The biosynthesis of polyphenols. In: Hillis W E (ed) Wood extractives. Academic Press, London, New York, 277–313
Swain T 1963 (ed) Chemical plant taxonomy. Academic Press, London, New York, 543 pp
Swain T 1965 The tannins. In: Bonner J, Varnet J E (eds) Plant biochemistry. Academic Press, London, New York, 552–580
Swan E P 1966 Chemical methods of differentiating the wood of several western conifers. For Prod J 16 (1): 51–54
Swan E P 1983 The forefront of sawmill technology. Forintek Rev 1 (3): 6–7
Swan E P, Jiang K S 1970 Formulation of heartwood extractives in western red cedar. Tappi 53: 844–846
Swan E P, Jiang K S, Gardner J A F 1969 The lignans of Thuja plicata and the sapwood-heartwood transformation. Phytochemistry 8: 345–351
Szopa P S, Garrett H E, McGinnes E A 1980 Growth-quality evaluation of black walnut wood. 1. Wood Sci 13: 95–98
Takahashi K, Ogiyama K 1985 Phenols of discolored sugi (Cryptomeria japonica) sapwood. II. Mokuzai Gakkaishi 31: 28–38
Talbot L 1983 Sandalwood. For Focus W Aust, 30 (Sept): 21–31
Talbot P H B 1977 The Sirex – Amylostereum-Pinus association. Ann Rev Phytopathol 15: 41–54
Tamblyn N 1937 Decay in timber with special reference to jarrah (Eucalyptus marginata). Aust For 2: 6–13
Tan H T, Leong T T 1975 Exploitation procedure for young mature rubber plantings to be felled early. Int Rubber Conf, Kuala Lumpur, Reprint: 11 pp
Tarkow H. Krueger J 1961 Distribution of hot-water soluble material in cell walls and cavities of redwood. For Prod J 11: 228–229
Teratini F, Haraguchi S, Kai Y 1976 Radial distribution of the extractives in the compression wood of guimatsu, Larix gmelinii. Mokuzai Gakkaishi 22: 133–136
Teratini F, Shimizu K, Miyazaki K 1969 (Purification of arabinogalactan from Japanese larch.) Mokuzai Gakkaishi 15: 266–269
ter Welle B J H 1976a On the occurence of silica grains in the secondary xylem of the Chrysobalanaceae. IAWA Bull N S 2: 19–29

ter Welle B J H 1976b Silica grams in woody plants of the neotropics especially Surinam. In: Baas P, Bolton A J, Catling D M (eds) Wood structure in biological and technological research. Leiden Bot Ser 3 Leiden Univ Press, Netherlands, 107–142

Thomas R J 1975 The effect of polyphenol extraction on enzyme degradation of bordered pit tori. Wood Fiber 7: 207–215

Timell T E 1986 Compression wood in gymnosperms. Springer, Berlin, Heidelberg, New York, 2210pp

Tindale M D, Roux D G 1974 An extended phytochemical survey of Australian species of Acacia. Phytochemistry 13: 829–839

Tippett J T 1982 Shedding of ephemeral roots in gymnosperms. Can J Bot 60: 2295–2302

Tippett J T 1986 Formation and fate of kino veins in Eucalyptus L'Herit. IAWA Bull N S 7: 137–143

Tippett J T, Bogle A L, Shigo A L 1982 Response of balsom fir and hemlock roots to injuries. Eur J For Pathol, 12: 357–364

Tippett J T, Shea S R, Hill T C, Shearer B L 1983 Development of lesions caused by Phytophthora cinnamomi in the secondary phloem of Eucalyptus marginata. Aust J Bot 31: 197–210

Tischler K 1976 Improvement of Eucalyptus camaldulensis Dehnh wood. PhD Thesis, Hebrew Univ, Jerusalem Israel, 129 pp

Trabut M 1901 Sur la manne de l'Olivier. C R Acad Sci (Paris) 32: 225–226

Trenard Y 1982 Making wood speak: an introduction to dendrochronology. For Prod Abstr 5: 267–298

Trendelenburg R, Mayer-Wegelin H 1955 Das Holz als Rohstoff. Hanser, München, 541 pp

Tschirch A, Stock E 1933–1936 Die Harze. 2 vols, 3rd edn. Borntraeger, Berlin, 1858 pp

Tsvetaeva I P 1959 Distribution of gums in the wood of dahurian larch. Chem Abstr 53: 13588

Turner C H, Balodis V, Dean G H 1983 Variation in pulping quality of E.globulus from Tasmanian provenances. Appita 36: 371–376

Uritani I, Muramatsu K 1952 Phytopathological chemistry of black rotted sweet potato. J Agr Chem Soc Jpn, 26: 289–295

Urquidi R L 1978 Gums for food uses – a review. Flavours Food Additives 9 (2): 73,75–76

Vanzetti B L 1936 (Constitution of olivil from olive resin). Atti V congr nazl chim pura appl 1935, Pt II, 932–938

Varner J E 1961 Biochemistry of senescence. Ann Rev Plant Physiol 12: 245–264

Vintila E 1939 Untersuchungen über Raumgewicht und Schwindmaß von Früh- und Spätholz bei Nadelhölzern.Holz Roh-Werkst 2: 345–357

Vité J P 1961 The influence of water supply on oleoresin exudation pressure and resistance to bark beetle attack on Pinus ponderosa. Contrib Boyce Thompson Inst 21 (2): 37–66

Wagner H, Wagner G J, Hrazdina G 1984 Endoplasmic reticulum as a site of phenylpropanoid and flavonoid metabolism in Hippeastrum. Plant Physiol 74: 901–906

Wangaard F F, Granados L A 1967 The effect of extractives on water-vapor sorption by wood. Wood Sci Technol 1: 253–277

Ward J C, Zeikus J G 1980 Bacteriological, chemical and physical properties of wetwood in living trees. In: Bauch J (ed) Natural variations of wood properties. Mitt Bundesforschungsanst Forst-Holzwirtsch 131: 133–166

Ward W C, Reid J S 1949 The properties and uses of rimu. N Z For Serv Inf Ser 2.

Wardell J F, Hart J H 1970 Early responses of sapwood of Quercus bicolor to mechanical injury. Can J Bot 48: 683–686

Wardell J F, Hart J H 1973 Radial gradients of elements in white oak wood. Wood Sci 5: 298–303

Wardrop A B, Cronshaw J 1962 Formation of phenolic substances in the ray parenchyma of angiosperms. Nature (London) 193: 90–92

Wardrop A B, Davies G W 1958 Some anatomical factor relating to the penetration of water into xylem of gymnosperms. Aust J Bot 6: 96–102

Waring R H, Gholz H L, Grier C C, Plummer M L 1977 Evaluating stem conducting tissue as an estimator of leaf area in four woody angosperms. Can J Bot 55: 1474–1477

Waring R H, Schroeder P E, Oren R 1982 Application of the pipe model theory to predict canopy leaf area. Can J For Res 12: 556–560

Webb J L, Saunders P R, Thienes C H 1949 Action of metabolic inhibitors on rat heart slice respiration. Arch Biochem Biophys 22: 458–466
Webb L J 1953 An occurrence of aluminium succinate in Cardwellia sublimis F.Muell. Nature (London) 171: 656
Webb L J 1954 Aluminium accumulation in Australian – New Guinea flora. Aust J Bot 2: 176–196
Wellwood R W 1955 Sapwood-heartwood relationships in second-growth Douglas fir. For Prod J 5: 108–111
Wellwood R W, Jurazs P E 1968 Variation in sapwood thickness, specific gravity, and tracheid length in western red cedar. For Prod J 18(12): 37–46
Werker E, Fahn A 1968 Site of resin synthesis in cells of Pinus halepensis. Nature (London) 218: 388–389
Westfelt L 1966 High-boiling neutral constituents from the wood of Pinus sylvestris. Acta Chem Scand 20: 2829–2840
Wheeler E A 1982 Ultrastructural characteristics of red maple (Acer rubrum) wood. Wood Fiber 14: 43–53
Wheeler E A 1983 Intervascular pit membranes in Ulmus and Celtis native to the United States. IAWA Bull N S 4: 79–88
Wheeler E A, Thomas R J 1981 Ultrastructural characteristics of mature wood of southern red oak (Quercus falcata) and white oak (Quercus alba). Wood Fiber 13: 169–181
Whistler R L 1973 Industrial gums. Academic Press, London, New York, 810 pp
Whistler R L 1982 Industrial gums from plants: guar and chia. Econ Bot 36,(2): 195–202
White T 1957 Tannins – their occurrence and significance. J. Sci Food Agric 8: 377–385
Whitehead D 1978 The estimation of foliage area from sapwood based area in Scots pine. Forestry 51: 137–149
Whitehead D, Edwards W R N, Jarvis P G 1984 Conducting sapwood area, foliage area, and permeability in mature trees of Picea sitchensis and Pinus contorta. Can J For Res 14: 940–947
Wiermann R 1981 Secondary plant products and cell and tissue differentiation. In: Conn E E, Stumpf P K (eds) The biochemistry of plants vol VII. Secondary plant products. Academic Press, London, New York, 85–116
Wilkinson R C 1980 Relationships between cortical monoterpenes and susceptibility of eastern white pine to white-pine weevil attack. For Sci 26: 581–589
Wolter K E 1977 Ethylene-potential alternative to bipyridilium herbicides for inducing lightwood in red pine. In: Esser M H (ed) Lightwood Res Coor Counc Proc Ann Meet, Atlantic Beach, FL, 90–99
Wolter K E, Zinkel D F 1976 Metabolic, hormonal, photosynthetic and lightering response to red pine treated with paraquat. In: Esser M H (ed) Lightwood Res Coor Counc Proc Ann Meet, Jacksonville FL, 31–39
Wolter K E, Zinkel D F 1984 Observations on the physiological mechanisms and chemical constituents of induced oleoresin synthesis in Pinus resinosa. Can J For Res 14: 452–458
Wooding F B F, Northcote D H 1965 Association of the endoplasmic reticulum and the plastids in Acer and Pinus. Am J Bot 52: 526–531
Wright T W, Will G M 1958 The nutrient content of Scots and Corsican pines growing on sand dunes. Forestry 31: 13–25
Wroblewska H, Connor A H, Rowe J W 1977 Lightwood formation in red pine treated with paraquat. Wood Sci 10: 1–5
Yamamoto K 1982 Yearly and seasonal process of maturation of ray parenchyma cells in Pinus species. Res Bull Exp For Hokkaido Univ 39: 245–296
Yamamoto K, Fukazawa K, Ishida S 1978 Study on the cell wall development of ray parenchyma in genus Pinus. 2. Res Bull Coll Exp For Hokkaido Univ 35: 99–108
Yamamoto K, Fukazawa K, Ishida S 1979 Study on the cell wall development of ray parenchyma in genus Pinus. 3. Res Bull Coll Exp For Hokkaido Univ 38: 451–458
Yamanaka K 1985 Ethylene production in Chamaecyparis obtusa phloem and xylem tissues in response to wounding. Mokuzai Gakkaishi 31: 703–710
Yang K C, Hazenberg G, Bradfield G E, Maze J R 1985 Vertical variation of sapwood thickness in Pinus banksiana and Larix laricina. Can J For Res 15: 822–828

Yang S F, Hoffman N E 1984 Ethylene biosynthesis and its regulation in higher plants. Ann Rev Plant Physiol 35: 155–189

Yang S F, Pratt H K 1978 The physiology of ethylene in wounded plant tissue. In: Biochemistry of wounded plant tissues. De Gruyter, Berlin, 595–622

Yazawa K 1960 The seasonal water content of sapwood and heartwood of broadleaved trees, especially of beech (Fagus crenata). Mokuzai Gakkaishi 6: 170–175

Yazawa K, Fukazawa K 1956 Studies on the relation between physical properties and growth condition for planted sugi in central district of Japan. I. Mokuzai Gakkaishi 2: 204–209

Yazawa K, Ishida S 1965a On the wet-heartwood of some broad-leaved trees grown in Japan. II. Seasonal moisture content by months. J Fac Agric Hokkaido Univ 54(2): 123–136

Yazawa K, Ishida S 1965b On the existence of the intermediate wood in some broad-leaved trees grown in Hokkaido, Japan. J Fac Agric Hokkaido Univ 54(2):137–150, 10 plates

Yazawa K, Ishida S, Miyajima H 1965 On the wet-heartwood of some broad-leaved trees grown in Japan. I. Mokuzai Gakkaishi 11: 71–76

Yazawa K, Ishida S, Ohtani J 1967 The study on the formation of the artificial heartwood. Res Bull Coll Exp For Hokkaido Univ 25 (1): 9–34

Yazdani R, Nilsson J E, Ericsson T 1985 Geographical variation in the relative proportion of monoterpenes in cortical oleoresin in Pinus sylvestris in Sweden. Silvae Genet 34: 201–205

Young H E, Guinn V P 1966 Chemical elements in complete mature trees of seven species in Maine. Tappi 49(5): 190–197

Zamski E 1970 The effect of microclimatic conditions on resin duct formation and resin secretion in Pinus halepensis. La–Yaaran 20 (1,2): 35–40

Zavarin E 1968 Chemistry of the genus Abies. II. Within tree variation of the terpenes in cortical oleoresin. Phytochemistry 7: 92–107

Zavarin E 1970 Qualitative and quantitative co-occurence of terpenoids as a tool for elucidation of their biosynthesis. Phytochemistry 9: 1049–1063

Zavarin E, Lin I-re, Zinkel D F 1977 Lightwood formation in Western pines. In: Esser M H (ed) Lightwood Res Coor Counc Proc Ann Meet, Atlantic Beach, FL, 57–61

Zavarin E, Snajberk K 1965 Chemotaxomomy of the genus Abies. I. Survey of the terpenes present in the Abies balsams. Phytochemistry 4: 141–148

Zelawski W 1960a Respiration intensity in oak wood in particular annual rings of sapwood. Bull Acad Pol Sci Cl II 8: 509–516

Zelawski W 1960b A contribution to the problem of temperature influence on respiratory processes in wood. Bull Acad Pol Sci C1 II 8: 517–519

Ziegler H 1957 Über den Gaswechsel verholzter Achsen. Flora (Jenna) 144: 229

Ziegler H 1964 Storage, mobilization and distribution of reserve material in trees. In: Zimmermann M H(ed) The formation of wood in forest trees. Academic Press, London, New York, 303–320

Ziegler H 1968 Biological aspects of heartwood formation. Holz Roh-Werkst 26: 61–68

Zinkel D F 1981 Turpentine, rosin and fatty acids from conifers. In: Goldstein I S (ed) Organic chemicals from biomass. CRS Press Boca Raton FL, 163–187

Zinkel D F, McKibben C R 1978 Chemistry of naval stores from pine light-wood – a critical review. In: Esser M H (ed) Lightwood Res Coor Counc Proc Ann Meet, Atlanta, GE, 133–156

Zycha H 1948 Über die Kernbildung und verwandte Vorgänge im Holz der Rotbuche. Forstwiss Cbl 67: 80–109

Subject Index

The scientific name of a wood species is followed in most cases by a common name in brackets.